Clinical Medicine and the Nervous System
Consulting Editor: Michael Swash

Syndromes that have an underlying neurological basis are common problems in many different specialties. Clinical Medicine and the Nervous System is a series of monographs concerned with the diagnosis and management of clinical problems due primarily to neurological disease, or to a neurological complication of another disorder. Thus the series is particularly concerned with those neurological syndromes that may present in different contexts, often to specialists without special expertise in neurology. Since the range of clinical practice embraced by neurologists is wide, the books in this series will appeal to many different specialists in addition to neurologists and neurosurgeons. It is the aim of the series to produce individual volumes that are succinct, informative and complete in themselves, and that provide sufficient practical discussion of the issues to prove useful in the diagnosis, investigation and management of patients. Important advances in basic mechanisms of disease are emphasized as they are relevant to clinical practice. In particular, individual volumes in the series will be useful especially to neurologists, neurosurgeons, physicians in internal medicine, oncologists, paediatricians, neuro-radiologists, rehabilitationists, otorhinolaryngologists and ophthalmologists, and to those in training in these specialties.

Michael Swash, MD, FRCP, MRCPath
The Royal London Hospital

Guillain–Barré Syndrome

Richard A.C. Hughes

With 88 Figures

Foreword by Professor P.K. Thomas

Springer-Verlag
London Berlin Heidelberg New York
Paris Tokyo Hong Kong

Richard A.C. Hughes, MD, FRCP
Professor of Neurology, Department of Neurology,
United Medical and Dental Schools of Guy's and
St Thomas's Hospitals, Guy's Hospital, London, UK

Consulting Editor

Michael Swash, MD, FRCP, MRCPath
Consultant Neurologist, Neurology Department, The Royal London Hospital,
Whitechapel, London E1 1BB, UK

The illustration on the cover shows a nerve fibre being demyelinated in
Guillain–Barré syndrome (Fig. 4.9)

ISBN-13:978-1-4471-3177-9   e-ISBN-13:978-1-4471-3175-5
DOI: 10.1007/978-1-4471-3175-5

British Library Cataloguing in Publication Data
Hughes, Richard *1942–*
Guillain–Barré syndrome
1. Man. Guillain–Barré syndrome
I. Title   II. Series
616.87
ISBN-13:978-1-4471-3177-9

Library of Congress Cataloging-in-Publication Data
Hughes, R.A.C. (Richard Anthony Cranmer)
Guillain–Barré syndrome / Richard A.C. Hughes.
p.   cm. – (Clinical medicine and the nervous system)
Includes bibliographical references.   Includes index.
ISBN-13:978-1-4471-3177-9
1. Polyradiculoneuritis.   I. Title.   II. Series.
[DNLM: 1. Polyradiculoneuritis.   WL 400 H894g]
RC416.H84   1990   616.8'7 – dc20
DNLM/DLC   90-10031
for Library of Congress   CIP

Typeset by Best-set Typesetter Ltd, Hong Kong

2128/3916-543210   Printed on acid-free paper

Dedicated to
Dr A.C.C. Hughes, MD, FRCP

Foreword

The period that followed World War II has witnessed a dramatic change in neurology. From being a discipline in which its participants were castigated for being interested solely in diagnosis, usually of disorders of unknown causation without effective therapy, neurology has evolved into a highly active treatment-orientated subject. This transition is clearly reflected in the approach to diseases of the peripheral nervous system, and to the Guillain–Barré syndrome (GBS) in particular. In a state-of-the-art review made in 1952, Elkington (1952) observed that no less than 56% of neuropathies remained undiagnosed, and amongst those of unknown causation he listed GBS. With intensive investigation and follow-up, the proportion of neuropathies seen at tertiary referral centres which elude diagnosis is now as little as 13% (McLeod et al. 1984). Overall, of course, the proportion is even less. This change is partly because of the introduction of new diagnostic techniques and partly because of the application of the great expansion in knowledge evident throughout medicine. In this book, Professor Richard Hughes has assembled current information on GBS and related disorders, including chronic inflammatory demyelinating polyneuropathy (CIDP), the existence of which was not appreciated until Austin's perspicacious study published in 1958.

In the Introduction, Professor Hughes gives an account of the way in which recognition of the GBS emerged and matured, and shows that it followed, pari passu, with the realisation that paralysis and sensory loss may result from peripheral nerve disorders. The descriptions of acute ascending paralysis by Landry in 1859 are clearly recognisable as the GBS, but it was not until the account by Guillain, Barré and Strohl in 1916 that the condition emerged as a clear entity, characterised by "albumino-cytological dissociation" in the cerebrospinal fluid. In the subsequent eponymous designations of the disease, poor Strohl's name is usually omitted. In this first chapter, Professor Hughes also addresses the definition of the GBS. It has to be accepted that definitions are still operational. It is not yet certain that it represents a single nosological entity or what are its precise boundaries. The same is even more true for CIDP.

The second chapter reviews relevant aspects of the biology of the peripheral nervous system, essential for understanding pathogenesis. This leads to Chapter 3 which is on experimental allergic neuritis. It is true to say that in no small measure our current knowledge as to the pathogenesis of the GBS and CIDP have depended on studies on this experimental model. Models for both the acute and chronic forms of the human disease are available. The animal studies have involved the identification of the responsible antigens, the detailed analysis of the pathological changes, and the dissection of disease mechanisms. In a later chapter, naturally occurring animal models of GBS are discussed. Chapter 4 considers the pathology and pathogenesis of GBS and it is here, in particular, that knowledge gained from the animal studies is providing dividends. Professor Hughes argues that the GBS may not represent a single pathological entity. It is clear that further clinicopathological studies are required, employing contemporary morphological and cell labelling techniques. The clinical heterogeneity which includes, in addition to the usual predominantly motor disorder, sensory forms and the ophthalmoplegic/ataxic Miller Fisher syndrome, also argues for pathological heterogeneity.

The electrophysiological changes, which are reviewed in Chapter 7, are helpful not only diagnostically but also prognostically, as was evident from observations made in the multicentre North American trial of plasma exchange. The treatment of GBS continues to be an area of active research. Although plasma exchange has been shown to be of benefit, it required a large-scale trial for this to be established. The studies so far have failed to demonstrate a beneficial effect of corticosteroids, but the result of Professor Hughes's trial on high dose methylprednisolone is keenly awaited, as is further evaluation of the use of intravenous human gamma globulin. But effective rational therapy, ultimately, will depend on a better knowledge of the pathogenesis.

CIDP is currently an enigmatic disorder, with phenotypic overlap with the GBS, with paraproteinaemic neuropathy and even perhaps with multiple sclerosis. Again it is unlikely to be a single pathological entity. Professor Hughes ends his book with a consideration of other inflammatory neuropathies including those related to paraproteinaemia, leprosy, sarcoidosis and vasculitis. The justification for this is the likelihood that overlapping disease mechanisms exist between these disorders.

This is a lively book on a lively subject. I warmly commend it.

Professor P.K. Thomas
Royal Free Hospital School of Medicine, London

References

Austin JG (1958) Recurrent polyneuropathies and their corticosteroid treatment – with five year observation of a placebo-controlled case treated with corticotrophin, cortisone and prednisolone. Brain 81:157–194

Elkington JStC (1952) Recent work on the peripheral neuropathies. Proc R Soc Med 45:661–664

Guillain G, Barré JA, Strohl A (1916) Sur une syndrome de radiculo-névrite avec hyperalbuminose du liquide cephalo-rachidien sans reaction cellulaire. Remarques sur les caractères cliniques et graphiques des reflexes tendineux. Bull Soc Med Hôp Paris 40:1462–1470

Landry O (1859) Note sur la paralysie ascendante aigue. Gaz Hebd Méd Chir 6:472–474

McLeod JG, Tuck RR, Pollard JD, Gameron J, Walsh JC (1984) Chronic polyneuropathy of undetermined cause. J Neurol Neurosurg Psychiat 47:530–535

Consulting Editor's Foreword

Guillain–Barré syndrome has proved a pivotal clinical problem in peripheral nerve disorders since it was first described nearly 100 years ago. Peripheral neuropathies are a common clinical problem and Guillain–Barré syndrome is particularly important not only because it is potentially a serious event, but because recovery is the usual outcome. The mechanisms of recovery in neurological disorders, of all kinds, remain poorly understood and therefore difficult to modify by medical intervention. Guillain–Barré syndrome presents in many different guises and so forms part of the differential diagnosis for clinicians in many different areas of expertise. Professor Hughes has provided a timely, succinct, and yet complete account of this disorder that will prove an invaluable source of reference. We welcome it to Clinical Medicine and the Nervous System, a series of short monographs intended to illuminate the importance of medicine in its generality to neurological problems, and vice versa.

London   Michael Swash

Preface

This book is both a historical statement of what is known about Guillain–Barré syndrome and related disorders and a look forward to the future. It is the product of a 15-year experience of looking after an unusually large number of patients with Guillain–Barré syndrome and an interest in immunological aspects of neurological disease. The choice of Guillain–Barré syndrome as an area of study was based on the prediction that it had an inflammatory and probably autoimmune pathogenesis which ought to be amenable to investigation and treatment. The lessons from a peripheral nerve inflammatory demyelinating disease should turn out to be relevant to the understanding and treatment of the neurologist's arch enemy, multiple sclerosis. Historically the understanding of simple nervous systems and the peripheral nerves has usually preceded the understanding of central nervous system mechanisms.

Guillain–Barré syndrome is the eponymous name given to a clinical syndrome of weakness with loss of the tendon reflexes progressing over one to three weeks. It affects about two per 100 000 of the population each year worldwide. According to some estimates 12% of patients die and 20% are left disabled and unable to work after a year. The pathogenesis and pathology have turned out to be heterogeneous and are still not completely understood. I hope that the information gathered in these pages will be a platform which will help future investigators define the relevant antigens and mechanisms of demyelination more clearly. Investigation of the cause of Guillain–Barré syndrome has required consideration of the immunology of the peripheral nervous system, which has an intrinsic interest constituting, like the central nervous system, an immunologically somewhat privileged site. This investigation has had to extend beyond experimentally induced neuritis in laboratory animals to veterinary conditions. Marek's disease, a herpes infection of chickens, and coonhound paralysis in dogs are both comparable with Guillain–Barré syndrome, being associated with lymphocytic infiltration and primary demyelination in peripheral nerves.

The book includes a section on treatment which should be of value to the intensive care physician. Shortly before I started writing this book, two large controlled trials clearly demonstrated that a form of immunosuppression, plasma exchange, is beneficial in Guillain–Barré syndrome. Other forms of immunological treatment are currently under investigation. For the neurologist the book covers not only the clinical features, diagnosis and treatment of Guillain–Barré syndrome but also related disorders. In particular Guillain–Barré syndrome is regarded as at the acute end of a spectrum which includes at its other end chronic progressive and chronic relapsing idiopathic demyelinating polyradiculoneuropathy. Different forms of paraproteinaemia are associated with demyelinating neuropathy, probably because the paraprotein is an antibody against neural antigens. These experiments of nature are teaching us much that is relevant to the understanding of acquired demyelinating polyradiculoneuropathy. Other forms of inflammatory neuropathy, vasculitis, sarcoidosis and leprosy, are included because of the contrasts which they provide with Guillain–Barré syndrome, generally producing axonal rather than primary myelin damage.

This book contains a mixture of clinical neurology, immunology, neurobiology, neurophysiology, neurochemistry and veterinary science. There have been many times while writing it when I have asked myself whether it should not have been a multiauthor volume. That I have managed to finish it is a tribute to many friends and I am deeply indebted to them. Their advice has been invaluable but responsibility for any misinformation rests firmly with me. Many colleagues have generously allowed me to reproduce their material which has been appropriately acknowledged in the figure and table legends, but I would like to thank them here. My own research has been supported by the UK Medical Research Council, Action Research for the Crippled Child, Moorgate Trust, Special Trustees of Guy's Hospital and Guillain–Barré Syndrome Support Group and I am grateful to all of them.

I would like particularly to thank Mr Ken Brady, Mr Ian Gray, Dr Norman Gregson, Dr Susan Hall, Dr Mick Kadlubowski, Professor Sidney Leibowitz, Mr Derek Lovell, Dr Henry Powell, Dr Ken Smith, Dr Bill Taylor, Dr John Winer, Mrs Penny Atkinson for her histological expertise, and Mrs Sheila Button for her indefatigable and cheerful secretarial help. The final acknowledgement must be to my family: I hope they will forgive me.

London   Richard A.C. Hughes
1990

Contents

Chapter 1

History and Definition

Landry's Acute Ascending Paralysis

During the first half of the nineteenth century the role of the peripheral nervous system was obscure. Anatomical dissections had revealed connections between the central nervous system and muscles and skin but pathological examinations had not revealed abnormalities of the peripheral nerves. The concept that peripheral nerve damage causes paralysis and sensory loss had not emerged. Case reports of progressive numbness and weakness occurring over several days and then recovering spontaneously were described in the medical journals without any understanding of their cause. In 1828 Chomel described an epidemic of cases of tingling, severe pain and numbness in the hands and feet and weakness of the limbs in Paris. This illness caused severe paralysis which was usually followed by recovery but was sometimes fatal. Although sometimes regarded as the first description of what was later to be called "Guillain–Barré syndrome", symptoms included severe distal pain, which prevented putting the feet to the ground, and a desquamating rash. These features are not consistent with a modern diagnosis of Guillain–Barré syndrome. In 1834 Wardrop described a man with progressive numbness and weakness confining him to his bed within ten days whose illness resembled Guillain–Barré syndrome more closely. Sensation and bowel function remained normal. The man remained paralysed apart from his head and toes for four months until Wardrop treated him with purgatives whereupon he gradually and completely recovered. Wardrop was impressed by the success of purgative treatment and attributed the paralysis to a disorder of the alimentary tract. Graves (1848) was the first to suggest that such paralysis arose "from disease commencing in the nervous system alone".

Striking examples of acute paralysis were published by Octave Landry (Fig. 1.1) in 1859 in a paper entitled "La Paralysie Ascendante Aiguë". To Landry's contemporaries acute ascending paralysis was a puzzling condition of uncertain cause: it was not even clear where in the nervous system the damage lay. With the clarity of hindsight his account is an unmistakable description of acute polyradiculoneuropathy. His analysis of five personal cases and five which he had discovered in the literature is worth quoting:

> The sensory and motor systems may be equally affected. However the main problem is usually a motor disorder characterised by a gradual diminution of muscular strength with flaccid limbs and without contractures, convulsions or reflex movements of any kind. In almost all cases micturition and defaecation remain normal. One does not observe any symptoms referable to the central nervous system, spinal pain or tenderness, headache or delirium. The intellectual faculties are preserved until the end. The onset of the paralysis can be preceded by a general feeling of

Fig. 1.1. Jean Baptiste Octave Landry de Thézillat (1826–1865). This elegant young neurologist had made several important contributions on the nature of sensation and paralysis before publishing the paper on acute ascending paralysis in 1859 which secured him posthumous fame. Landry never published any more papers possibly because he had just married a distinguished, aristocratic but impecunious beauty. Landry took charge of a hydrotherapy practice which proved extremely successful. His accomplishments included dancing, singing, playing the cello, hunting, mountaineering and crystallography. His elegance made him the talk of the salons. He caught cholera while looking after the destitute in Paris and is said to have died with Charcot at his bedside at the age of 39. This is a photograph of the portrait by Courbet. (From Haymaker (1953), with permission.)

weakness, pins and needles and even slight cramps. Alternatively the illness may begin suddenly and unexpectedly. In both cases the weakness spreads rapidly from the lower to the upper parts of the body with a universal tendency to become generalised. The first symptoms always affect the extremities of the limbs and the lower limbs particularly. When the whole body becomes affected the order of progression is more or less constant: 1. toe and foot muscles, then the hamstrings and glutei, and finally the anterior and adductor muscles of the thigh; 2. finger and hand, arm and then shoulder muscles; 3. trunk muscles; 4. respiratory muscles, tongue, pharynx, oesophagus, etc. The paralysis then becomes generalised but more severe in the distal parts of the extremities. The progression can be more or less rapid. It was eight days in one and fifteen days in another case which I believe can be classified as acute. More often it is scarcely two or three days and sometimes only a few hours. When the paralysis reaches its maximum intensity the danger of asphyxia is always imminent. However in eight out of ten cases death was avoided either by skilful professional intervention or a spontaneous remission of this phase of the illness. In two cases death occurred at this stage ... When the paralysis recedes it demonstrates the reverse of the phenomenon which signalled its development. The upper parts of the body, the last to be affected, are the first to recover their mobility which then returns from above downwards.

Landry's description is immediately recognisable to a twentieth-century neurologist as the disorder which is now most commonly called Guillain–Barré syndrome. His rather lengthy account of one patient contains some astute clinical observations. For instance he recognised the presence of diaphragmatic weakness from the presence of paradoxical movement of the abdomen: "on the day before dying of asphyxia the patient was lying quietly on his back and there was hollowing of the abdomen during inspiration and outward movement during expiration". This paradoxical movement was much less evident when the patient was sitting up. He also recognised the flaccid paralysis of the limbs and absence of central nervous system and sphincter involvement. The absence of reflex movements which he describes probably refers to absence of movement following cutaneous stimuli since the tendon reflexes were first recognised and introduced into neurological examination by Westphal and Erb in 1875 (McHenry 1969). It would be left to Guillain, Barré and Strohl nearly 70 years later to stress the loss of tendon reflexes. It is notable that Landry's paper has only one author, a welcome contrast with the multiauthor papers of today. The chief under whose care that patient had been admitted merely adds a note in which he pays tribute to Landry who had predicted a fatal outcome at an early stage. The consultant originally thought that Landry's patient had a hysterical illness, a mistake which is still sometimes made today. The consultant concerned was Monsieur Gubler who is known to modern medical students because his name was given to a form of focal pontine infarction called the Millard–Gubler syndrome. Landry himself noted that no abnormality was observed at autopsy either in his own case or another in the literature and declined to express an opinion on the mechanism of the paralysis. However, Gubler discussed the possibility that there was a close connection between Landry's cases and the paralysis which follows diphtheria. This was a prophetic comparison since both conditions were eventually shown to be due to a demyelinating neuropathy albeit, as we shall see, with different underlying pathogenetic mechanisms.

During the later decades of the nineteenth and early years of the twentieth centuries many more cases of acute paralysis corresponding to Landry's paralysis were reported. Levi (1865) acknowledged the significance of Landry's paper. Westphal (1876) described four cases of paralysis (one a post-diphtheritic neuropathy) in which autopsy did not reveal any pathological lesion and first applied the eponymous title Landry's ascending paralysis. The site of the pathology in Landry's acute ascending paralysis became the subject of debate and confusion (Hun 1891). Gowers (1888) called it "a mysterious disease", considering that "multiple neuritis" (polyneuropathy in modern terminology) and myelitis could mimic the clinical picture. He discussed toxic influences on the upper motor neuron pathway as an explanation for the predominantly motor disorder without any wasting. As time went by the view that acute ascending paralysis could be produced either by myelitis or multiple neuritis was frequently expressed (Eisenlohr 1890). In the cases in which myelitis was demonstrated to be the cause there were features which differed from the description of Landry, such as bladder involvement and prominent reflex movements. The nosological limits of "acute ascending paralysis" continued to puzzle neurologists during the first half of the twentieth century. In particular it was unclear what its relationship was to "acute febrile polyneuritis".

Acute Febrile Polyneuritis

In 1892 Osler writing the first edition of his *Principles and Practice of Medicine* classified polyneuropathy into:

1. Acute febrile polyneuritis
2. Recurring multiple neuritis
3. Alcoholic neuritis
4. Multiple neuritis in the infectious diseases
5. Arsenical and saturnine (lead) neuritis
6. Endemic neuritis (Beri Beri)

He regarded many cases of Landry's ascending paralysis as having acute febrile polyneuritis, others as having myelitis. Osler's description of acute febrile polyneuritis is of an illness starting with a temperature rapidly rising to 103°F and causing aching limbs and back, tingling and ascending or descending paralysis, with respiratory involvement and muscle wasting in severe cases. Sensory involvement was variable. The tendon reflexes were not mentioned. Some patients died of respiratory involvement or "paralysis of the heart" within a week to ten days, others remained stable for five to six weeks and then slowly recovered. In the *British Medical Journal* of July 1917 Gordon Holmes described a similar clinical picture based on 12 cases which he had encountered in the winter of 1916. He stressed the onset with fever and rapidly progressive motor paralysis. He also included facial paralysis, absent tendon reflexes and frequent sphincter involvement. What distinguished his cases from other forms of neuritis was the early onset of improvement provided the patient survived the acute episode. It is puzzling to know what modern disease Osler and Holmes were describing since absence of fever is the rule in Guillain–Barré syndrome (GBS). It is possible that the prior infection which so often precedes GBS was regarded, in those cases with a short delay, as being part of the neuritis. Alternatively those cases with acute fulminating neuropathy may have developed fever due to secondary chest infection.

Acute Infective Polyneuritis

Next in the argument, though not in time, comes a paper by Bradford et al. (1918) which described more clearly the syndrome which the authors regarded as synonymous with acute febrile polyneuritis. However, they made clear that the febrile episode, if it occurred, took place and recovered before the onset of the neuropathic symptoms, i.e. the modern concept of Guillain–Barré syndrome. These authors injected cerebrospinal fluid (CSF) from patients into monkeys and transferred a paralysing illness which could then be transferred to further monkeys. They reported inflammatory changes in the central nervous

SUR UN SYNDROME DE RADICULO-NÉVRITE AVEC HYPERALBUMINOSE DU LIQUIDE
CÉPHALO-RACHIDIEN SANS RÉACTION CELLULAIRE. REMARQUES SUR LES
CARACTÈRES CLINIQUES ET GRAPHIQUES DES RÉFLEXES TENDINEUX,

par MM. Georges Guillain, J.-A. Barré et A. Strohl.

Fig. 1.2. Title of the original paper by Guillain, Barré and Strohl published in the *Bulletins et Memoirs de la Société Médicale des Hôpitaux de Paris* in 1916. A translation of part of the paper was published in *Archives of Neurology*, 1968, 18:450–452.

system of the monkeys. They proposed that the condition was caused by a virus. Although a viral aetiology has been frequently proposed for Guillain–Barré syndrome, subsequent attempts to grow viruses from the CSF and body tissues or excretions have not proved successful except in rare single case reports.

A Syndrome of Radiculoneuritis with Albumino-cytological Dissociation

The next landmark in the history of Guillain–Barré syndrome was the paper of Guillain, Barré and Strohl which was presented to a meeting of the hospitals of

Fig. 1.3. Georges Guillain (1876–1961). Guillain worked as intern for Raymond and then Pierre Marie whom he succeeded as Professor of Neurology at the Salpêtrière Hospital in Paris in 1923. His publications covered a wide range but none was more famous than the description of what became known as Guillain–Barré syndrome. He retired in 1947 and his position was taken by Alajouanine (From Bonduelle (1977, 1978), with permission.)

Fig. 1.4. Jean-Alexandre Barré (1880–1967). Barré came from Brittany, worked as an intern in Paris for Babinski and joined forces with Guillain in the neurological centre of the Sixth Army during World War I. He was Professor of Neurology in Strasbourg from 1919 until 1950. A colourful description of his career was given in 1984 by Professor Rohmer, to whom the author is indebted for this photograph.

Paris in October 1916 (Fig. 1.2). It is remarkable that three doctors (Figs. 1.3, 1.4 and 1.5) working under the conditions which must have prevailed in the neurological centre of the French Sixth Army in 1916, just after the Battle of the Somme, should have had the *sang froid* to publish this "short note" on two paralysed soldiers whom they encountered within the same month. Their first patient was a hussar who developed progressive pins and needles and weakness of his limbs without any preceding illness. Examination a month after the onset showed profound predominantly distal weakness of the limbs, absent tendon reflexes and only slight sensory loss. The sphincters were not affected. Within a month he had improved so that he was able to walk for an hour and was sent to the rear to convalesce. Their second case was rather more acute: an infantryman was brought to their centre on the fourth day of an illness which had rendered him so weak that he fell over backwards when he put on his backpack and could not pull himself to his feet. He had improved slightly when he was evacuated a month later. Their graphic records of the knee and ankle reflexes showed that the reflex muscle contraction was delayed to almost twice the normal latency and reduced in amplitude. They deduced that nerve conduction or the central part of the reflex must be severely and predominantly affected. However, they also observed that the response to direct percussion of the muscle was diminished which led them to suppose that the muscle was also affected. Only their first deduction was later confirmed. The major point of Guillain, Barré and Strohl's

Fig. 1.5. A photograph of Professor André Strohl. Strohl was born in Poitiers in 1887 and became interested in physical medicine of which he was professor in Paris in 1925. He contributed more than 200 publications including several on reflexes and nerve conduction. He retired in 1957. (From Green (1962), with permission.)

paper was to draw attention to the increased cerebrospinal fluid (CSF) protein without any cellular reaction in this form of "radiculoneuritis". This observation was made possible by the introduction in 1891 of the technique of lumbar puncture by Quincke, professor of medicine in Bern, who, in searching for a method of draining hydrocephalus, hit upon the idea of inserting a needle between the lumbar vertebrae (McHenry 1969). The first of Guillain, Barré and Strohl's cases had a CSF albumin of 2.5 g/l and the second 0.85 g/l but the CSF cell count was normal in both. They stressed the importance of this finding which had not previously been reported in pure nerve root or peripheral nerve disease. However, such albumino-cytological dissolution had already often been described in association with spinal cord compression, Pott's disease (tuberculosis of the spine) and syphilis affecting the central nervous system. The last example seems rather surprising today since active neurosyphilis is usually associated with a raised CSF cell count. Neurosyphilis was extremely common and could mimic almost every neurological syndrome. Landry and Guillain, Barré and Strohl went to great pains to emphasise that their patients denied venereal disease.

Guillain, Barré and Strohl did not consider that they could be precise about the pathogenesis of the syndrome but wondered whether it might be due to an intoxication or infection. They did state that the prognosis was not grave, a view to which Guillain was to adhere rather vehemently for the remainder of his career.

Guillain–Barré Syndrome or Landry's Ascending Paralysis

After 1917 neurologists had to contend with three imprecisely defined but similar conditions, Landry's ascending paralysis, acute febrile polyneuritis and the condition described by Guillain, Barré and Strohl. Draganescu and Claudian (1927) used the term Guillain–Barré syndrome for the first time to describe a case of radiculoneuritis after staphylococcal osteomyelitis. Their presentation to a meeting of the Société de Neurologie de Paris was introduced by Barré himself and Strohl's name was omitted both from the eponymous title and the list of authors in the reference citation to the famous 1916 paper on polyradiculoneuritis by Guillain, Barré and Strohl. It has been suggested that the omission of Strohl's name was due to his leaving neurology for a career in physical medicine (Green 1962). The omission does seem unfair though now hallowed by tradition and the least we can do is to republish his portrait in acknowledgement of his contribution (Fig. 1.4).

With additional experience Guillain recognised that the cranial nerves, particularly the facial nerves, were often involved and that difficulty with micturition and altered bladder sensation also occurred. He stressed the importance of both the clinical features and the CSF findings in defining the syndrome. Thus in poliomyelitis the raised CSF cell count, as well as the more acute clinical picture and absent sensory disturbance, would make the diagnosis clear. Albuminocytological dissociation might occur in diphtheria but that condition could be diagnosed by the pseudomembranous sore throat and onset with paralysis of the palate and accommodation. Guillain (1953) resented a suggestion made by Baker (1943) that his syndrome was none other than the acute febrile polyneuritis already recognised by Osler (1892) and Holmes (1917).

These points of view were powerfully expressed by Guillain (1938) when summing up a meeting in Belgium devoted to the "Syndrome de Guillain–Barré" in 1937 and published the following year. Guillain stressed that in their syndrome there was no fever and that Osler could not have diagnosed their syndrome because lumbar puncture had not been introduced and the albuminocytological dissociation was an essential part of the definition. He also considered that Landry's cases represented a mixed bag of diagnoses which might have included other causes of ascending paralysis including poliomyelitis and acute encephalomyelitis. However, as we have seen, Landry clearly described sensory disturbances which excluded poliomyelitis and flaccid limbs and normal sphincter function which made encephalomyelitis unlikely.

Haymaker and Kernohan (1949) took a broad view of the definition of GBS including cases with normal CSF protein and high CSF cell counts and related the story of the debate which had taken place since Guillain, Barré and Strohl's paper. They concluded that Landry's clinical description could not be differentiated from that of Guillain and Barré's patients and published their paper under the title "Landry–Guillain–Barré Syndrome". This view enraged Guillain (1953) who reiterated his narrower definition of the condition. Now that the protagonists are no longer on the stage we can peacefully follow modern usage and use the simpler title "Guillain–Barré syndrome". It is only fair to acknowledge here that Landry's precise clinical description probably did represent the same clinical entity. We should also remember that Strohl's contribution, like that of many a hard-working registrar or resident, has been forgotten.

Later chapters will explore just how heterogeneous an entity this "syndrome" will turn out to be. Guillain (1953) had already begun to distinguish the following subdivisions:

1. Spinal
2. Spinal and brainstem
3. Brainstem
4. Polyradiculoneuritis with mental involvement.

The pure brainstem form represents what has come to be called the Miller Fisher syndrome in which ophthalmoplegia, depressed tendon reflexes and ataxia are attributed to a demyelinating polyradiculoneuropathy. The mental involvement consisted of "psychopathic symptoms" coinciding with polyradiculoneuritis: these probably consisted of somnolence and agitation which are well recognised today in patients with incipient respiratory failure due to Guillain–Barré syndrome. Other features recognised by Guillain at this time included ataxia, occasional papilloedema, autonomic disturbances including diffuse sweating and painful extremities.

It is interesting to see how far the developments of the last quarter of a century were presaged by Guillain's last paper in 1953. Although he continued to stress the benign prognosis in most cases, one of the 19 patients whom he reported in this last paper had died. The autopsy revealed extensive damage to the peripheral nervous system without any sign of inflammation. Guillain continued to regard an infectious cause as the most likely explanation of his syndrome. He did not think that the viruses suggested up to that time were responsible, these including Coxsackie virus, hepatitis and infectious mononucleosis. Collaborative attempts with Pierre Mollaret, head of the Institut Pasteur, to induce disease by injecting CSF from a GBS patient into mice, guinea pigs and monkeys were unsuccessful. He dismissed as fanciful suggestions, already promulgated by Bannwarth (1943), that the disease was due to allergy. He considered that the generally good prognosis of the syndrome scarcely justified the use of adrenocorticotrophic hormone which had just been introduced. However, he strongly supported the use of measures to guard against inhalation and pulmonary collapse, including stopping oral fluids in favour of intravenous infusion, pharyngeal suction, chest physiotherapy with postural drainage, large doses of antibiotics, oxygen and even the use of the iron lung. Guillain had lived through an era from the discovery of the albumino-cytological dissociation to the first descriptions of the pathology, modern sounding discussions of the pathogenesis and the beginning of modern intensive care.

Pathology

Peripheral nerve damage as a cause of paralysis and sensory loss had been predicted by Graves (1848). Pathological changes in peripheral nerves were first described in 1864 by Dumenil in a patient who had accumulated multiple individual nerve lesions and died after 5 months; the clinical picture resembled

multiple mononeuropathy rather than GBS. Microscopic examination of un-
stained specimens of nerve revealed "atrophy of the medullary substance of the
peripheral nerve fibres" (Dumenil 1866). Dumenil considered that Graves had
originated the concept of peripheral nerve disease and did not mention Landry.
Landry (1859) had himself undertaken a detailed pathological study of the
brain, spinal cord and nerve roots, including microscopy, without finding any
abnormalities. Degenerative changes in the peripheral nerves of individual
cases, sometimes proximal and sometimes distal, were described by several
authors during the late nineteenth and early twentieth centuries (reviewed by
Haymaker and Kernohan 1949). Eichhorst (1877) described the microscopic
appearance of perivascular cuffing and myelin breakdown in a woman whose
clinical illness he regarded as Landry's acute ascending paralysis. The detailed
description of stepwise accumulation of individual nerve lesions with oedema
and proteinuria suggests a modern diagnosis of multiple mononeuropathy due
to vasculitis. The tendency of many authors to add the suffix "-itis" to mean
"disease of" rather than inflammation makes it difficult to discover who was
the first to describe a truly inflammatory peripheral neuropathy. Leyden (1880)
probably deserves the credit for describing a clearly recognisable microscopic
perivascular inflammation in a case with a clinical course consistent with the
modern concept of GBS (Fig. 1.6). In a long review Leyden distinguished poliomyelitis and myelitis from multiple neuritis and argued that multiple neuritis
alone could be the substrate of Landry's ascending paralysis. He studied a pa-

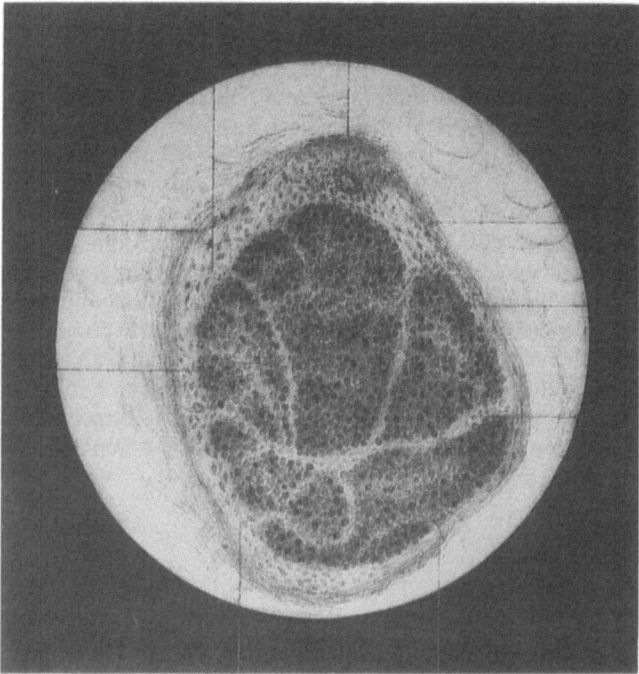

Fig. 1.6. Drawing of subperineurial inflammatory cell infiltrate in a case of neuritis resembling
GBS by Leyden (1880).

tient who died eight weeks after the onset of an acute illness with paralysis, sensory impairment and loss of tendon reflexes. At autopsy microscopic examination of the spinal cord was normal but the peripheral nerves showed florid infiltration by cells around the endoneurial blood vessels. The cells were "partly small lymphoid cells, the usual products of inflammation, partly larger flat protoplasmic cells, round or oval with yellowish granular contents".

Thereafter many cases of inflammatory changes in the nerves of patients dying from acute ascending paralysis or Guillain–Barré syndrome were reported. Sigmund Freud (1886) reported a fatal case of ascending paralysis which resembled the cases described by Leyden but also had brisk reflexes suggesting cord involvement. An early English language account is that of Putnam (1889). Schweiger (1909) described inflammation of the peripheral nerves and dorsal root ganglia without any spinal cord affection in a case of typical Landry's ascending paralysis. Krücke (1955) reviewed these early accounts and described seven personal cases. He stressed that cellular and predominantly lymphocytic infiltration of the peripheral nerves was the first change, present from the first days of the disease. However, Haymaker and Kernohan (1949) in a lengthy review of the clinical and pathological features put forward the view that lymphocytic infiltration was an inconstant finding. When lymphocytic infiltration did occur it was a late event and probably a response to the nerve fibre damage rather than its cause. Haymaker and Kernohan regarded oedema of the spinal roots as the most important early finding. Their paper being in the English language was more readily accessible to the Anglo-Saxon world and deflected attention for a time from the important inflammatory component of the pathological changes.

By the 1950s the idea of autoallergic causes for inflammatory diseases was becoming popular. It was first considered as an explanation of the acute encephalomyelitis which sometimes followed rabies vaccine (Hurst 1932). Rivers and colleagues had reproduced the clinical syndrome of acute encephalomyelitis by giving monkeys prolonged courses of injections of rabbit brain (Rivers and Schwentker 1935; Rivers et al. 1933). Freund made such experiments much easier by showing that if the nervous tissue was emulsified in oil with dead tubercle bacilli (Freund's adjuvant) a single injection would produce encephalomyelitis in guinea pigs. Rivers et al. (1933) showed incidentally that the spinal roots were also involved. Waksman and Adams in the 1950s described how rabbits, guinea pigs and mice immunised with peripheral nerve tissue in Freund's adjuvant developed an acute paralysing illness followed by gradual recovery (Waksman and Adams 1955, 1956). Pathologically this disease was characterised by multifocal infiltration of peripheral nerves, spinal roots and root ganglia with lymphocytes and histocytes. Waksman and Adams pointed out that this experimental allergic neuritis (EAN) provided a model for GBS, which gave substance to previous suggestions that GBS might have an autoallergic basis (Pette and Kornyey 1930; Bannwarth 1943; Pette and Pette 1956).

In 1969 Asbury, Arnason and Adams, in an important clinicopathological study supporting the idea that GBS has an autoallergic basis, pointed out the discrepancy in the literature between the bland, degenerative or oedematous pathology favoured by Haymaker and Kernohan (1949) as the basis of GBS and the inflammatory changes in many other reports. They described 19 personal cases of GBS in all of which they found lymphocytic infiltration, particularly in the very early stages of the peripheral nervous system. Their series included

Table 1.1. Synonyms for Guillain–Barré syndrome

Landry's acute ascending paralysis
Acute febrile polyneuritis
Polyradiculoneuritis with albuminocytological dissociation
Acute infective polyneuritis
Guillain–Barré syndrome
Landry–Guillain–Barré syndrome
Landry–Guillain–Barré–Strohl syndrome
Idiopathic polyneuritis
Acute inflammatory demyelinating polyradiculoneuropathy

all the clinical cases of "idiopathic polyneuritis", as they then called GBS, which they had encountered at the Massachusetts General Hospital during the previous 14 years. This classical paper has dominated thinking about the pathology and pathogenesis of GBS until the present time. Further discussion of these problems and how to account for the cases of GBS lacking similar lymphocytic infiltration belongs to the present rather than to history.

Definition

The difficulties clinicians have had in defining the diagnostic limits of GBS are readily understood against the historical background of evolving clinical practice and pathological understanding. The difficulties are indicated by the enormous list of names which have been applied, only some of which have been mentioned (Table 1.1). For clinical practice and epidemiological studies a definition of the core syndrome is essential and in the absence of a laboratory test to apply during life or opportunities for pathological confirmation of the diagnosis this definition has to be arbitrary. In 1960 Osler and Sidell proposed rather strict diagnostic criteria to prevent the label Guillain–Barré syndrome being too widely applied.

Table 1.2. Diagnostic criteria for Guillain–Barré syndrome: 1

Required criteria	Supportive criteria
Progressive weakness of more than one limb due to neuropathy	*Clinical* Relatively symmetrical weakness
Areflexia	Relatively mild sensory signs
Duration of progress less than 4 weeks[a]	Cranial nerve involvement, especially facial nerve

Absence of other cause of acute neuropathy e.g. porphyria
    toxin exposure
    diphtheria
	Autonomic dysfunction: vasomotor instability

	Absence of fever with neuropathic symptoms
	CSF (after the first week): protein concentration increased–cell count more or less normal
	Neurophysiological: Slowing of nerve conduction suggestive of demyelination

[a] Modified from Asbury et al. (1978). In the original version the duration of progression was only a supportive criterion but this arbitrary boundary has been applied to distinguish GBS from chronic idiopathic demyelinating polyradiculoneuropathy.

Table 1.3. Diagnostic criteria for the Guillain–Barré syndrome: 2. Variants

Clinical
1. Fever at onset of neuropathic symptoms
2. Severe sensory loss with pain
3. Progression beyond four weeks[a]
4. Cessation of progression without recovery or with major permanent residual deficit
5. Transient bladder paralysis
6. Central nervous system involvement. Severe ataxia interpretable as cerebellar in origin, dysarthria, extensor plantar responses and ill-defined sensory levels need not exclude the diagnosis if other features are typical

CSF
1. Normal CSF protein for 1–10 weeks
2. CSF mononuclear leucocytes 11–50/mm^3

After Asbury et al. (1978).
[a] Such patients will be regarded as having a subacute form of chronic idiopathic demyelinating polyradiculoneuropathy in this book.

Table 1.4. Diagnostic criteria for the Guillain–Barré syndrome: 3. Features casting doubt on the diagnosis

1. Marked persistent asymmetry of weakness
2. Persistent bladder or bowel dysfunction
3. Bladder or bowel dysfunction at onset
4. More than 50 mononuclear leucocytes/mm^3 in the CSF
5. Polymorphonuclear leucocytes in the CSF
6. Sharp sensory level

After Asbury et al. (1978).

Inclusion of cases with severe sensory loss, sphincter involvement, optic nerve involvement and CSF pleocytosis seemed to them to make the syndrome meaningless. Each subsequent reviewer proposed slightly different criteria. When an epidemic of cases of GBS occurred in the USA in 1976 following the swine influenza vaccination programme, the National Institute of Neurological and Communicative Disorders and Stroke charged a committee to produce an ad hoc definition to permit epidemiological studies (Asbury et al. 1978). This definition laid down required and supportive criteria which have been adopted in subsequent treatment trials. An abbreviated and slightly modified version is listed in Table 1.2. In the original version variants (Table 1.3) and features casting doubt on but not excluding the diagnosis (Table 1.4) were recognised.

Table 1.5. Diagnostic criteria proposed for chronic idiopathic demyelinating polyradiculoneuropathy

---

Inclusion criteria

Progressive weakness of two or more limbs due to polyradiculoneuropathy

Loss or diminution of tendon reflexes

Progression for more than four weeks or recurrence or relapse (deterioration of at least one disability grade[a] following improvement of at least one disability grade for at least one week)

Fulfilment of neurophysiological criteria of demyelination[b]

Exclusion criteria

Intoxication by drugs or environmental agents

Family history of similar polyradiculoneuropathy

Neuropathy attributable to metabolic causes including diabetes mellitus, vitamin deficiency, liver or renal failure

Systemic vasculitis, systemic lupus erythematosus, polyarteritis nodosa, neoplasm, paraproteinaemia

---

[a] See Chapter 6 for definition of disability grade.
[b] See Chapter 10 for neurophysiological criteria for demyelination.
After Hughes et al. (1987).

Chronic Idiopathic Demyelinating Polyradiculoneuropathy

The occurrence of recurrent attacks of neuritis was recognised by Osler in 1892. The earlier descriptions are difficult to analyse in modern terms because they lack important clinical details. The earliest case accepted by Austin in his review of the early literature (Austin 1958) was a woman of 39 who had had three attacks in 20 years each lasting several months (Targowla 1894). From his review of 32 recurrent cases including two of his own, Austin drew a representative clinical picture of a young adult with a progressive, more or less symmetrical, predominantly motor, mainly distal, polyradiculoneuropathy. The facial, bulbar and ocular motor nerves were occasionally involved but not the sphincters. The cerebrospinal fluid protein concentration was usually increased. Two-thirds of the cases were males. Symptoms usually increased over about five months and took rather longer to recover. The interval between bouts was about 4 years and the average number of bouts three per patient. However, the picture was very variable. Thickening of the nerves was noted in a third of cases and Austin questioned whether recurrent attacks might not cause "hypertrophic polyneuropathy". He noted in particular the beneficial effect of steroid treatment. From a patient who had 20 separate bouts in 5 years treated with ACTH, cortisone, prednisone or placebo, he concluded that steroids in appropriate doses would suppress the disease but that if withdrawn too quickly the disease would re-emerge. It seemed to Austin that the tempo of the disease could be modulated with hormones and that acute "Landry–Guillain–Barré syndrome", as he called it, and more chronic polyneuropathies might represent a spectrum.

Later neurophysiological and pathological studies showed that the predominant changes in the nerves were demyelination and conduction block and that this was a common denominator to both GBS and the patients with recurrent attacks of neuropathy such as those described by Austin. Difficulties arose,

Table 1.6. Subgroups of idiopathic demyelinating polyradiculoneuropathy

---

1. Acute Guillain–Barré syndrome
   Progressive phase 4 weeks or less
2. Subacute
   Progressive phase more than 4 weeks but less than 12 weeks
3. Chronic progressive
   Progressive phase 12 weeks or more
4. Chronic relapsing
   At least two attacks[a] with progressive phase more than 4 weeks in at least one
5. Recurrent acute
   At least two attacks[a] with progressive phase lasting 4 weeks or less in each

---

[a] An attack must last at least 4 weeks. A second attack must be preceded by a remission in which the patient improves by at least one disability grade (see Chapter 6) for at least one week.

and remain, over what was meant by "acute" versus "chronic" and "relapsing" versus "recurrent". The difficulties were somewhat compounded when Dyck et al. (1975) and Prineas and McLeod (1976) conjoined both chronic progressive and chronic relapsing cases of demyelinating neuropathy under the titles "chronic inflammatory polyradiculoneuropathy" and "chronic relapsing polyneuritis" respectively. The word "inflammatory" and suffix "itis" carry connotations about pathogenesis which are rarely established by investigation in individual cases and I prefer the neutral expression chronic idiopathic demyelinating polyradiculoneuropathy. The abbreviation CIDP does allow translation of the "I", by those so inclined, into "inflammatory". However, the word idiopathic does remind us that this is a disorder *sui generis* whose cause has still to be discovered.

Although no agreed definition of CIDP exists an arbitrary definition is clearly needed for epidemiological and investigative purposes. The definition in Table 1.5 has been modelled on that proposed for Guillain–Barré syndrome and attempts to formalise previous criteria (Dyck et al. 1975; Prineas and McLeod 1976). The definition is deliberately based on clinical and neurophysiological criteria (Hughes et al. 1987). The cut-off between acute and chronic cases with a progressive course lasting less than or more than four weeks is arbitrary but based on the conclusion of Prineas (1970) that those with a shorter course tend to recover while those with a longer course are more likely to progress and/or be left with deficit. Most patients with CIDP have a progressive course evolving in the initial attack over at least six months (Dyck and Arnason 1984). However, some patients with a progressive course lasting 5–6 weeks improve and eventually recover in the manner of GBS with or without treatment, whereas others pursue a progressive or relapsing and remitting course suggestive of CIDP. It is desirable to distinguish those with a "subacute" monophasic course. I therefore propose a subdivision of idiopathic demyelinating polyradiculoneuropathy according to its pattern of evolution into the groups defined in Table 1.6. These groups should identify more homogeneous nosological entities than the broader term CIDP and so help to identify the underlying pathological substrate and pathogenetic mechanisms. Pathological and immunological studies of these groups should define whether there really is a spectrum of disease from GBS to CIDP with a common pathogenesis or whether different pathogenetic mechanisms can produce the clinical picture of acute or chronic acquired demyelinating neuropathy.

Guillain–Barré Syndrome in History and the News

GBS has occasionally hit the headlines. From October 1976 to January 1977 several cases of GBS were detected following vaccination with influenza A/ New Jersey influenza ("swine flu") vaccine. This vaccine had been produced in response to an expected epidemic of swine flu. It was not until a quarter of the population of the USA had been immunised that it was appreciated that there was a probable real increase in incidence of GBS during the six weeks following vaccination (Chapter 5). The vaccine was abruptly withdrawn and the abandonment of this massive vaccination programme caused some embarrassment to the administration with a possible adverse effect on President Ford's campaign for re-election. That event is perhaps the closest GBS has come to altering world history.

The most famous doctor to have developed GBS was Harvey Cushing, the doyen of neurosurgery. He experienced typical symptoms of GBS following a flu-like illness at the end of World War I. His illness made walking difficult but not impossible and he was back at work within a few weeks. During this time Cushing was known to have been at a neurosurgical meeting also attended by Guillain but the diagnosis was not made until a scholarly article by Reich 70 years later (Reich 1987). The nearest GBS has come to a place in literature is a personal account of his own illness by Heller (Heller and Vogel 1986), the author of *Catch 22*. The book, *No Laughing Matter*, published in 1986, gives an insider view of Heller's illness which threatened his ventilation and left him with residual symptoms at the time of writing. Although the book tells us more about the man than his disease it is instructive for the physician to share the frustration of paralysis with an introspective and articulate patient.

Summary

Guillain–Barré syndrome is defined in clinical terms as an illness causing progressive weakness of the limbs with diminished or absent tendon reflexes due to polyradiculoneuropathy and not having a toxic or other recognised cause. Such a clinical picture was described by Landry in 1859 but Landry's acute ascending paralysis was often confused with myelitis. Guillain, Barré and Strohl in 1916 described the raised CSF albumin concentration and normal cell content. The pathological basis of the Guillain–Barré syndrome was described as an inflammatory polyradiculoneuropathy by Asbury, Arnason and Adams in 1969. An illness resembling Guillain–Barré syndrome but pursuing a more chronic course was described by Austin in 1958. This illness is often called chronic inflammatory demyelinating polyradiculoneuropathy but since the inflammation is often difficult to find the term chronic idiopathic demyelinating polyradiculoneuropathy is preferred. Diagnostic criteria have been proposed which define chronic idiopathic demyelinating polyradiculoneuropathy in clinical and neurophysiological terms.

References

Asbury AK, Arnason BG, Adams RD (1969) The inflammatory lesion in idiopathic polyneuritis. Its role in pathogenesis. Medicine 48:173–215

Asbury AK, Arnason BGW, Karp HR, McFarlin DF (1978) Criteria for diagnosis of Guillain–Barré syndrome. Ann Neurol 3:565–566

Austin JH (1958) Recurrent polyneuropathies and their corticosteroid treatment. Brain 81:157–192

Baker AB (1943) Guillain–Barré's disease (encephalo-myelo-radiculitis). A review of 83 cases. The Journal Lancet 63:384–398

Bannwarth A (1943) Die entzündliche Polyneuritis mit dem Liquor-syndrome von Guillain–Barré (polyradiculitis) in Rahmen einer biologischen Krankheit-Betrachtung. Arch Psychiatr Nerven-Krankheiten 115:566–672

Bonduelle M (1977) George Guillain. Rev Neurol 133:661–666

Bonduelle M (1978) George Guillain: 1876–1961. In: den Hartog Jager WA, Bruyn GW, Heijstee APG (eds) Proceedings of the 11th World Congress of Neurology. Excerpta Medica, Amsterdam, pp 18–21

Bradford JB, Bashford EF, Wilson JA (1918) Acute infective polyneuritis. Q J Med 12:88–126

Chomel J (1828) De l'épidémie actuellement régnante à Paris. J Hebdomadaire Méd 1:332–339

Draganescu S, Claudian J (1927) Sur un cas de radiculo-névrite curable (syndrome de Guillain–Barré) apparue au cours d'une osteomyelite du bras. Rev Neurol 2:517–519

Dumenil L (1864) Paralysie peripherique du mouvement et du sentiment portant sur les quatre membres – atrophie des rameaux nerveux des parties paralysees. Gazette Hebdomadaire Méd Chir 1:203–207

Dumenil L (1866) Contributions pour servir à l'histoire des paralysies périphériques et spécialement de la névrite. Gazette Hebdomadaire Méd Chir 54–89

Dyck PJ, Arnason BGW (1984) Chronic inflammatory demyelinating polyradiculoneuropathy. In: Dyck PJ, Thomas PK, Lambert EH, Bunge R (eds) Peripheral neuropathy, WB Saunders, Philadelphia, pp 210–214

Dyck PJ, Lais AC, Ohta M, Bastron JA, Okazaki H, Groover RV (1975) Chronic inflammatory polyradiculoneuropathy. Mayo Clin Proc 50:621–651

Eichhorst H (1877) Virchow's Archiv für pathologische Anatomie und Physiologie und Klinische Medizin 69:265–285

Eisenlohr C (1890) Ueber Landry'sche Paralyse. Dtsch Med Wochenschr 16:841–844

Freud S (1886) Akute multiple Neuritis der Spinalen und Hirnnerven. Wien Med Wochenschr 36:167–171

Gowers WK (1888) A manual of diseases of the nervous system. Blakiston, Philadelphia

Graves RJ (1848) Clinical lectures in the practice of medicine. ed 2, New Sydenham Society, London

Green D (1962) Infectious polyneuritis and Professor André Strohl – a historical note. N Engl J Med 267:821–822

Guillain G (1938) Synthese générale de la discussion. J Belge Neurol Psychiatr 38:323–329

Guillain G (1953) Considérations sur le syndrome de Guillain et Barré. Ann Med 54:81–92

Guillain G, Barré JA, Strohl A (1916) Sur un syndrome de radiculo-névrite avec hyperalbuminose du liquide cephalorachidien sans réaction cellulaire. Remarques sur les caracteres cliniques et graphiques des reflexes tendineux. Bull Soc Méd Hôp Paris 40:1462–1470

Haymaker WE, Jean Baptist Octave Landry de Thezillat (1953) The founders of neurology. 133 biographical sketches. CC Thomas, Springfield, Illinois

Haymaker W, Kernohan JW (1949) The Landry–Guillain–Barré syndrome: a clinicopathologic report of fifty fatal cases and a critique of the literature. Medicine 28:59–141

Heller J, Vogel S (1986) No laughing matter. Jonathan Cape, London

Holmes G (1917) Acute febrile polyneuritis. Br Med J 2:37–41

Hughes RAC, Sanders EACM, Winer JB (1987) Guillain–Barré syndrome and chronic idiopathic demyelinating polyradiculoneuropathy. Prog Clin Neurosci 1:143–155

Hun H (1891) The pathology of acute ascending (Landry's) paralysis. NY Med J 53:609–615

Hurst EW (1932) The effects of the injection of normal brain emulsion into rabbits with special reference to the aetiology of the paralytic accidents of automatic treatment. J Hyg (Lond) 32:3–12

Krücke W (1955) Die primär entzündliche Polyneuritis unbekannter Ursache. In: Lubasch O, Henke F, Rossle G (eds) Handbuch der speziellen pathologischen Anatomie und Histologie. Vol 13 Erkrankungen der peripheren Nerven. Springer, Berlin, Gottingen, Heidelberg, pp 164–182

Landry O (1859) Note sur la paralysie ascendante aiguë. Gazette Hebdomadaire Méd Chir 6: 472–488

Levi P (1865) Contribution a l'étude de la paralysie ascendante aiguë ou extenso progressive aigue. Arch Gen Med 5:129–147

Leyden E (1880) Ueber poliomyelitis und neuritis. Z Klin Med 1:387–427

McHenry LC (1969) Garrison's history of neurology. CC Thomas, Springfield, Illinois

Osler LD, Sidell AD (1960) The Guillain–Barré syndrome: the need for exact diagnostic criteria. N Eng J Med 262:964–969

Osler W (1892) Acute ascending (Landry's) paralysis. In: Young J (ed) Principles and practice of medicine. Pentland, Edinburgh and London

Pette E, Pette H (1956) Zur Atiopathogenese der Entmarkungsencephalomyelitis (einschliesslich der akuten Multiplen Sklerose) und der Polyneuritis. Klin Wochenschr 34:718–720

Pette H, Kornyey S (1930) Zur Histologie und Pathogenese der akut entzündlichen Formen der Landryschen Paralyse. Z Ges Neurol Psychiatr 128:390–397

Prineas J (1970) Polyneuropathies of undetermined cause. Acta Neurol Scand 46 (suppl 44):1–72

Prineas JW, McLeod JG (1976) Chronic relapsing polyneuritis. J Neurol Sci 24:365–377

Putnam JF (1889) A case of acute fatal neuritis of infectious origin with post-mortem examination. Boston Med Surg J 120:159–190

Reich SG (1987) Harvey Cushing's Guillain–Barré syndrome: an historical diagnosis. Neurosurgery 21:135–141

Rivers TM, Schwentker FF (1935) Encephalomyelitis accompanied by myelin destruction experimentally produced in monkeys. J Exp Med 61:689–703

Rivers TM, Sprunt DH, Berry GP (1933) Observations on attempts to produce acute disseminated encephalomyelitis in monkeys. J Exp Med 58:39–52

Rohmer F (1984) Jean-Alexandre Barré (1880–1967) L'homme, sa vie et son oeuvre. Conférences de l'Institut d'Histoire de la Médicine de Lyon Cycle 1983–1984:185–207

Schweiger L (1909) Ueber Veränderungen der Spinal-ganglion in einem Fall von Landryscher Paralyse (mit status hypoplasticus). Dtsch Z Nervenheilkünde 37:35–48

Targowla J (1894) Polynévrite recidivante, envahissement des nerfs craniens et diplégie faciale. Rev Neurol 2:465–472

Waksman BH, Adams RD (1955) Allergic neuritis: experimental disease of rabbits induced by the injection of peripheral nervous tissue and adjuvants. J Exp Med 102:213–225

Waksman BH, Adams RD (1956) A comparative study of EAN in the rabbit, guinea-pig and mouse. J Neuropathol Exp Neurol 15:293–310

Wardrop J (1834) Clinical observations on various diseases. Lancet i:380–382

Westphal C (1876) Ueber einige Fälle von acuter tödlicher Spinallähmung (sogenannter acute aufsteigender Paralyse). Arch F Psychiat 6:765–822

Chapter 2

Immunobiology of the Peripheral Nervous System

Introduction

Peripheral nerves have unique properties which govern the outcome of inflammatory and immunological disorders affecting them. Nerve cells are in danger of being affected by inflammatory disorders anywhere along their extremely lengthy processes. Some motor neurons and dorsal root ganglion cell axons are more than a metre long. Fortunately the axons are wrapped and protected by myelin or Schwann cell cytoplasm. The integrity of the nerve terminals depends on uninterrupted axonal transport which occurs both at slow and fast (over 400 mm/day) rates. RNA is transcribed and protein and membrane materials are synthesised in the cell body and then transported down the axons. However, the traffic is two-way and for instance IgG (Fabian and Petroff 1987) and viruses, including rabies virus, may be carried from the periphery to the central nervous system via a retrograde axonal transport system. Inflammatory disorders have a particular tendency to affect the proximal ventral and dorsal roots. Vulnerable sites are the spinal cord entry zones, the dorsal root ganglia, and the junction of the ventral and dorsal roots where they become tightly invested by collagenous tissue. This junction is a point at which the nerve fibres might be expected to be particularly prone to compression by an oedematous inflammatory process.

The nerve fibres are protected from inflammation and infection in the surrounding tissues by the connective tissue of the *epineurium* and particularly by the circumferentially arranged layers of cells constituting the *perineurium* (Figs. 2.1 and 2.2). These cells abut on each other and are tightly joined together by zonnulae occludentes. A basement membrane invests the inner and outer aspects of these cells) (Fig. 2.3). These anatomical arrangements provide a passive diffusion barrier but an active transport mechanism is probably also present since the perineurial cells are rich in phosphorylating enzymes. These cells also contain numerous pinocytotic vesicles which are often seen opening onto the inner or outer surfaces. After nerve damage myelin debris may be observed within the cells. The perineurium is morphologically equipped to serve as a supporting and protective barrier and to transport material out of the endoneurium (see Thomas and Olsson (1984) for review).

Within the perineurium lies the *endoneurium*, which contains fascicles of myelinated and non-myelinated nerve fibres supported by a matrix of collagen fibres bathed in endoneurial fluid. The fluid is maintained at a pressure slightly higher than that of the epineurial connective tissue, producing a gradient which helps to prevent the endoneurium being contaminated by toxic substances in the neighbouring environment (Powell et al. 1979). The endoneurial fluid has the

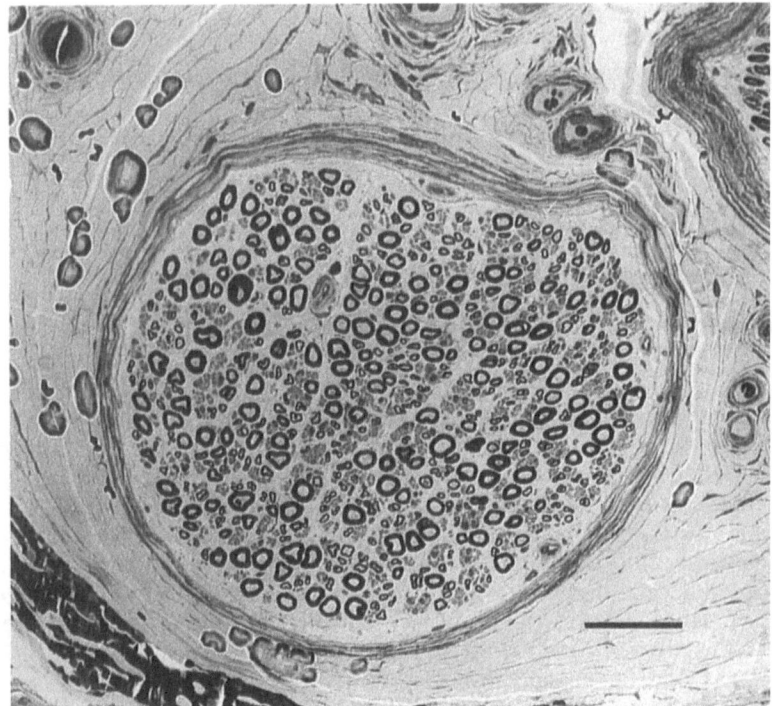

a

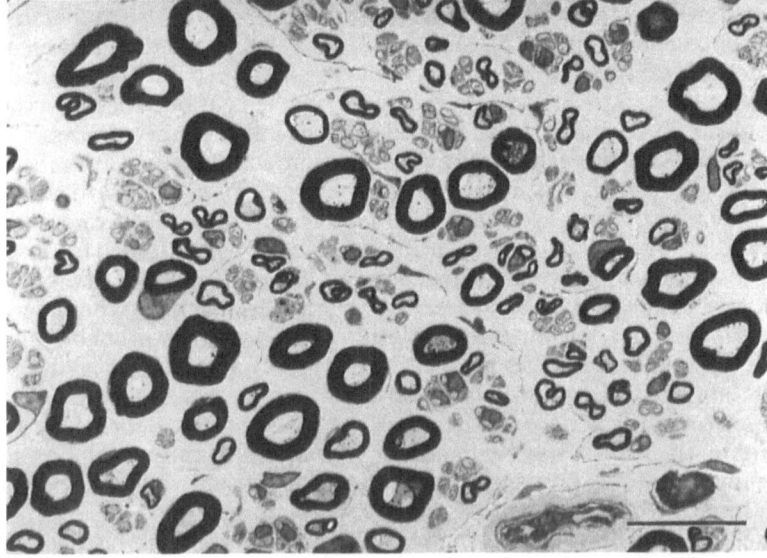

b

Fig. 2.1. Photomicrographs of a fascicle from a sural nerve obtained at biopsy from a normal male volunteer aged 45; 1 μm epon embedded section stained with thionin and acridine orange. **a** Bar = 50 μm. **b** Higher power view, bar = 20 μm. Figures 2.2–2.7 are from this nerve.

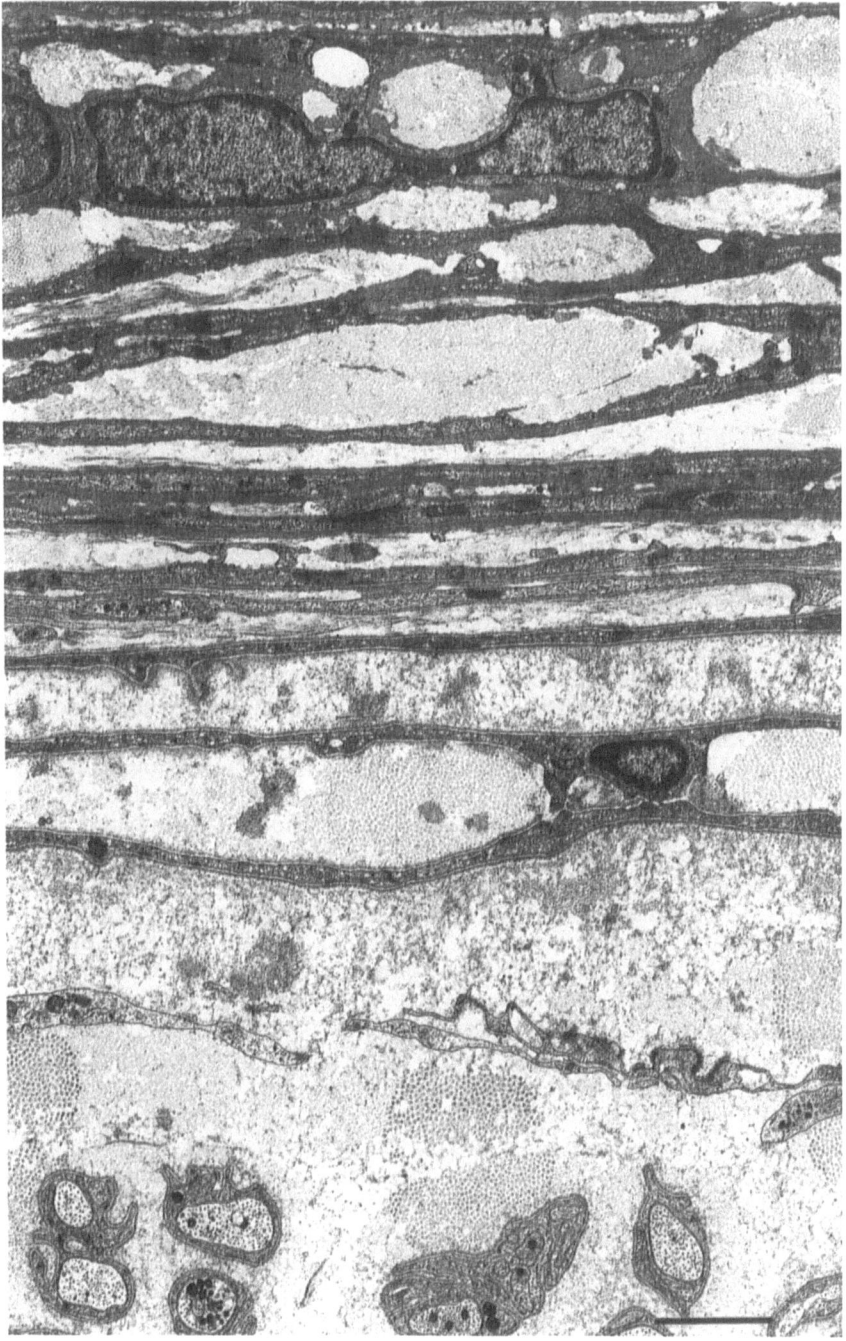

Fig. 2.2a. Electron micrograph of perineurium from the sural nerve. Bar = 2 μm.

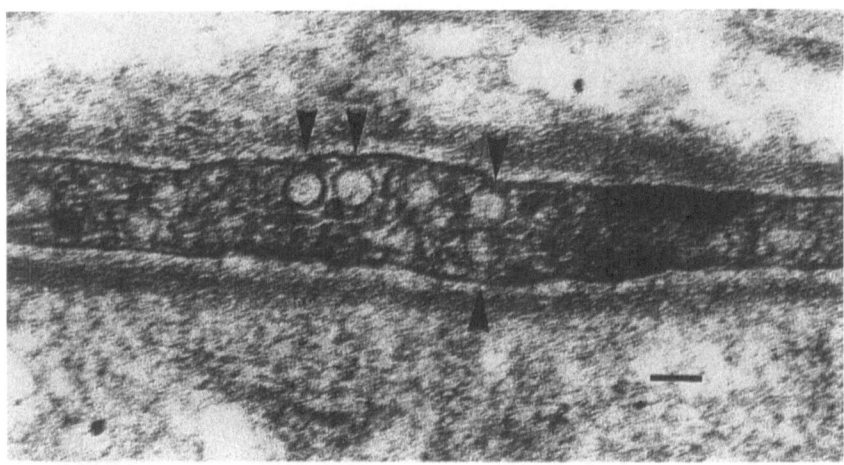

b

Fig. 2.2b. Electron micrograph of perineurium from the sural nerve. Higher magnification to show caveolae (*arrowheads*) Bar = 0.5 μm.

electrolyte characteristics of the fluid bathing the extracellular connective tissue (Low 1984) and is important in maintaining the milieu for nerve conduction. The protein and immunoglobulin content are also very low. The normal endo-neurium contains a limited number of cell types, including myelinating and non-myelinating Schwann cells, fibroblasts, mast cells, and mononuclear cells which resemble tissue histiocytes and are potential phagocytes. The endoneurium has a rich blood supply and in most regions the blood vessels have walls which are impermeable to macromolecular molecules under normal circumstances and constitute the blood–nerve barrier (Fig. 2.3).

In this privileged and protected endoneurial environment lie the nerve fibres themselves. The myelinated nerve fibres can be clearly seen by light microscopy (Fig. 2.1) and vary in diameter from about 1 to 12 μm. The fibre diameter distribution is bimodal (Fig. 2.4). The myelin sheath of each internode consists of the compacted membranes of a single Schwann cell. A thin strip of cytoplasm, the mesaxon of the Schwann cell, is often preserved at the adaxonal and abaxonal surfaces of the myelin sheath (Fig. 2.5). In particular a column of cytoplasm persists at the tip of the external mesaxon, which may mark the point of origin of the original wrapping of Schwann cell cytoplasm when the myelin sheath was formed. Cuffs of cytoplasm also persist at the paranodes, and in the Schmidt–Lantermann clefts, which are obliquely arranged series of expansions of the myelin lamellae of uncertain function along the length of the internodes. A basement membrane surrounds the outer aspect of the Schwann cell cytoplasm. Many demyelinating insults initiate demyelination at the paranodal region or at the Schmidt–Lantermann clefts. Macrophages invade the myelin sheath usually by penetrating the external mesaxon, splitting the myelin sheath at the intra-period line representing the apposed external surfaces of the original Schwann cell surface membranes. The unmyelinated nerve fibres are wrapped by Schwann cells singly or in groups but myelin is not elaborated (Fig. 2.6). The diameter of the axons is unimodal ranging from about 0.5 to 1.5 μm (Fig. 2.7).

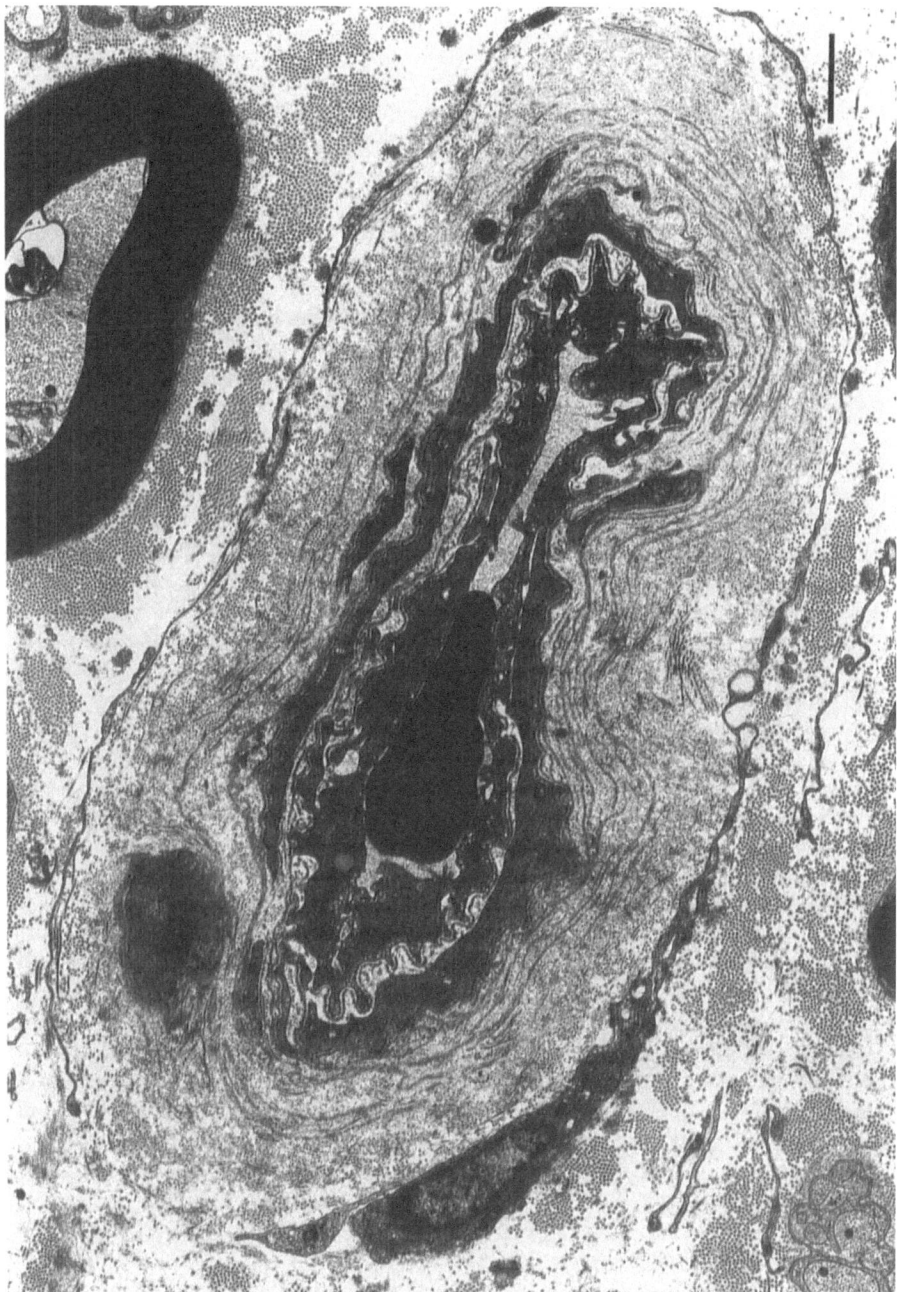

Fig. 2.3a. Endoneurial blood vessel containing an erythrocyte. Bar = 2 μm.

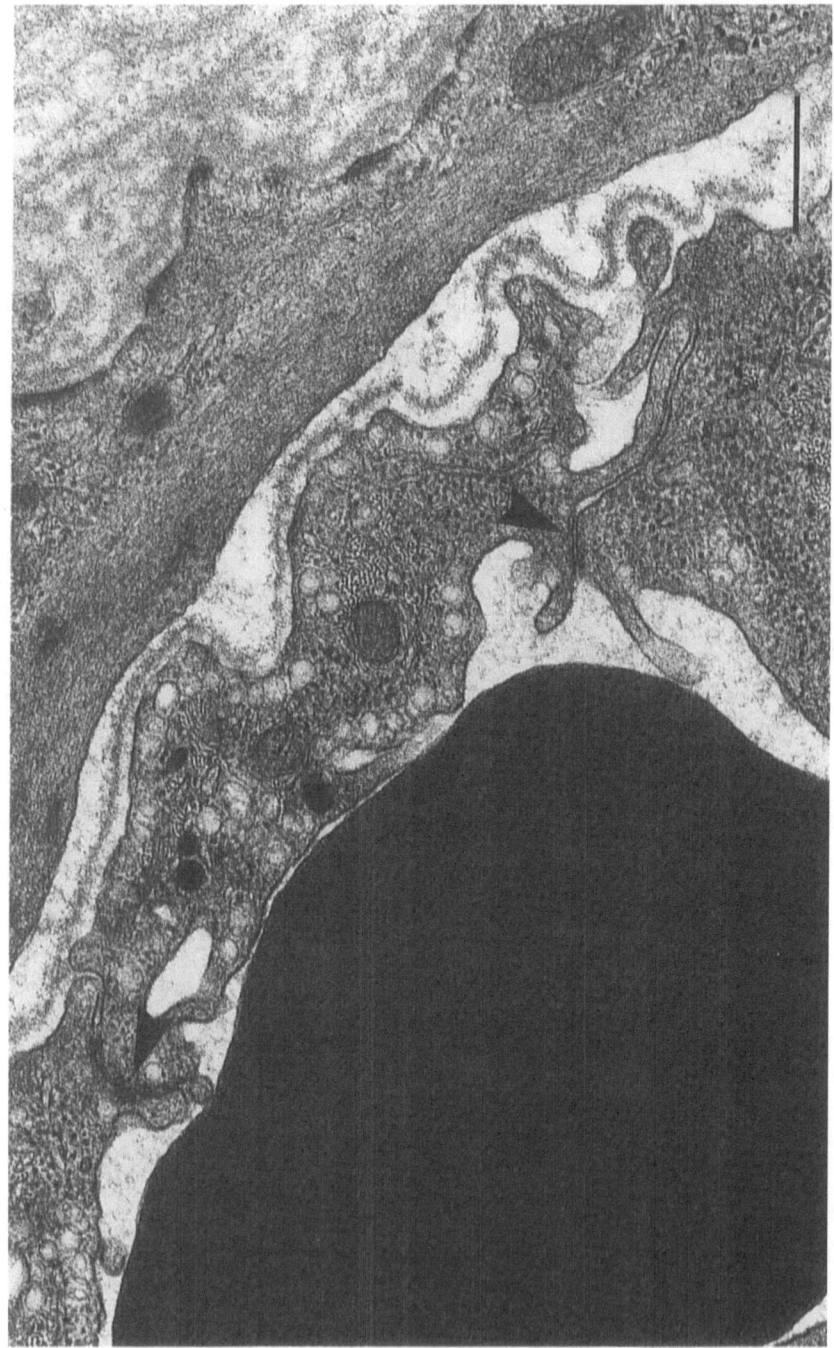

Fig. 2.3b. Enlargement of Fig 2.3a showing tight junctions (*arrowheads*) between endothelial cells. Bar = 0.5 μm.

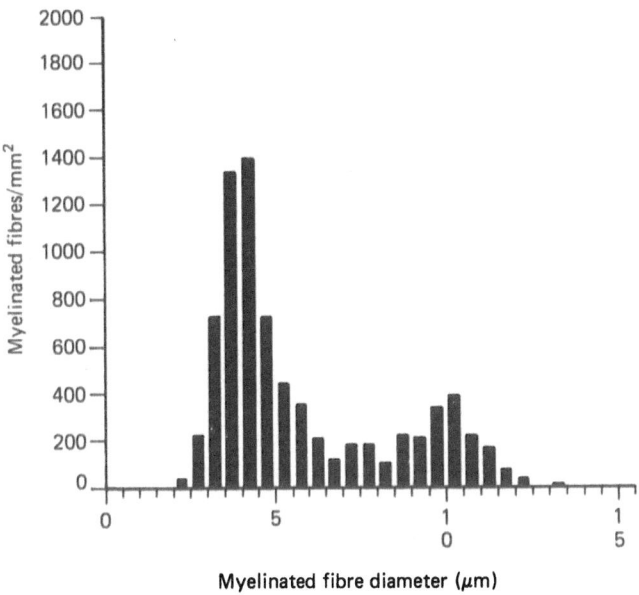

Fig. 2.4. Distribution of the myelinated nerve fibre diameters of the normal human sural nerve.

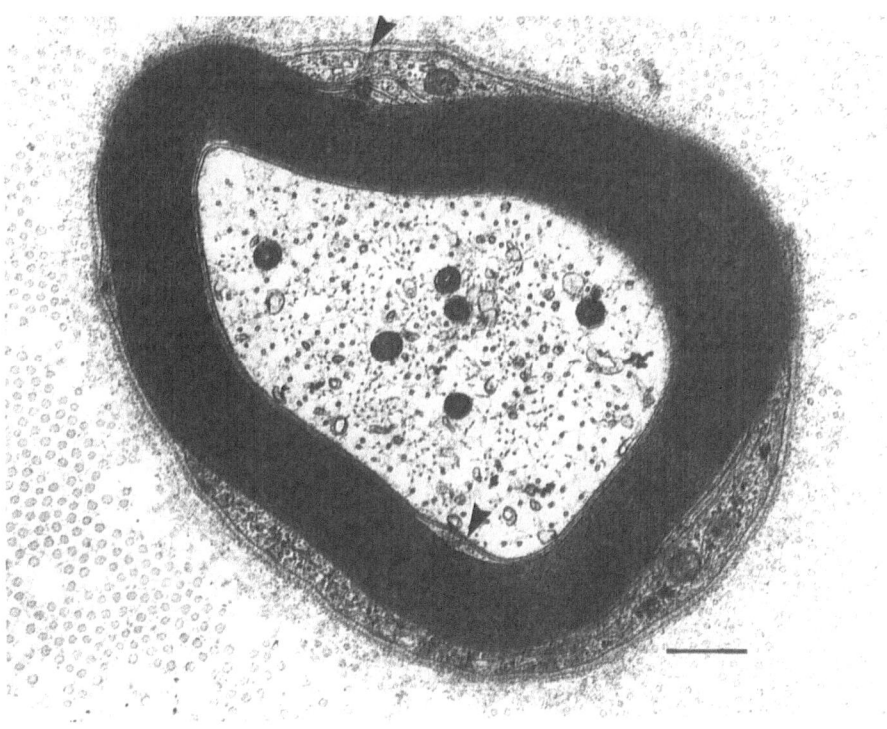

Fig. 2.5. Electron micrograph picture of a transverse section of a human myelinated nerve fibre. Note the external and internal mesaxons (*arrowheads*). Bar = 0.5 μm.

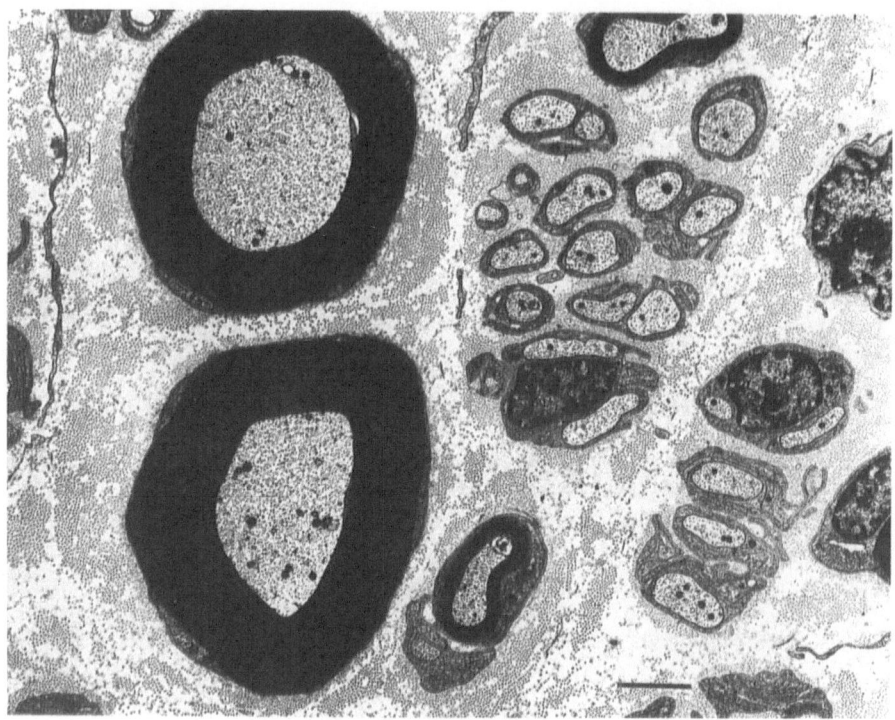

Fig. 2.6. Electron micrograph of myelinated and unmyelinated nerve fibres. Bar = 2 μm.

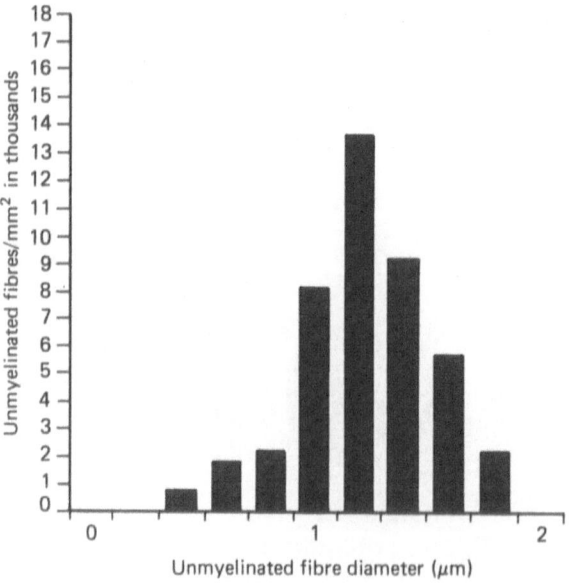

Fig. 2.7. Axon diameter distribution of unmyelinated nerve axons from normal human sural nerve.

Wallerian Degeneration Versus Primary Demyelination

If a myelinated nerve fibre is transected, the distal portion of the axon degenerates and the myelin sheath breaks down into a line of "ovoids" of myelin debris. This debris is phagocytosed and removed, a process called secondary demyelination, which must be differentiated from primary demyelination in which the axon is preserved. The Schwann cells proliferate within the original basal lamina tube. Regenerating axon sprouts grow down the tube and become ensheathed and eventually remyelinated by the Schwann cells. In the early stages of degeneration the axon either appears shrunken so that its organelles appear more tightly packed than normal or swollen and watery. The anterior horn cell or dorsal root ganglion cell of axons which have undergone Wallerian degeneration have to switch on RNA transcription and protein synthesis to meet the increased metabolic demand associated with regeneration. These changes correspond morphologically to the migration of the Nissl substance from a single clump by the nucleus to a peripheral position around the perikaryon. The regenerative process following Wallerian degeneration is slow; although commonly reported to be about 1 mm/day, in clinical practice the regeneration often proceeds at only half that speed and is often incomplete. According to a single report (Schwartz et al. 1982), antibodies to myelin gangliosides and basic proteins are formed following nerve transection and during regeneration. This observation needs to be repeated and the possible significance of these antibodies in delaying regeneration should be explored.

By contrast with axonal degeneration and secondary demyelination, primary demyelination in the peripheral nervous system (PNS) is a relatively rapidly reversible process. Following demyelination by a toxin, such as diphtheria toxin, or an inflammatory process, as in experimental allergic neuritis, the integrity of the axon may be preserved. The demyelinated segments may become wrapped by two or three lamellae of myelin within a few days after demyelination. About 10 wraps of myelin would probably be sufficient to provide electrical insulation and restore saltatory conduction to an otherwise normal fibre (KJ Smith, personal communication). For remyelination to proceed properly the debris of the myelin sheath must first be removed by phagocytes, mostly macrophages (Beuche and Friede 1984). Schwann cells may also participate in the phagocytic process. The macrophages migrate towards the endoneurial capillaries and the perineurium and clear the debris. Following demyelination the Schwann cells proliferate: several Schwann cells become related to each old internode and two or three short new internodes form in place of the single old one. Each new internode is remyelinated by only one axon-associated Schwann cell, and any supernumerary Schwann cells degenerate or migrate away. Under some circumstances, especially following repeated cycles of demyelination and remyelination, the supernumerary Schwann cells persist, overlapping each other in a whorl of variable thickness which has been likened to the cut section of an onion. The concentric layers of Schwann cell processes in these "onion bulbs" are separated from each other by longitudinal arrays of collagen fibres.

Myelin

The peripheral nerve myelin sheath is continuous with and a specialisation of the Schwann cell plasma membrane. It is extremely rich in lipids (Table 2.1) which constitute 70%–80% of its dry weight and produce the glistening white appearance of myelinated nerves. Some of the constituent lipids differ from those elsewhere in the body. For instance galactocerebroside is found almost only in myelin sheaths and oligodendrocyte and Schwann cell membranes. It is immunogenic in rabbits and antibodies to galactocerebroside react with its galactose portion and produce demyelination. Accumulation of galactocerebroside occurs in the autosomal recessive condition globoid cell leucodystrophy (Krabbe's disease) in which demyelinating neuropathy is associated with cerebral leucodystrophy. Sulphatide is present in smaller quantities in myelin, the myelin-producing cells and possibly other glial cells. Antibodies to sulphatide can be produced but the evidence as to whether they have a demyelinating effect is conflicting (see below). Accumulation of sulphatide causes a demyelinating neuropathy and cerebral leucodystrophy in the autosomal recessive condition metachromatic leucodystrophy.

Gangliosides represent only 0.3%–0.7% of total myelin lipids. They are acidic glycosphingolipids containing one or more residues of sialic acid in asso-

Table 2.1. Peripheral nerve myelin lipids

	Dry weight (%)	Structure
Total lipid	71	
Cholesterol	23	
Total galactolipid	22	
Galactocerebroside[a]		ceramide—Glc—Gal—Gal—Gal
Sulphatide[a]		ceramide—Glc—Gal—Gal—Gal—SO$_4$
Ganglioside	<1	

GM$_1$[b]		ceramide—Glc—Gal—NAN
		            |
		          GalNAc
		            |
		          Gal

LM$_1$[b]		ceramide—Glc—Gal—GlcNAc—Gal—NAN

GD1b[b]		ceramide—Glc—Gal—NAN—NAN
		            |
		          GalNAc
		            |
		          Gal

Total phospholipids	55	
Ethanolamine phosphoglycerides	19	
Sphingomyelin	19	ceramide—phosphorylcholine

[a] Most of the galactolipid is galactocerebroside.
[b] LM$_1$ is the major ganglioside: GM$_1$ and GD1b are only present in small amounts and may be derived from the axon.
Glc, glucose; Gal, galactose; NAN, N-acetylneuraminic acid; GalNAc, N-acetylgalactosamine; GlcNAc, N-acetylglucosamine.

ciation with a variety of neutral sugars which generate a family of antigenically distinct molecules (Table 2.1). There are marked species differences in the proportions of gangliosides in myelin and also differences between the PNS and CNS. In rat CNS myelin ganglioside GM_1 is predominant whereas in human CNS myelin GM_4 and GD_{1b} are the major components. In PNS myelin LM_1 is the dominant form of ganglioside (Norton and Cammer 1984; Norton 1985; Quarles et al. 1986)). There is a growing literature concerning an association between antibodies to gangliosides, especially the carbohydrate determinants on GM_1, and lower motor neuronopathy or chronic idiopathic demyelinating poly-radiculoneuropathy with multifocal conduction block (Baba et al. 1989; Marcus et al. 1989). In experiments in which antiserum to ganglioside GM_1 was injected into rat sciatic nerve, demyelination was not produced (Hughes et al. 1985). However, recent evidence raises the possibility that antibodies to minor gang-liosides, or neutral glycolipids sharing epitopes with gangliosides, induce demyelination. Sera from 65% of patients with IgM paraproteinaemia and demyelinating neuropathy react with a glycoprotein called myelin-associated glycoprotein (see below). Such sera usually also react with a recently discovered acidic glycolipid, sulphate-3-glucuronyl-lacto-N-neotetraosylceramide (SGN). SGN lacks sialic acid and is not classified as a ganglioside but resembles the major peripheral nerve ganglioside LM_1, differing only in having a terminal glucuronic acid sulphate group instead of sialic acid (Quarles et al. 1986). About half the patients with neuropathy and IgM paraproteinaemia who do not have antibodies to SGN have antibodies to other myelin glycolipids demonstrable by immuno-overlay on a thin-layer chromatogram of nerve or brain glycolipids.

About 20%–30% of the dry weight of myelin is protein. The most abundant protein in peripheral nerve myelin is a glycoprotein designated P_0, molecular weight 28 000, which is a transmembrane protein whose carbohydrate moiety lies in the myelin intraperiod line. P_0 is not found in central nervous system myelin where the equivalent abundant protein is proteolipid protein. P_0 forms about 50% of peripheral nerve myelin protein and probably has an important structural role in maintaining the integrity of the myelin sheath. P_0 is the only myelin protein currently investigated which is thought to be truly specific for PNS myelin, a point relevant to the consideration of candidate autoantigens in GBS. Injection of rabbit antiserum to P_0 into rat sciatic nerves has been reported to induce demyelination consistent with the idea that antibody to antigen ex-pressed on the surface of the Schwann cell or myelin will induce demyelination (Hughes et al. 1985).

Peripheral nerve myelin contains two main basic proteins, myelin basic pro-tein and P_2. Myelin basic protein (MBP), sometimes called P_1, is also present in CNS myelin and its major form has a molecular weight of 18 500. It has been intensively studied as the major antigen responsible for experimental allergic encephalomyelitis (EAE). The sequence of MBP has been determined and the gene cloned. Some species have duplicated genes with partial deletions so that additional smaller myelin basic proteins are produced, for instance a protein with a molecular weight of 14 000 in mice (Norton 1985). The basic properties of the protein are accounted for by its high proportion of the basic amino acids, lysine, arginine and histidine. The sequence of amino acid residues in MBP is highly conserved from species to species. However, the identity of the encephalitogenic sequence varies from species to species and within strains of the same species (Chapter 3). The reason for these differences might lie in the differing availability of T cell receptors capable of recognising these epitopes or in differing ability to suppress T cell responses. The disease produced by immunisation with MBP, EAE, predominantly affects the CNS. However, lesions do also occur in the nerve roots and are easily explained since MBP is present in peripheral nerve myelin, constituting 2%–16% of the total protein of peripheral nerve myelin in different species (Brostoff 1984).

The second and smaller basic protein in peripheral nerve myelin, P_2, molecular weight 14 500, is the major antigen responsible for experimental allergic neuritis. Bovine, rabbit, and human P_2 proteins have been sequenced: there are 131 amino acid residues which again are highly conserved from species to species (Chapter 3). P_2 protein has an intermolecular disulphide bridge. In aqueous solution it forms an ordered β pleated sheet structure which may differ from the structure in its normal lipid environment. X-ray crystallography studies of the purified protein suggest that the β pleated sheet is arranged like a barrel with its ends partially occluded by short alpha helical chains. This structure and the amino acid sequence show that P_2 is related to a family of homologous cytoplasmic proteins which bind lipid ligands. The X-ray crystallography studies show that the P_2 molecule contains an electron-dense area in the centre of its barrel which may represent fatty acid. It has been proposed that P_2 may be involved in transport of fatty acid into Schwann cells (Jones et al. 1988). Such an explanation does not provide a *raison d'être* for P_2 in compact myelin, where the opportunity for binding between lipid and protein is more likely to have a structural function. P_2 protein forms a variable percentage of peripheral nerve myelin protein ranging from <2% in guinea pigs to 15% in bovine tissue. This percentage also varies in different regions of the PNS being greater in the ventral than the dorsal spinal roots. There are small amounts of P_2 in the central nervous system especially in the spinal cord (Kadlubowski et al. 1984). It is not yet clear whether this is merely due to a few myelin sheaths within the dorsal and ventral root entry zones which are myelinated by Schwann cells or whether oligodendrocytes make small amounts of P_2. A protein isolated from bovine spinal cord by acid extraction and named spinal cord protein by McPherson was subsequently identified as P_2 (Deibler et al. 1978). The amino acid sequence and determinants of P_2 which induce experimented allergic neuritis will be described in Chapter 3.

Myelin-associated glycoprotein (MAG) is a minor glycoprotein component of peripheral and central nervous system myelin. A helpful review of its properties and immunology is provided by Quarles (1984). It contributes less than 1% of myelin protein but plays an important part in Schwann cell– or oligodendrocyte–axon interactions. MAG is a large molecule, molecular weight 100 000, consisting of a cytoplasmic C-terminus, a transmembrane portion and oligosaccharide components. These components form 30% of the molecular weight and project on the extracellular surface. MAG is related in structure to the immunoglobulin supergene family whose other members include the intercellular adhesion molecules, ICAM-1 and NCAM, and P_0. Immunohistochemical studies have localised MAG to the interface between the axon and the myelin-producing cell, either Schwann cell or oligodendrocyte. It is not present in compact myelin but is present at Schmidt–Lantermann incisures, the outer mesaxon and paranodal cytoplasm. These are all situations where the Schwann cell meets the axon or another Schwann cell membrane. It has been proposed that the oligosaccharide tail preserves the 12–14 nm separation between axon and Schwann cell, or

Schwann cell and Schwann cell, while the C-terminal portion prevents the compaction of Schwann cell cytoplasm (Quarles 1984). On this hypothesis MAG must be removed or catabolised when myelin compaction occurs. There are minor differences in biochemistry and antigenicity of MAG between species and some antibodies to human MAG do not cross-react with rat MAG. MAG shares antigenic epitopes not only with the peripheral nerve acidic glycolipids already mentioned, but also with some CNS neurons (Gregson and Leibowitz 1985) and an epitope on human natural killer cells defined by a monoclonal antibody called HNK1 (Quarles 1984).

Blood–Nerve Barrier

The endoneurium is protected from toxic or infective agents in the circulation by a barrier similar to that which guards the CNS. The blood–nerve barrier is formed by the specialised properties of the endoneurial capillaries and perineurium (Olsson 1984). The capillaries have continuous non-fenestrated endothelium with tight intercellular junctions and a surrounding basement membrane. The perineurium has several layers of cells also surrounded by basement membrane and connected to their neighbours by tight junctions. If large molecules such as horseradish peroxidase are injected intravenously they are retarded by these barriers and do not enter the endoneurium as readily as they enter the extracellular space in the epineurium and other body tissues. However, horseradish peroxidase does gradually enter the endoneurium where it is ingested by macrophages which may be regarded as performing a rearguard action mopping up marauding materials/infective agents which penetrate the front line defences. The blood–nerve barrier is very much less effective in the dorsal root and autonomic ganglia (Jacobs et al. 1976). Evans blue injected intravenously stains dorsal root ganglia but not the endoneurium of the peripheral nerves (or the spinal cord and brain parenchyma). Macromolecular tracers have been shown to reach the endoneurium of small peripheral intramuscular nerve branches. This might be because the peripheral ends of the perineurium are open ended or because the distal perineurium or capillaries have different properties. Thus the blood–nerve barrier, though important in protecting the endoneurial environment, is only relative, and dorsal root ganglia and distal parts of peripheral nerves theoretically might be more prone to damage by immune reactions in the blood stream. Under normal circumstances the blood–nerve barrier separates the endoneurial mast cells from circulating substances which cause mast cell degranulation. These mast cells lurk like mines ready to discharge their granules loaded with vasoactive amines. The discharge can be triggered experimentally by intraneural injection of an ionophore (Powell et al. 1980). In this experiment degranulation of the mast cells was associated with disruption of the blood–nerve barrier, endoneurial oedema, and increased endoneurial fluid pressure. The blood–nerve barrier also prevents substances which have penetrated the endoneurium from leaving. This escape route is provided in other tissues by the lymphatic system which the endoneurium lacks. Instead the endoneurial macrophages phagocytose myelin debris and extraneous material and migrate towards the endoneurial capillaries and perineurium. The perineurial cells also have a phagocytic capability and play a role in clearing debris. The tight investment by perineurium and slowness of transport across the blood–nerve barrier contribute to the increase in endoneurial fluid pressure which follows injury to peripheral nerves. Vascular permeability and endoneurial fluid pressure increase throughout a nerve segment where axons are undergoing Wallerian degeneration following a proximal lesion (Myers et al. 1981).

In addition to lacking a lymphatic drainage it is possible the normal endoneurium has little, if any, lymphocyte traffic. Most tissues are thought to permit an immunological surveillance mechanism whereby T-helper cells in particular can encounter antigens and recruit other lymphocytes and macrophages to the site of the encounter. Although some mononuclear cells are found by light or electron microscopy in the normal endoneurium, their morphological characteristics are more suggestive of monocytes or macrophages than lymphocytes. In immunohistochemical studies of frozen sections of normal rat sciatic nerve or spinal root endoneurium, we have not been able to find T cells with the monoclonal antibody OX19 (Hughes et al. 1987). In similar studies of human nerve biopsies, T cells have not been found except in pathological situations where inflammatory cells would be expected. Under normal conditions lymphocyte traffic must be a rare event but if the blood–nerve barrier is breached lymphocytes can accumulate rapidly. It has been proposed that activated lymphocytes are able to cross the blood–brain barrier whereas resting lymphocytes are not. The evidence in support of this is that radiolabelled activated T lymphocyte line cells directed against an irrelevant antigen such as ovalbumin injected intravenously can be detected by radioautography in the brain of the recipient rats (Wekerle et al. 1986a,b). Such a mechanism provides a convenient and economical explanation permitting only those lymphocytes which are ready for action to perform this surveillance role.

Antigen Presentation in the Peripheral Nervous System

It is now accepted immunological dogma that cell-mediated immune reactions have to be initiated by the presentation of antigen in the context of immunologically recognisable self markers called major histocompatability (MHC) antigens. There are two classes of these. MHC class I antigens are present on all nucleated cells in the body and are recognised by cytotoxic T cells. MHC class II antigens are present on a more limited range of cells including some macrophages and specialised tissue dendritic cells. Activated T lymphocytes and B cells also express MHC class II antigen. In the rat, low levels of MHC class I antigen expression by all endoneurial and perineurial cells can be demonstrated with a monoclonal antibody and an avidin–biotin–peroxidase labelling system. Some endoneurial cells, probably resident macrophages, and some perineurial cells constitutively express more MHC class I antigen. MHC class I antigen expression is also evident on the endothelial lining of small and large blood vessels. In normal rats MHC class II antigen expression is shown with the appropriate monoclonal antibody to be much more restricted: staining is observed on only a few endoneurial and perineurial cells. The positively labelled endoneurial cells are again probably resident macrophages. The endothelial cells and Schwann cells probably do not express MHC class II. These observations indicate that under normal conditions T-helper cells will not be able to recognise and respond to antigen in the endoneurium unless they escape from the vascular compartment and encounter their relevant antigen on the surface of one of the MHC class II antigen-bearing endoneurial cells. On the other hand cytotoxic T cells do have access to their appropriate MHC antigen, the class I antigen, in the surface of the endothelial cells. If, as has been proposed, but not proved, endoneurial, especially myelin, antigens are presented at the endothelial cell surface, the stage is set for cytotoxic T cells to initiate an immune reaction which might disrupt the blood–nerve barrier, stimulate the secretion of lymphokines and trigger the inflammatory cascade.

The nature of the endoneurial cells in the rat which express MHC class II antigens has not been unequivocally established but they are likely to be endoneurial macrophages. In culture a few Schwann cells can be stimulated with γ-interferon to express MHC class II antigen weakly and to present an antigen, MBP, to a T cell line. However, they present antigens less efficiently than professional antigen presenting cells (Wekerle et al. 1986b). Although Schwann cells are more numerous, the resident endoneurial macrophages are probably more important in presenting antigen and triggering the development of peripheral nerve autoimmune disease.

Final Common Pathway of Immunologically Mediated Demyelination

In inflammatory demyelinating lesions of peripheral nerves the final common pathway is macrophage invasion of the Schwann cell basal lamina, penetration of myelin lamellae, phagocytosis of myelin and debris, and stripping of the axon. It is difficult to tell whether demyelination has arisen from direct disruption of the myelin sheath or as a secondary consequence of damage to the myelin-producing cell. The macrophage invasion, phagocytosis and stripping occur rapidly and appear to be aimed directly at the myelin. It is usual to see healthy Schwann cell cytoplasm within the basal lamina of demyelinated axons but this does not rule out the previous occurrence of Schwann cell necrosis since Schwann cells hypertrophy and proliferate rapidly in response to demyelination. Schwann cell cytoplasm sometimes appears watery, but this change may be reversible. Sometimes necrotic cells in the position of Schwann cells are observed, but the very fact that they are necrotic makes their identification problematical.

The experimental injection of an ionophore into mouse or rat sciatic nerve selectively opened calcium channels and provided an informative model in which myelin disruption occurred without the inevitable death of Schwann cells (Smith and Hall 1988). Within an hour vesicular disruption of myelin had occurred at the paranodes and Schmidt–Lantermann incisures. The disruption spread within 24 hours to involve the whole thickness of the myelin sheath while the Schwann cells showed only minor reversible changes and the axon remained intact (Figs. 2.8 and 2.9). Concurrently some fibres were invaded by macrophages. Within a week the invasion of fibres by macrophages was extensive. By

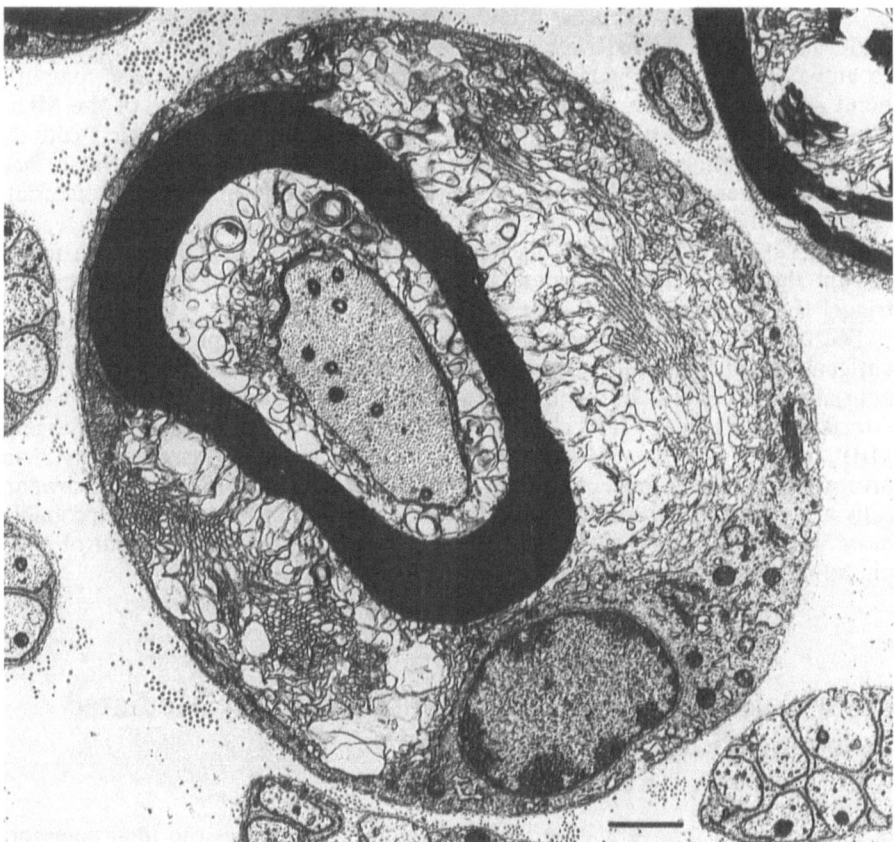

Fig. 2.8. Myelin vesicular change 24 hours following injection of a calcium ionophore. Bar = 1 μm.
(From Smith and Hall (1988) with permission.)

two weeks the macrophages had phagocytosed the myelin debris and migrated
out of the basal lamina into the endoneurium. In this model the initial myelin
vesiculation might be a consequence of a non-lethal rise in Schwann cell calcium
concentration or the direct activation of myelin lipases or proteases. When
activated, phospholipase A_2 breaks down myelin phospholipid into lysophosphatidylcholine which has detergent properties. The calcium ionophore model
induced demyelination which closely resembled that previously demonstrated
following intraneural injection of lysophosphatidylcholine (Hall and Gregson
1971). Westland and Pollard (1987) have produced a similar rapid demyelination
by intraneural injection of proteinase K. Myelin itself contains a neutral
protease which could break down myelin basic protein.

   Agents which kill Schwann cells also cause demyelination but more slowly
than the agents which disrupt myelin directly. Tunicamycin inhibits protein
glycosylation and intraneural injection induces demyelination after 3 days. The
chelating agents mitomycin C and doxorubicin inhibit DNA-directed RNA synthesis: their intraneural injection causes Schwann cell death and, after six to ten
days, demyelination (Westland and Pollard 1987).

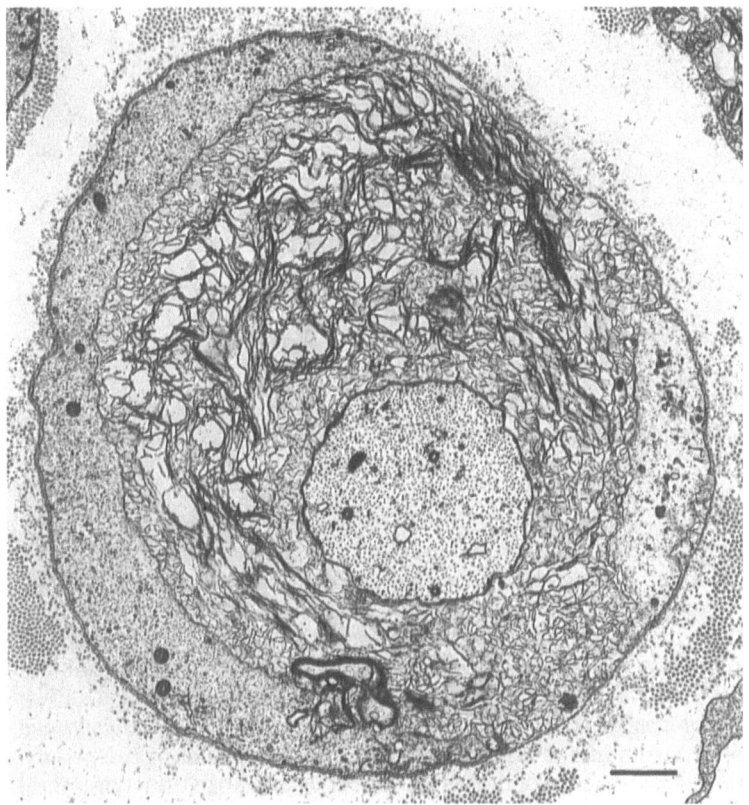

Fig. 2.9. Myelin vesicular change 48 hours following injection of a calcium ionophore. Bar = 1 μm. (From Smith and Hall (1988) with permission.)

Antibody-Mediated Demyelination

For many years classical antibody-mediated complement dependent tissue damage has been invoked as a mechanism for demyelination (Fig. 2.10). The evidence that demyelinating antibodies exist in human diseases such as multiple sclerosis remains scanty but serum from animals with EAE or EAN has been shown to induce demyelination in tissue cultures or following intrathecal or intraneural injection (Seil 1977). The identity of the antigens responsible for the demyelinating effect in these experimental antisera has not yet been fully clarified. Theoretically any antigen present on the Schwann cell membrane might provide an adequate target for the attachment of complement-fixing antibodies and fixation of the first component of complement. This will activate the classical complement pathway, with the successive activation of C4, C2 and C3 and then the formation of the C5–C9 membrane attack complex which has the property of forming a pore in the cell membrane. Such a pore will be larger than that produced by the selective calcium ionophore in the experiments described above and may cause Schwann cell lysis or myelin damage by osmotic disruption as

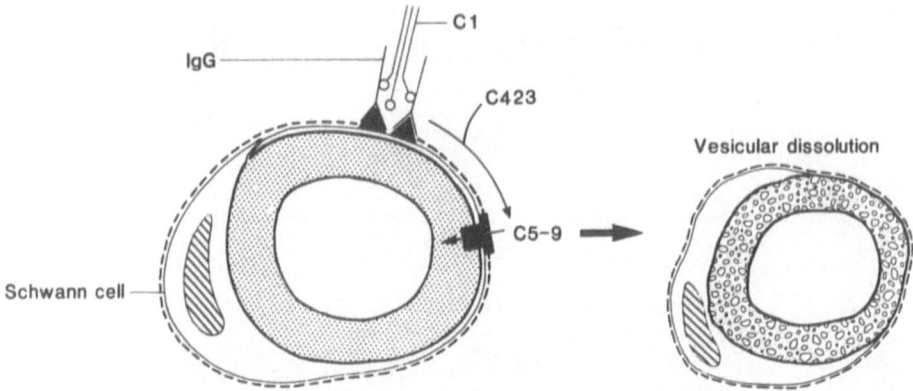

Fig. 2.10. Diagram of complement-fixing antibody mediated demyelination. Complement-fixation can be achieved by two IgG molecules as shown or by a single IgM molecule.

sodium ions follow their concentration gradient. The antibodies which have been shown to produce complement-dependent demyelination are all directed against antigens expressed on the cell surface (Table 2.2).

It is possible that Schwann cells and myelin are not passive participants in immunological reactions but may possess receptors which permit the attachment of immunologically active molecules. For instance Vedeler (1987) has shown that Schwann cells have receptors for the Fc portion of IgG molecules which permit the non-specific binding of antibodies to their surfaces, a possibility which has to be considered since the blood–nerve barrier is not absolute. Peripheral nerve myelin does not bind the C1 component of complement (although CNS myelin does) but will activate the alternative pathway in vitro (Koski et al. 1985). The P_0 molecule is probably responsible. In addition Schwann cells

Table 2.2. Antibody-induced complement dependent demyelination

Antibody directed against	Demyelination produced in	
	CNS	PNS
Galactocerebroside[a]	+	+
Sulphatide[a,b]		0/+
Ganglioside GM_1[a]		0
Myelin associated glycoprotein[c]	0	+
Myelin/oligodendrocyte glycoprotein (CNS only)	+	0
P_0[a] (PNS only)	0	+
Myelin basic protein	0	0
P_2 (mostly PNS)	0	0

All these antigens are present in CNS and PNS myelin except where shown.
[a] Present at the surface of the Schwann cell and/or oligodendrocyte.
[b] No demyelination in the rat, demyelination in the mouse (see text).
[c] Demyelination associated with antibodies to myelin associated glycoprotein (MAG) which is present in CNS and PNS myelin may be due to a cross-reaction with a glycolipid which is confined to peripheral nerve (see text).

possess CR1 complement receptors to which the activated complement component C3b will attach. These properties provide a mechanism for myelin or myelin breakdown products to initiate complement fixation and so inflammation in the absence of specific antibodies to myelin or Schwann cell antigens.

Demyelination induced by antibodies to galactocerebroside has been more thoroughly studied than that induced by antibodies to other antigens. In 1979 Saida et al. showed that rabbits immunised repeatedly with galactocerebroside would develop weakness after several months which was shown by electrophysiological techniques to be due to multifocal conduction block. Pathologically the rabbits had multifocal demyelinating lesions in the spinal roots, especially close to the dorsal root ganglia, and to a lesser extent in the distal peripheral nerves (Saida et al. 1981). Rabbits with galactocerebroside neuritis had high titres of circulating antibody to galactocerebroside. Rabbits are much better at producing anti-galactocerebroside antibodies than other species studied so far and attempts to produce galactocerebroside neuritis in rats, guinea pigs and monkeys have failed. In one unconfirmed report repeated injections of galactocerebroside into monkeys induced a paralytic illness with histological evidence of axonal degeneration (Yonezawa et al. 1981). Curiously, although CNS myelin is also at risk, galactocerebroside immunisation does not induce CNS demyelination. Although active immunisation with galactocerebroside does not induce demyelination in the CNS, injection of anti-galactocerebroside serum into cat, rat or guinea pig optic nerves does (Carroll et al. 1984; KJ Smith personal communication 1990; Sergott et al. 1984). This discrepancy could be accounted for by the relative leakiness of the blood–nerve barrier, especially in the region of the dorsal root ganglia, compared with the blood–brain barrier. It has been shown that complement-fixing antibodies against galactocerebroside are largely directed against the galactose portion of the molecule (Gregson et al. 1971). Injection of anti-galactocerebroside serum into rat nerves induces demyelination (Saida et al. 1979; Powell et al. 1984; Hughes et al. 1985).

Glucocerebroside is identical to galactocerebroside except that the galactose is substituted with glucose. However, antibodies to glucocerebroside did not induce demyelination following intraneural injection in the rat (Hughes et al. 1985) but did in the mouse (Gregson NA and Hall SM, personal communication). Intraneural injection of anti-galactocerebroside serum causes morphological changes and sometimes necrosis of Schwann cells as well as myelin damage (Saida et al. 1979, 1981; Powell et al. 1984; Hughes et al. 1985). Since both myelin and Schwann cell membrane contain galactocerebroside it is likely that both would be damaged simultaneously by anti-galactocerebroside antibodies. However, the rapid time course of anti-galactocerebroside antibody-induced demyelination resembles that induced by calcium ionophores rather than Schwann cell poisons.

The other myelin glycolipids in Table 2.2 have been less intensively studied than galactocerebroside. Direct intraneural injections of rabbit antisera to sulphatide or ganglioside GM_1 did not induce demyelination in the rat (Hughes et al. 1985). Anti-sulphatide antibody injected into the mouse sciatic nerve did induce demyelination but anti-ganglioside GM_1 did not (Gregson NA and Hall SM, personal communication). Nagai et al. (1976) reported that active immunisation with gangliosides and lecithin in complete Freund's adjuvant will cause a paralytic illness characterised histologically by extensive Wallerian degeneration in peripheral nerves. However, a characteristic autoimmune disease induced by gangliosides has not yet been described. Rabbits produce antibodies to glycolipids more readily than mice, rats or guinea pigs. Experiments in which lipids are used to immunise rabbits have to be interpreted with caution since immunisation of rabbits with Freund's complete adjuvant alone has been shown to cause endoneurial oedema and demyelination in the dorsal root ganglion (Mizisin et al. 1987). There are no reports of experimental autoimmune disease following immunisation with sulphatides.

The conclusion that antibodies to myelin associated glycoprotein (MAG) cause demyelination largely depends on the circumstances that most patients with an IgM paraprotein directed against this antigen have a demyelinating sensory and motor neuropathy. Although most patients with this syndrome have a postural tremor, overt clinical signs of CNS demyelination do not occur. MAG is present in both CNS and PNS myelin but shares epitopes with a glycolipid which occurs only in PNS and not CNS myelin. Reactivity with this glycolipid would explain why the clinical syndrome is a predominantly demyelinating peripheral neuropathy sparing the CNS. Experimental demonstration of the demyelinating effect of anti-MAG serum has been difficult to provide because antibodies to human MAG do not cross-react with rodent myelin. Recently Hays et al. (1987) demonstrated that intraneural injections of anti-MAG serum induce demyelination in cat sciatic nerve. In many patients with anti-MAG antibodies, IgM paraprotein and demyelinating neuropathy, there is a characteristic myelin ultrastructure in which there is a widening of the intraperiod line, due to failure of compaction of the external Schwann cell cytoplasmic surfaces (Chapter 12). A similar ultrastructural appearance is observed in EAE and EAN and also produced in tissue culture by anti-myelin antibodies in the absence of complement (Raine and Bornstein 1979; Raine et al. 1981).

Recent experiments have demonstrated that antibodies to a CNS-specific oligodendrocyte surface glycoprotein, MOG (myelin-oligodendrocyte glycoprotein), will induce CNS demyelination. Monoclonal antibodies to this antigen administered intravenously to Lewis rats with passive EAE induced by a myelin basic protein specific cell line produced extensive demyelination, and more severe clinical disease, than the cell line alone, which produced cellular infiltration but not demyelination (Lassmann et al. 1988; Linington et al. 1988). Although not directly relevant to the PNS this model serves to support the concept that complement-dependent antibody mediated attack on myelin or myelin-forming cells affects the PNS or CNS individually according to the specificity of the antibody.

Antibodies to the major peripheral nerve glycoprotein, P_0, produced demyelination following intraneural injection into rat sciatic nerves (Hughes et al. 1985) (Fig. 2.11). Since the oligosaccharide portion of the P_0 molecule is present at the external surface of the Schwann cell this observation is consistent with the notion that antibodies to myelin antigens will induce demyelination if their epitopes are present on the surface of a myelin-producing cell. However P_0 is difficult to purify and no other work on the demyelinating effect of anti-P_0 serum has been published except for the observation that active immunisation of Lewis rats with P_0 will induce an EAN-like disease (Milner et al. 1987). Antibodies against myelin basic protein and P_2, both of which are present in the major dense line of myelin, and are not expressed on the surface of the myelin producing cell, do not induce demyelination (Seil et al. 1981) (Fig. 2.12). Myelin basic protein is the major antigen responsible for inducing experimental allergic encephalo-

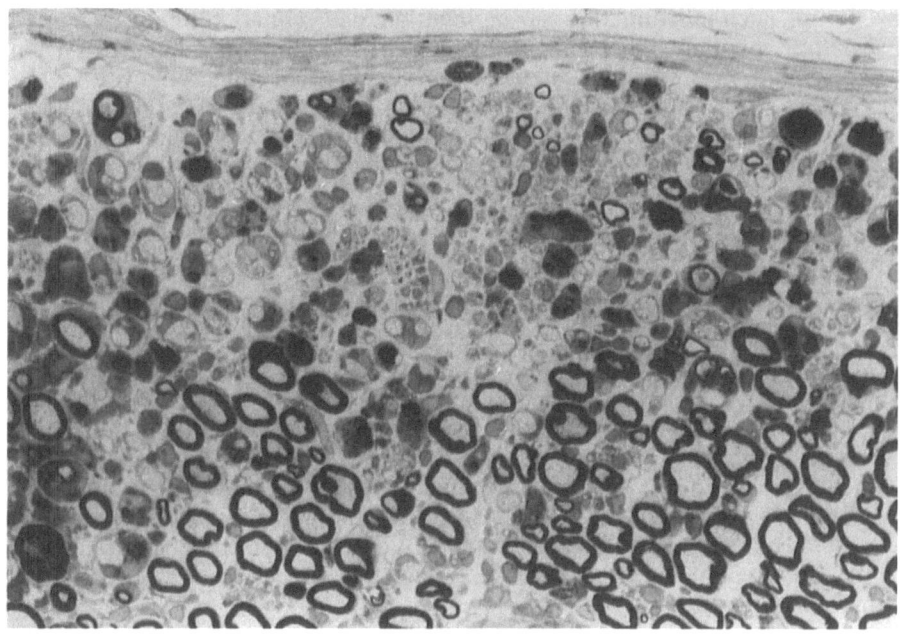

Fig. 2.11. Light micrograph of rat sciatic nerve following intraneural injection of rabbit anti-P_0 serum. One micrometre paraphenylene-diamine stained section. (From Hughes et al. (1985) with permission.)

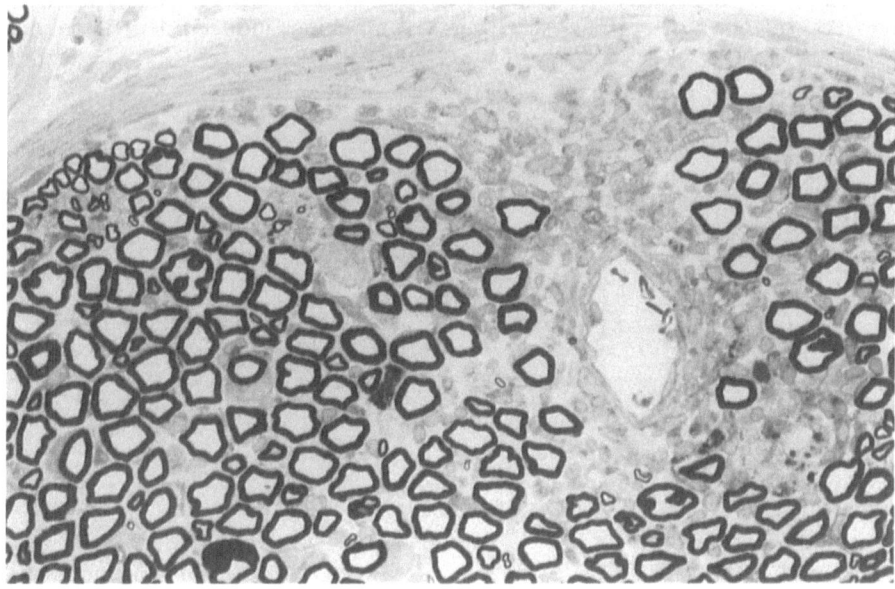

Fig. 2.12. Light micrograph of rat sciatic nerve following intraneural injection of rabbit anti-P_2 serum. One micrometre paraphenylene-diamine stained section. (From Hughes et al. (1985) with permission.)

myelitis in which inflammation is prominent but demyelination relatively limited unless the inoculum is boosted with lipids. In the peripheral nervous system active immunisation with P_2 induced prominent demyelination even in the absence of lipids (Hughes and Powell 1984; Rostami et al. 1984). Passive immunisation with a P_2-specific T cell line (Linington et al. 1984; Izumo et al. 1985) also induces demyelination which is probably cell- rather than antibody-mediated.

Antibody-Dependent Cell-Mediated Demyelination

Antibodies are known to be able to target macrophages or K-cells to kill cells bearing a relevant antigen on their surface, a process called antibody-dependent cell-mediated toxicity (ADCC) (Fig. 2.13). The antibody attaches itself to the antigen as usual by its $F(ab)_2$ binding sites. Fc receptors on the macrophage or K-cell surface then bind the Fc portion of the immunoglobulin molecule. The cells which induce ADCC bear activated C3 receptors and the presence of C3b on the target enhances the cytotoxic reaction although it is not essential (Perlmann et al. 1975).

Experiments with the rabbit eye model have implicated ADCC in CNS demyelination. Brosnan et al. (1977) took advantage of the fact that substances injected into the vitreous have direct access to myelinated nerve fibres on the retinal surface of the rabbit eye. Injection of the supernatant from activated (concanavalin A stimulated) lymphocytes into the vitreous induced mononuclear cell infiltration but not demyelination. Injection of anti-myelin serum alone had no effect. Injection of both antiserum and lymphokines induced demyelination

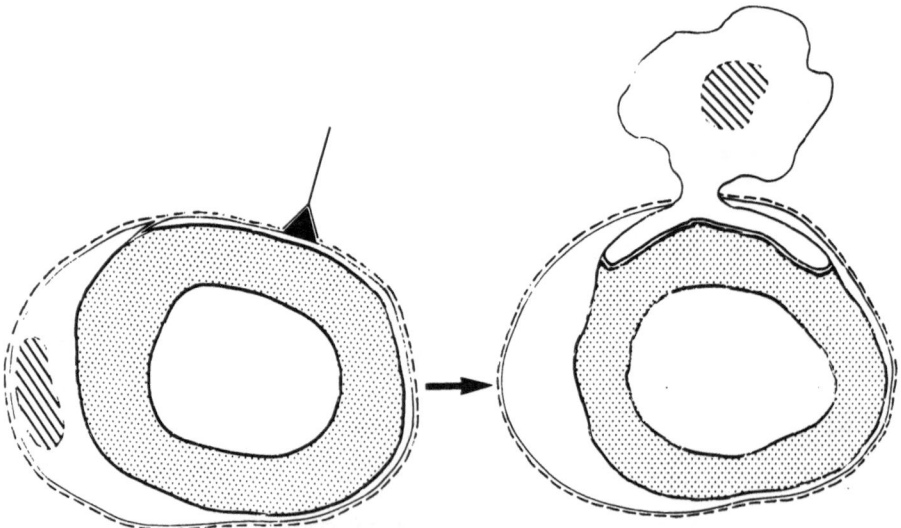

Fig. 2.13.   Diagram of antibody dependent cell-mediated demyelination.

as well as cell infiltration. This demyelination was probably induced by ADCC since the activated cells were not sensitised to a myelin antigen and the same experiment performed with an antiserum to an irrelevant antigen did not cause demyelination. In Lewis rats the intraneural injection of activated peritoneal exudate cells and heat-inactivated anti-galactocerebroside serum has been described as inducing demyelination whereas the injection of cells alone or antisera to MBP, P_2, or irrelevant antigens did not induce demyelination (Saida et al. 1987). The conclusion was that the demyelination required both the participation of cells and antibody and a mechanism akin to ADCC was invoked. This hypothetical mechanism has not been sufficiently studied but would provide a convenient explanation for the morphological findings in human disease.

Cell-Mediated Cytotoxicity Causing Demyelination

Some T cells, usually those expressing the CD8 surface antigen, have the property of killing the cell bearing the antigen which they recognise (Fig. 2.14). This killer function requires the simultaneous presentation of the class I major histocompatability antigen (MHC class I) and the relevant antigen by the target cell. The MHC class I antigen is expressed by all the nucleated cells in the body but the level of expression may be up-regulated by lymphokines and particularly interferon. Such up-regulation is therefore usual in any inflammatory disorder. The existence of cytotoxic T cell mediated demyelination has been hypothesised to explain human and experimental demyelinating disease but has never been proved to play a role. The precise mechanisms by which cytotoxic T cells kill

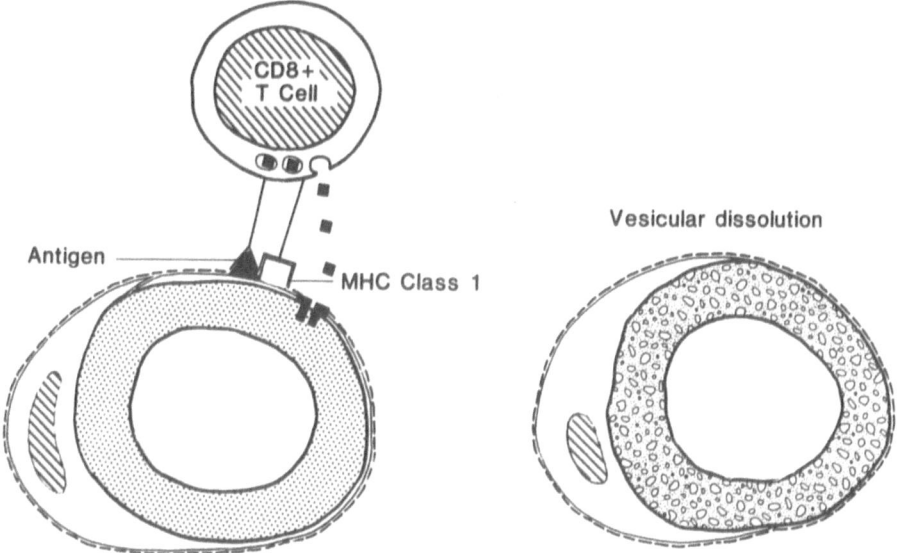

Fig. 2.14. Diagram of cell-mediated cytotoxicity causing demyelination.

target cells is not fully understood and may be multiple (Young and Liu 1988). One favoured mechanism is the calcium-dependent release of pore-forming proteins (perforins) from granules in the cytotoxic T cells. Three or four perforin molecules inserted into the membrane of the target cell produce a homopolymer with porelike properties sufficient to cause osmotic damage to the cell. The aggregation of more perforin molecules forms even larger pores which have internal diameters as large as 5–20 nm visible with the electron microscope. The perforin-induced pore has many similarities to that produced by the membrane attack complex of complement, and perforin and complement component C9 have considerable immunological cross-reactivity. There is also a suggestion of similarity with ADCC in that a monoclonal antibody against membranes damaged by cytotoxic cells will cross-react with the complement membrane attack complex and block ADCC (Ward and Lachmann 1985). Since lympho-cytes can exert cytotoxic effects in the absence of perforin or calcium other mechanisms must exist. These may include the release of tumour necrosis factor and the activation of endogenous endonucleases. Whatever the mechanism of action the exquisite specificity of cytotoxic T lymphocytes for their target antigen should stimulate further investigation of their role in primary demyelination.

Delayed Hypersensitivity to Myelin Antigens Causing Demyelination

The major mechanism responsible for demyelination in EAN has often been considered to be delayed hypersensitivity. The exact means by which delayed hypersensitivity to a myelin antigen might induce demyelination have not been elucidated. EAN will be discussed in detail in the next chapter and here it is only pertinent to explain that primary demyelination has been produced in Lewis rats by intravenous injection of an anti-P_2 T lymphocyte line bearing the CD4 surface marker which characterises T helper cells (Linington et al. 1984; Izumo et al. 1985). There is an interval of 6–7 days before the disease begins which gives ample time for the transferred T helper cells to recruit host immune responses. These might be any of those already mentioned, B cell production of antibodies to P_2 (although these have not been shown to be demyelinating), antibody-dependent cell-mediated immunity, or cytotoxic T cells. In addition activated T helper cells secrete lymphokines such as tumour necrosis factor which might damage myelin directly (Brosnan et al. 1988) or which localise and activate macrophages. The accumulation of macrophages is an obvious feature of delayed hypersensitivity reactions in general and EAN in particular. However, the mere presence of activated macrophages in the endoneurium is not sufficient to cause demyelination *per se* and there is no known mechanism whereby T helper cells against a myelin antigen can confer such exquisite sensitivity on a macrophage that it will invade the Schwann cell base membrane and invade the external mesaxon of the myelin sheath as has been frequently demonstrated in EAN and GBS. There is either some specific means of transferring specificity from T helper cells to macrophages which we do not yet understand or one of the other more specific mechanisms must be operating.

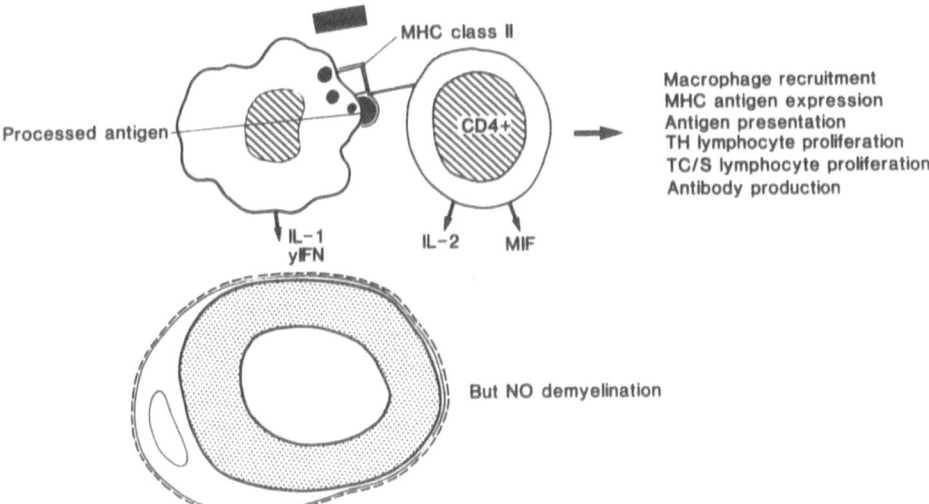

Macrophage recruitment
MHC antigen expression
Antigen presentation
TH lymphocyte proliferation
TC/S lymphocyte proliferation
Antibody production

But NO demyelination

Fig. 2.15. Diagram of immune response to an extraneous antigen introduced into peripheral nerve which produces inflammation and may cause axonal damage but does not cause primary demyelination.

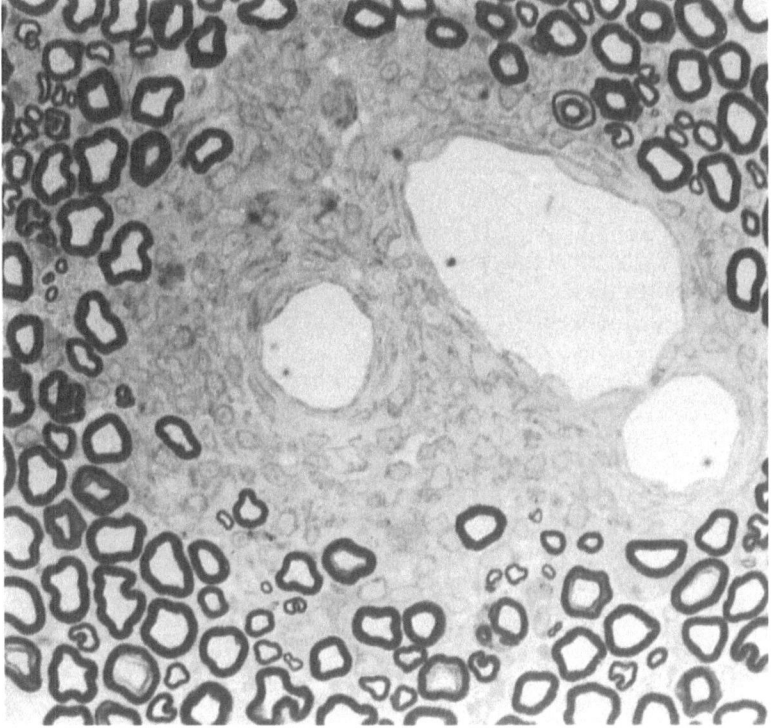

Fig. 2.16. Tuberculin reaction in the sciatic nerve of a rat not immunised with galactocerebroside. (From Powell et al. (1984) with permission.)

Bystander Demyelination

Theoretically demyelination might be induced by the release of proteases or
other factors in the vicinity of an inflammatory reaction or an immune reaction
against a foreign antigen which has been introduced into the peripheral nerve
(Fig. 2.15). Such a mechanism was postulated by Wisniewski and Bloom (1975)
who detected myelin destruction in the vicinity of tuberculin injected into the
brain, spinal cord or peripheral nerves of guinea pigs previously sensitised by
immunisation with tuberculin and Freund's complete adjuvant. Such "bystander
demyelination" is frequently invoked as a theoretical mechanism which would
explain human demyelinating disease. Nevertheless it is not absolutely clear that
the myelin destruction surrounding the tuberculin reactions of Wisniewski and
Bloom (1975) was due to primary demyelination. Much of the myelin loss which
they illustrated was due to axonal damage and secondary myelin breakdown.
Subsequent similar experiments by the same authors have failed to confirm the
occurrence of primary demyelination in association with immune reactions
against foreign antigens introduced into the nervous system: an inflammatory
reaction induced by activated lymphocyte supernatant introduced into the rabbit
eye vitreous did not cause primary demyelination (Stoner et al. 1977). Tuber-

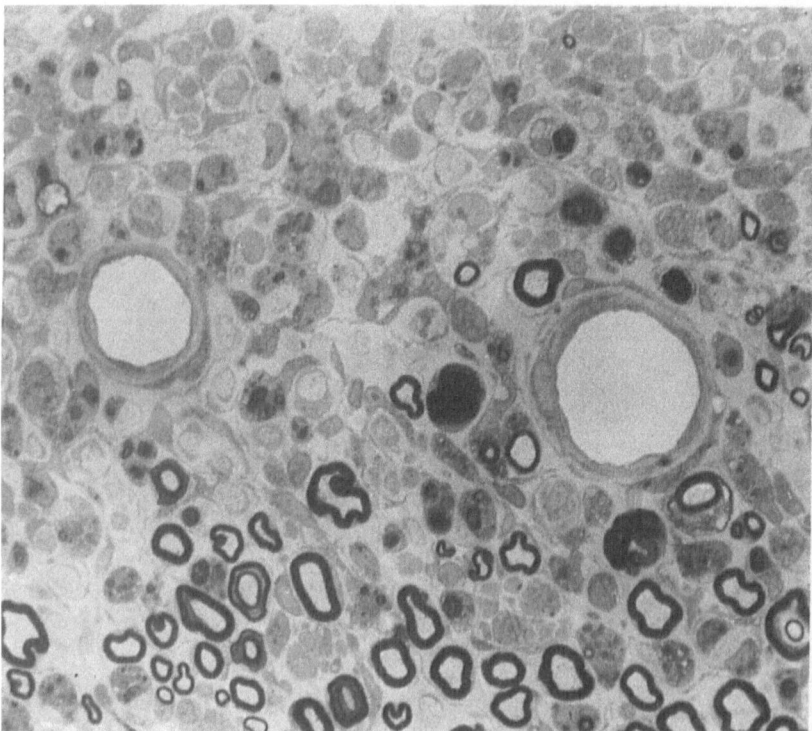

Fig. 2.17. Tuberculin reaction in the sciatic nerve of a rat immunised with galactocerebroside.
(From Powell et al. (1984) with permission.)

culin reactions in the nerves of rats, guinea pigs or rabbits also did not elicit primary demyelination although Wallerian degeneration and secondary myelin breakdown were seen (Powell et al. 1984; Goban et al. 1986; Powell and Hughes 1987) (Fig. 2.16). An Arthus reaction induced by injecting bovine albumin into the nerves of previously immunised Lewis rats caused the expected inflammatory reaction and oedema but no demyelination (Powell et al. 1984). There is also much circumstantial evidence against an important role for bystander demyelination. It has been frequently observed that vigorous cellular reactions occur in acute EAE which are accompanied by very little demyelination and in viral encephalitis inflammation and cell destruction are common but primary demyelination is rare and where it occurs autoimmune reactions are usually masked. For primary demyelination to occur, inflammatory reactions in the nervous system must be superimposed on a background of autoimmunity to myelin antigens (Fig. 2.17).

Summary

Peripheral nerve fibres are protected from circulating immune reactions by a specialised blood–nerve barrier and the perineurium. Damage to myelin or Schwann cells alone causes primary demyelination but the axon survives and remyelination restores function rapidly. Myelin destruction, following axon damage, i.e., secondary demyelination, is followed by a much slower process of axonal regeneration and remyelination with delayed recovery. Endoneurial macrophages are the main antigen presenting cells in the peripheral nervous system. Schwann cells can express major histocomptability antigen in vitro but function poorly as presenters of antigen compared with macrophages. Primary demyelination can be induced by complement-fixing antibodies against epitopes on glycoproteins and glycolipids which are present on the surface of Schwann cells and myelin. T-helper cell mediated reactions against the myelin protein, P_2, which is buried in the myelin sheath, induce demyelination by an unknown mechanism. Primary demyelination does not occur as a bystander effect of immune responses to foreign antigens introduced into the endoneurium. T cell cytotoxic cell-mediated demyelination and antibody-dependent cell-mediated demyelination have not been demonstrated experimentally although either would explain the morphological findings in EAN and GBS.

References

Baba H, Daune GC, Ilyas AA et al. (1989) Anti GM1 ganglioside antibodies with differing fine specificities in patients with multifocal motor neuropathy. J Neuroimmunol 25:143–150
Beuche W, Friede RL (1984) The role of non-resident cells in Wallerian degeneration. J Neurocytol 13:767–796
Brosnan CF, Stoner GL, Bloom BR, Wisniewski HM (1977) Studies in demyelination by activated lymphocytes in the rabbit eye. J Immunol 118:2103–2110

Brosnan CF, Selmaj K, Raine CS (1988) Hypothesis: a role for tumor necrosis factor in immune-mediated demyelination and its relevance to multiple sclerosis. J Neuroimmunol 18:87–94

Brostoff SW (1984) Antigens of peripheral nervous system myelin. In: Dyck PJ, Thomas PK, Lambert EH, Bunge R (eds) Peripheral neuropathy. WB Saunders, Philadelphia, pp 562–576

Carroll WM, Jennings AR, Mastaglia FL (1984) Experimental demyelinating optic neuropathy induced by intra-neural injection of galactocerebroside antiserum. J Neurol Sci 85:125–135

Deibler GE, Driscoll BF, Kies MW (1978) Immunochemical and biochemical studies demonstrating the identity of a bovine spinal cord protein (SCP) and a basic protein of bovine peripheral nerve myelin (BF). J Neurochem 30:401–412

Fabian RH, Petroff G (1987) Intraneuronal IgG in the central nervous system: uptake by retrograde axonal transport. Neurology 37:1780–1784

Goban Y, Saida T, Saida K, Nishitkani H, Kameyana M (1986) Role of non-specific myelin destruction by delayed type hypersensitivity in primary demyelination. J Neurol Sci 74:97–109

Gregson NA, Leibowitz S (1985) IgM paraproteinaemia, polyneuropathy and MAG. Neuropathol Appl Neurobiol 11:329–347

Gregson NA, Kennedy MC, Leibowitz S (1971) Immunological reactions with lysolecithin-solubilised myelin. Immunology 20:501–512

Hall SM, Gregson NA (1971) The in vivo and ultrastructural effects of injection of lysophosphatidyl choline into myelinated peripheral nerve fibres. J Cell Sci 9:769–789

Hays AP, Latov N, Takatsu M, Sherman WH (1987) Experimental demyelination of nerve induced by serum of patients with neuropathy and an anti MAG M protein. Neurology 37:242–246

Hughes RAC, Powell HC (1984) Experimental allergic neuritis: demyelination induced by P_2 alone and non-specific enhancement by cerebroside. J Neuropathol Exp Neurol 43:154–161

Hughes RAC, Powell HC, Braheny SL, Brostoff SW (1985) Endoneurial injection of antisera to myelin antigens. Muscle Nerve 8:516–522

Hughes RAC, Atkinson PF, Gray IA, Taylor WA (1987) Major histocompatibility antigens and lymphocyte subsets during experimental allergic neuritis in the Lewis rat. J Neurol 234:390–395

Izumo S, Linington C, Wekerle H, Meyermann R (1985) Morphological study on EAN mediated by T-cell line specific for bovine P_2 protein in Lewis rats. Lab Invest 53:209–218

Jacobs JM, MacFarlane RM, Cavanagh JB (1976) Vascular leakage in the dorsal root ganglia of the rat, studied with horseradish peroxidase. J Neurol Sci 29:95–107

Jones TA, Bergfors T, Sedzik J, Unge T (1988) The three-dimensional structure of P_2 myelin protein. EMBO 7:1597–1604

Kadlubowski M, Hughes RAC, Gregson NA (1984) Spontaneous and experimental neuritis and the distribution of the myelin protein P_2 in the nervous system. J Neurochem 42:123–129

Koski CL, Vanguri P, Shin ML (1985) Activation of the alternative pathway of complement by human peripheral nerve myelin. J Immunol 134:1810–1814

Lassmann H, Brunner C, Bradl M, Linington C (1988) Experimental allergic encephalomyelitis: the balance between encephalitogenic T lymphocytes and demyelinating antibodies determines the size and structure of demyelinated lesions. Acta Neuropathol 75:566–576

Linington C, Izumo S, Suzuki M, Uyemura M, Meyermann R, Wekerle H (1984) A permanent rat T cell line that mediates experimental allergic neuritis in the rat in vitro. J Immunol 133:1946–1950

Linington C, Bradl M, Lassmann H, Brunner C, Vass K (1988) Augmentation of demyelination in rats: acute allergic encephalomyelitis directed against a myelin/oligodendrocyte glycoprotein. Am J Pathol 130:443–454

Low PA (1984) Endoneurial fluid pressure and microenvironment of the nerve, In: Dyck PJ, Thomas PK, Lambert EH, Bunge R (eds) Peripheral neuropathy. WB Saunders, Philadelphia, pp 599–617

Marcus DM, Latov N, Hsi BP, Gillard BK (1989) Measurement and significance of antibodies against GM1 ganglioside. Report of a Workshop, 18 April 1989, Chicago, IL, USA. J Neuro-immunol 25:255–259

Milner P, Lovelidge CA, Taylor WA, Hughes RAC (1987) P_0 myelin protein produces experimental allergic neuritis in Lewis rats. J Neurol Sci 79:275–285

Mizisin AP, Wiley CA, Hughes RAC, Powell HC (1987) Peripheral nerve demyelination in rabbits after inoculation with Freund's complete adjuvant alone or in combination with lipid hapten. J Neuroimmunol 16:381–395

Myers RR, Powell HC, Heckman HM, Costello ML, Katz J (1981) Biophysical and pathological effects of cryogenic nerve lesion. Ann Neurol 10:478–485

Nagai Y, Momoi T, Saito M, Mitsuzawa E, Ohtani S (1976) Ganglioside syndrome, a new autoimmune neurologic disorder, experimentally induced with brain gangliosides. Neurosci Lett 2:107–111

Norton WT, Cammer W (1984) Isolation and characterisation of myelin. In: Morell P (ed) Myelin, 2nd edn. Plenum Press, New York

Norton WT (1985) Recent advances in myelin biochemistry. Ann NY Acad Sci 436:5–10

Olsson Y (1984) Vascular permeability in the peripheral nervous system. In: Dyck PJ, Thomas PK, Lambert EH, Bunge R (eds) Peripheral neuropathy. WB Saunders, Philadelphia, pp 579–597

Perlmann P, Perlmann H, Muller-Eberhard HJ (1975) Cytolytic lymphocytic cells with complement receptor in human blood. Induction of cytolysis by Igh antibody but not by target cell-bound C3. J Exp Med 141:287–296

Powell HC, Hughes RAC (1987) Role of non-specific myelin destruction by delayed type hypersensitivity in primary demyelination. J Neurol Sci 74:97

Powell HC, Myers RR, Costello ML, Lampert PW (1979) Endoneural fluid pressure in Wallerian degeneration. Ann Neurol 5:550–573

Powell HC, Myers RR, Costello ML (1980) Increased endoneurial fluid pressure following injection of histamine and compound 48/80 into rat peripheral nerves. Lab Invest 43:564–573

Powell HC, Braheny SL, Hughes RAC, Lampert PW (1984) Antigen-specific demyelination and significance of the bystander effect in peripheral nerves. Am J Pathol 114:443–453

Quarles RH (1984) Myelin-associated glycoprotein in development and disease. Dev Neurosci 6:285–303

Quarles RH, Ilyas AA, Willison HJ (1986) Antibodies to glycolipids in demyelinating diseases of the human peripheral nervous system. Chem Phys Lipids 42:235–248

Raine CS, Bornstein MB (1979) Experimental allergic neuritis – ultrastructure of serum induced myelin aberrations in peripheral nervous system cultures. Lab Invest 40:423–432

Raine CS, Johnson AB, Marcus DM, Suzuki A, Bornstein MB (1981) Demyelination in vitro – absorption studies demonstrate that galactocerebroside is a major target. J Neurol Sci 52:117–131

Rostami A, Brown MJ, Lisak RP, Sumner AJ, Zweiman B, Pleasure DE (1984) The role of myelin P_2 protein in the production of experimental allergic neuritis. Ann Neurol 16:680–685

Saida T, Saida K, Dorfman SH (1979) Experimental allergic neuritis induced by sensitisation with galactocerebroside. Science 204:1103–1106

Saida T, Saida K, Silberberg DH, Brown MK (1981) Experimental allergic neuritis induced by galactocerebroside. Ann Neurol 9 suppl:87–101

Saida T, Saida K, Olawa K, Goban Y, Kawanishi T (1987) Experimental models of immune-mediated demyelination in peripheral nerve. In: Aarli JA, Behan WMH, Behan PO (eds) *Clinical Neuroimmunology*. Blackwell Scientific Publications, Oxford, pp 102–113

Schwartz M, Sela BA, Eshhar N (1982) Antibodies to gangliosides and myelin autoantigens are produced in mice following sciatic nerve injury. J Neurochem 38:1192–1195

Seil FJ (1977) Tissue culture studies of demyelinating disease: a critical review. Ann Neurol 2:345–355

Seil FJ, Kies MW, Bacon ML (1981) A comparison of demyelinating and myelination inhibiting factor induction by whole peripheral nerve tissue and P_2 protein. Brain Res 210:441–448

Sergott RC, Brown MJ, Silberberg DH, Lisak RP (1984) Antigalactocerebroside serum demyelinates optic nerve in vitro. J Neurol Sci 64:297–303

Smith KJ, Hall SM (1988) Peripheral demyelination and remyelination initiated by the calcium-selective ionophore ionomycin: in vivo observations. J Neurol Sci 83:37–53

Stoner GL, Brosnan CF, Wisniewski HM, Bloom BR (1977) Studies on demyelination by activated lymphocytes in the rabbit eye. Effects of a monuclear cell infiltrate induced by products of activated lymphocytes. J Immunol 118:2094–2102

Thomas PK, Olsson Y (1984) Microscopic anatomy and function of the connective tissue components of peripheral nerve. In: Dyck PJ, Thomas PK, Lambert EH, Bunge R (eds) Peripheral neuropathy. WB Saunders, Philadelphia, pp 97–120

Vedeler CA (1987) Demonstration of Fc gamma receptors on human peripheral nerve fibres. J Neuroimmunol 15:207–216

Ward RHR, Lachmann PJ (1985) Monoclonal antibodies which react with lymphocyte-lysed target cells and which cross-react with complement-lysed ghosts. Immunology 56:179–188

Wekerle H, Linington C, Lassmann H, Meyermann R (1986a) Cellular immune reactivity within the CNS. Trends in Neurol Sci 9:271–277

Wekerle H, Schwab M, Linington C, Meyermann R (1986b) Antigen presentation in the peripheral nervous system: Schwann cells present endogenous myelin autoantigens to lymphocytes. Eur J Immunol 16:1551–1557

Westland K, Pollard JD (1987) Proteinase induced demyelination. An electrophysiological and histological study. J Neurol Sci 82:41–53

Wisniewski HM, Bloom BR (1975) Primary demyelination as a non-specific consequence of

circulating immunocytes in Guillan–Barré syndrome. A cell-mediated immune reaction. J Exp
    Med 141:346–359
Yonezawa T, Hasegawa M, Arizona N, Okabe H (1981) Antigenicity of galactocerebroside in
    experimental allergic demyelinating diseases. Acta Neuropathol suppl. 7:162–164
Young JD-E, Liu C-C (1988) Multiple mechanisms of lymphocyte-mediated toxicity. Immunol
    Today 9:140–144

Chapter 3

Experimental Allergic Neuritis

Experimental Allergic Encephalomyelitis

Experimental allergic neuritis (EAN), an experimental model of GBS, developed as a by-product of research into experimental allergic encephalomyelitis (EAE). In 1895 Louis Pasteur had prepared the first rabies vaccines from dried formalin-treated rabbit brain. The vaccine provided some protection from rabies but also occasionally caused encephalomyelitis. Rivers and Schwentker (1935) showed that repeated injections of rabbit brain into monkeys would produce encephalomyelitis and Freund in 1949 showed that the same result could be achieved with a single injection provided that the nervous tissue was emulsified with an adjuvant. Early experiments were directed towards identifying the antigen. The encephalitogenic component of nervous tissue was found to reside largely in the major myelin basic protein (*see* Leibowitz and Hughes 1983 for review). More recently it has been demonstrated that highly purified proteolipid protein will also induce EAE (Yoshimura et al. 1985; Endoh et al. 1986). There is no theoretical reason why other CNS myelin antigens should not be capable of inducing EAE. Myelin basic protein (MBP), molecular weight 18 500, is rich in basic aminoacids, and its amino acid sequence is highly conserved from one species to another (Hashim 1980). However, there is great species variation in the part of the protein which is responsible for inducing EAE. For example, the tryptophan-containing sequence:

H—Phe—Ser—Trp—Gly—Ala—Glu—Gly—Gln—Lys—OH
114   122

is encephalitogenic in the guinea pig but not in the rat. The encephalitogenic determinants for the rat, rabbit, guinea pig and monkey are all different and located in different parts of the parent molecule (Fig. 3.1). MBP is located in the major dense line of myelin which represents the apposed cytoplasmic surfaces of the oligodendrocyte or Schwann cell. It remains a mystery how lymphocytes gain access to a protein at that site in order to produce EAE. The explanation may be that minor damage involved in trivial trauma to the CNS or PNS disrupts the myelin sufficiently to expose or release small amounts of antigen. For obvious reasons the identity of the encephalitogenic sequence for man has not been identified. However, reports of encephalomyelitis following injection of brain emulsions for vaccination against rabies, quack cures, or experimental treatment of multiple sclerosis indicate that human encephalitogenic determinants do exist.

The immunology of EAE has also been investigated in great detail. Acute EAE can be transferred from one animal to another of the same strain with T

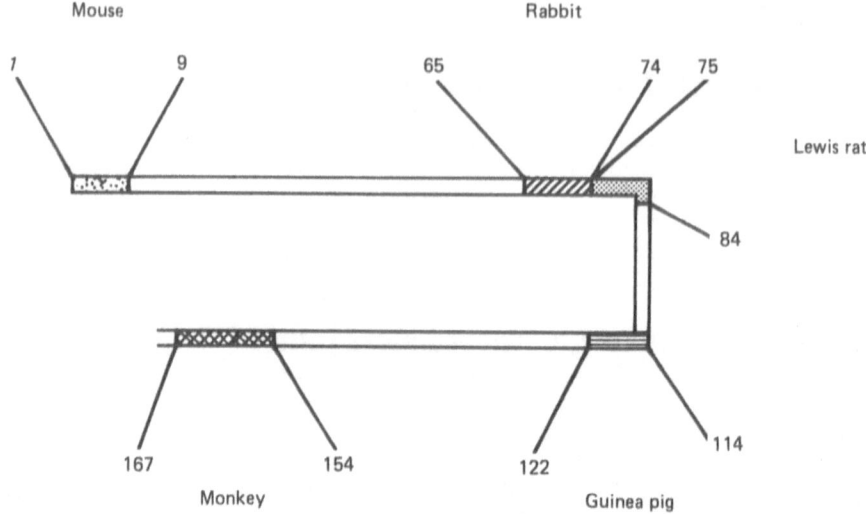

Fig. 3.1. Diagram of myelin basic protein showing the different sequences which induce EAE in different species. (Data from Zamvil et al. (1986) for PL/J × SJL (H-2^s) mice and review by Hashim (1980) for the other species.)

cells but not with serum. The responsible cells are T cells bearing the surface marker CD4+ and respond to antigen in the context of major histocompatability class II antigen. The T cells can be cultured as cell lines which retain their encephalitogenic potency. In recent experiments it was shown that individual CD4+ T cell clones from these cell lines can have different fine specificities from each other, reacting with slightly different epitopes within the encephalitogenic region (Happ et al. 1988). Nevertheless only a small proportion of the T cell receptor genes are used by T cell clones directed against MBP (Acha-Orbea et al. 1988). In particular most anti MBP clones in PL/J mice used the Vβ8 gene: this observation carries the hope that therapeutic deletion of the T cells expressing this gene would permit specific immunosuppression (Heber-Katz and Acha-Orbea 1989).

The EAE is normally an acute monophasic disease and the mechanisms involved in switching off the inflammatory process are incompletely understood. One possibility is that antibody inhibits the cell-mediated immune response. It has long been known that injections of convalescent serum will protect rats from developing EAE (Paterson and Harwin 1963; Hughes 1974). A more popular idea is that suppressor T cells inhibit the encephalitogenic action of the helper T cells. Recently both Ellerman et al. (1988) and Cohen and Weiner (1988) have reported that T cell lines having the suppressor phenotype will inhibit the proliferation of a myelin basic protein responsive helper T cell line in response to specific stimulation and prevent the helper line cells transferring EAE to naive recipients.

The problem with EAE as a model for human demyelinating disease is that histologically it consists of a marked inflammatory process in the brain and spinal cord with relatively little demyelination. This problem has been overcome in two ways. First, a relapsing model has been developed by injecting juvenile guinea pigs or susceptible mouse or rat strains with large amounts of CNS tissue. In these models there is extensive demyelination which bears comparison with multiple sclerosis (Lassmann 1983). Second, Linington and colleagues have injected a myelin basic protein T cell line into inbred Lewis rats and simultaneously injected antibody to a CNS glycoprotein called myelin/ oligodendrocyte glycoprotein (Lassmann et al. 1988; Linington et al. 1988). The combined injections induced widespread demyelination and inflammation. This supports the popular idea that antibody to myelin antigens may be important in inducing CNS demyelination whereas cell-mediated immunity to myelin antigens breaks down the blood–brain barrier and initiates a sequence of inflammatory changes which permit antibody to reach the CNS parenchyma.

Pathology of EAN

Waksman and Adams (1955) described how rabbits immunised with rabbit or bovine cerebral white matter would develop inflammatory lesions in the CNS and PNS whereas rabbits immunised with peripheral nerve tissue from several species (rabbit, bovine, human, dog, guinea pig) would develop lesions in the

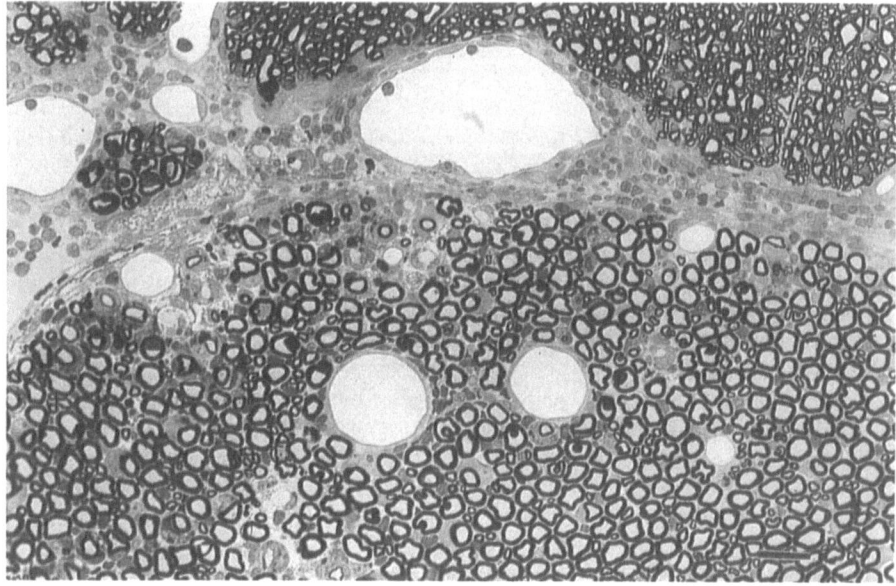

Fig. 3.2. Cauda equina of a Lewis rat with EAN. **a** 11 days after immunisation with myelin. There is a cellular infiltrate in the subarachnoid space and nerve root (*below*) but not in the spinal cord (*above*). Some demyelinated axons are present in the nerve root. Paraphenylene diamine. Bar = 20 μm.

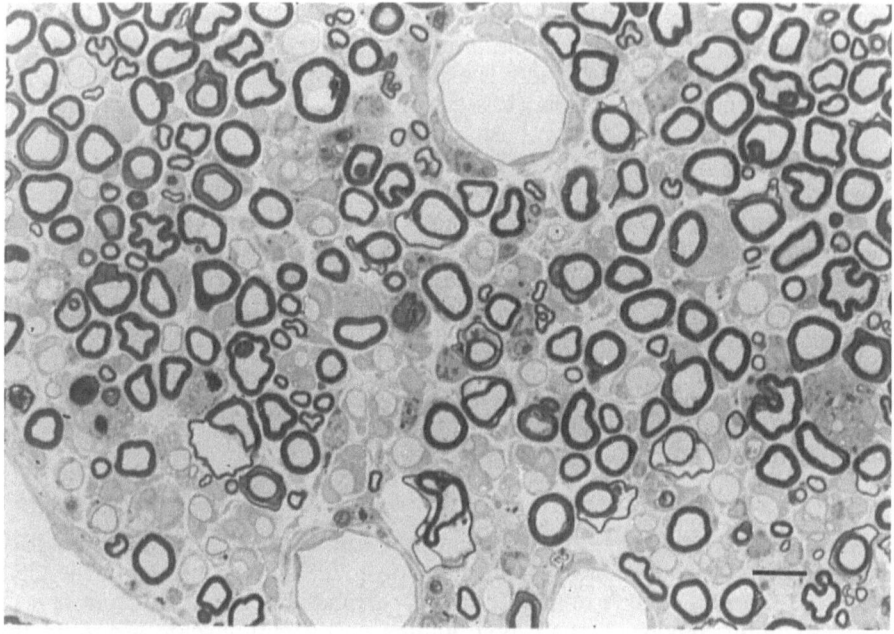

b

Fig. 3.2b.   21 days after immunisation with P_2 protein. Many demyelinated fibres and some debris-laden macrophages are present in a nerve root. Some surviving myelin sheaths show ballooning of lamellae. Bar = 10 μm.

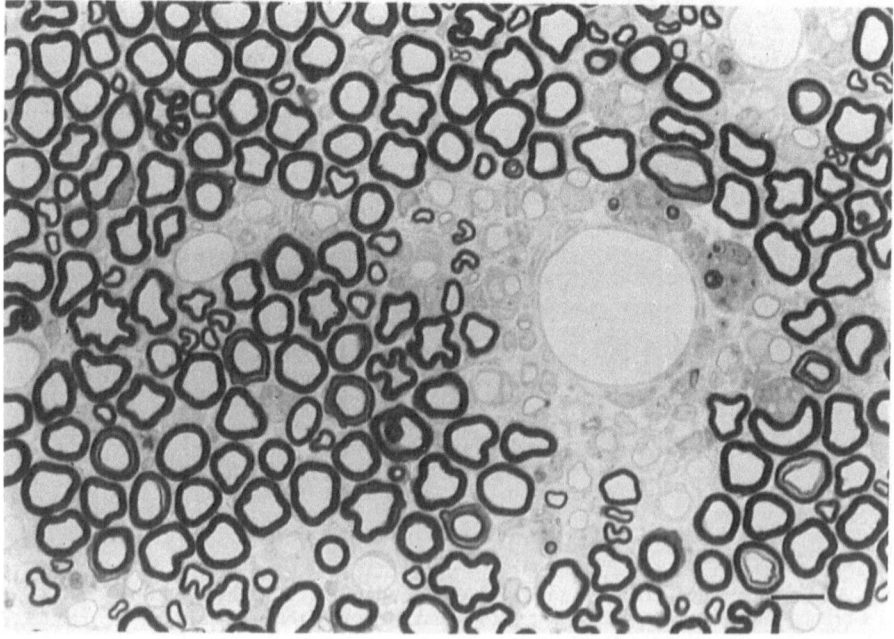

c

Fig. 3.2c.   21 days after immunisation with myelin. A cuff of demyelinated axons surround a blood vessel. Bar = 10 μm. (From Hughes and Powell (1984), with permission.)

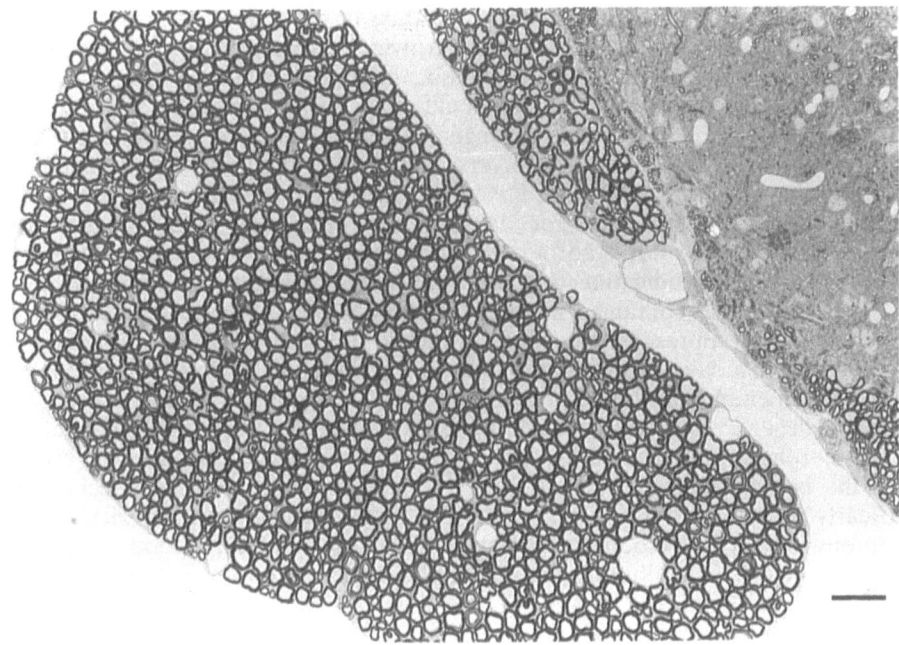

Fig. 3.3.   Cauda equina of a normal Lewis rat. Paraphenylene diamine. Bar = 20 μm.

PNS alone. The inflammatory lesions consisted of multifocal infiltrates of lymphocytes and macrophages around capillaries and vessels, and sometimes spreading diffusely within the endoneurium. Close to the inflammation, the myelin was destroyed, but axons remained intact (Figs. 3.2 and 3.3). Waksman and Adams pointed out that this was similar to the findings in GBS. They also compared EAN in mice and guinea pigs with the disease in rabbits (Waksman and Adams 1956). Following immunisation with peripheral nerve, mice developed disease similar in distribution but milder in severity than that in rabbits with lesions preferentially and predominantly in the spinal roots and dorsal root ganglia. On the other hand guinea pigs commonly developed CNS lesions as well. They argued convincingly that the disease for which they introduced the term "experimental allergic neuritis" was not due to any of the several known animal infections and was likely to be due to an allergic process. They also correctly deduced that the antigen responsible for EAE must be different from EAN. They noted that the inflammatory lesions might destroy the whole neural parenchyma or cause loss of myelin leaving the axons intact. Subsequent studies of teased fibres (Cragg and Thomas 1964) showed segmental demyelination with preservation of proximal and distal internodes, an appearance which had already been demonstrated in lead and diphtheritic neuropathy. The minimal lesion was a retraction of the myelin sheath from the node causing widening of the nodal gap.

Identifying the very earliest morphological lesion in EAN is not an easy matter because the onset is explosive. Animals which are healthy one day may become severely paralysed within 24 to 48 hours. Furthermore, because of the vagaries of sampling and the multifocal nature of the lesions, it is very difficult to be sure that a change, such as oedema, which appears to be acellular in one section, is not the consequence of cellular infiltration which is nearby but out of the plane of the section. A similar difficulty and controversy exist in EAE in which it is not clear whether leakage of immunoglobulins and oedema occur before cellular infiltration, or vice versa, or whether both occur simultaneously (Leibowitz and Hughes 1983). Powell et al. (1983) reported oedema in the sciatic nerves of Lewis rats ten days after immunisation with human sciatic nerve. At that time horseradish peroxidase injected intravenously appeared in the endoneurium indicating that the blood–nerve barrier had become leaky. The oedema fluid contained immunoglobulin. The endoneurial fluid pressure, measured with a pressure transducer attached to a micropipette, was shown to start rising ten days after immunisation and reach a peak two to six days later. The rise in endoneurial fluid pressure may be a contributory factor, causing the axonal degeneration which occurs in EAN as well as primary demyelination. The morphological observations so far do not explain what initiates the breach of the blood–nerve barrier. It might be a humoral or cellular mechanism. Clearly everything happens very fast. The most detailed morphological observations suggest that blood–nerve barrier breakdown, immunoglobulin leakage, and perivascular lymphocyte and macrophage infiltration occur simultaneously (Hahn et al. 1985).

Mast cells are abundant in the sciatic nerves of rats though very sparse in the spinal roots and dorsal root ganglia. They decrease in number and degranulate in the early stages of EAN (Brosnan et al. 1985). Mast cells have been proposed to play an important role in initiating and facilitating delayed hypersensitivity reactions (Askenase et al. 1980). Mast cells might make a contribution to the oedema and increase in endoneurial fluid pressure in the peripheral nerves. However, the absence of mast cells from the spinal roots and dorsal root ganglia of rats, where the morphological changes of EAN are most conspicuous, suggests that mast cells play only a minor role in its pathogenesis.

The morphological sequence of demyelination has been studied at the ultrastructural level. The popular view is that demyelination occurs by a process of macrophage invasion of the myelin sheath followed by stripping of the lamellae or vesicular dissolution, and phagocytosis while the Schwann cells remain intact. In the first ultrastructural study, Ballin and Thomas (1968) stressed that the earliest changes were separation of the terminal myelin loops from the axon at the nodes of Ranvier accompanied by vesicular dissolution. They did not observe the insertion of macrophage processes between myelin lamellae and stripping of myelin by macrophages which had already been reported as a mechanism of demyelination in EAE (Lampert 1967). Rather they noted that most myelinated fibres were either intact or completely demyelinated and that intermediate forms in the process of demyelination were rare. Ballin and Thomas used guinea pigs immunised with rabbit sciatic nerve, and the disease in the affected animals was probably more severe than in the rat. Although only 20 of 38 animals were affected, five of them died. Furthermore, the lesions in guinea pigs are more widespread throughout the nervous system and may affect the sciatic nerve more severely than in rats (Waksman and Adams 1956). Ballin and Thomas (1968) described how myelin was lost by primary segmental demyelination, in which one complete internode was demyelinated while internodes proximal and distal to this segment remained intact.

Lampert (1969) published a beautifully illustrated paper claiming that demyelination in EAN only occurred in relation to mononuclear cell invasion. His experiments were conducted on Lewis rats immunised with human sciatic nerve myelin. The earliest changes, perivascular infiltrates of mononuclear cells in the spinal roots, sciatic nerves and dorsal root ganglia, were observed 11 days after immunisation. This is exactly consistent wiith results in Lewis rats immunised with bovine spinal root myelin (compare Figs. 3.2 and 3.3) (Hughes and Powell 1984). In very early severe lesions there is leakage of erythrocytes, polymorphonuclear leucocytes and fibrinous exudates into the endoneurium.

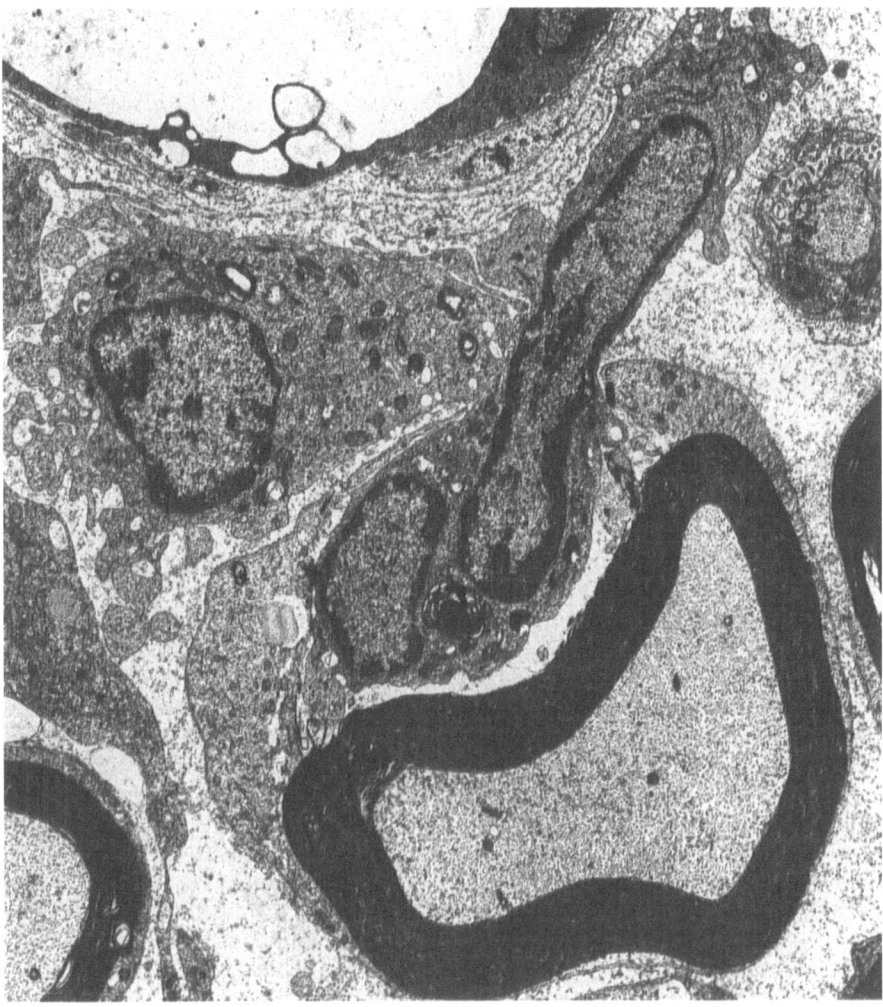

Fig. 3.4. Early experimental allergic neuritis. Electron micrograph from the cauda equina of a Lewis rat. Note that one macrophage has penetrated the basal lamina, pushed aside the Schwann cell cytoplasm and begun to digest myelin debris. Another macrophage nearby contains debris. The capillary endothelium shows reactive changes. (From Leibowitz and Hughes (1983), with permission. Electron micrograph prepared by Dr C Meier.)

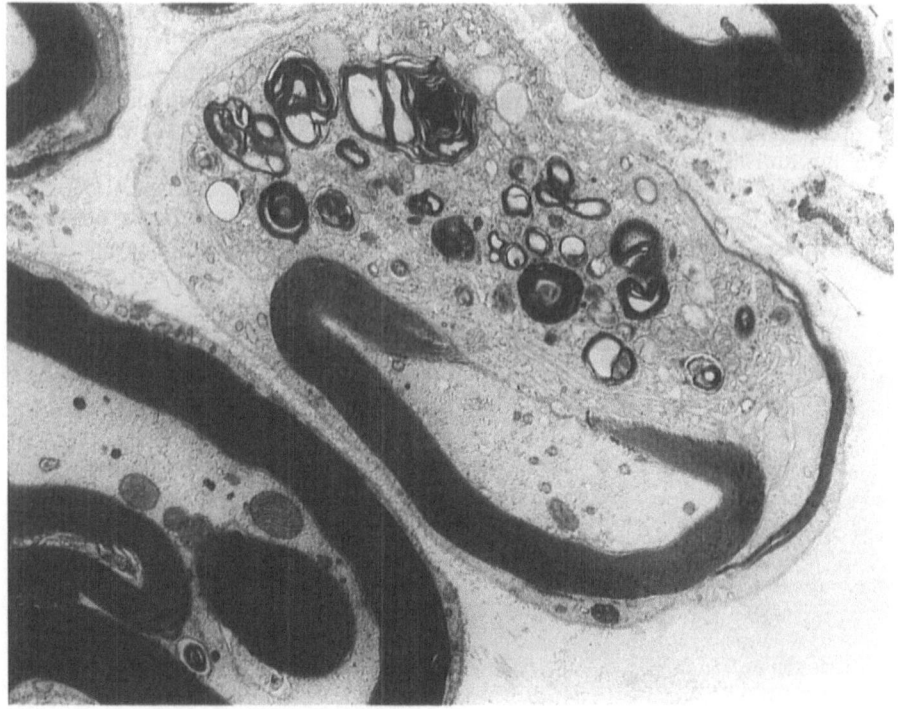

Fig. 3.5. Macrophage-mediated demyelination in EAN. A macrophage has digested half the myelin sheath. Electron micrograph.

Lampert described the penetration of processes of mononuclear cell cytoplasm through the basal lamina and between the lips of cytoplasm of the external mesaxon (Fig. 3.4). The mononuclear cell then pushed its way into and under the myelin sheath, separating it from its Schwann cell. The myelin either underwent vesicular dissolution or was ingested by the phagocytosing cell (Fig. 3.5). The end result was the stripping of the whole sheath from one internode to the next. In large lesions, particularly where there was severe inflammation, some axons were seen to be undergoing Wallerian degeneration. Throughout the process of demyelination the Schwann cells were considered to remain intact which led Lampert to propose that the myelin sheath was the target of an "allergic" reaction. He stressed that demyelination only occurred in the presence of infiltrating mononuclear cells, but the origin and identity of these cells was unclear. Identical conclusions were reached concerning the mechanism of demyelination in the spinal roots of rabbits with EAE induced with bovine white matter (Wisniewski et al. 1969), and with EAN induced with human sciatic nerve (Allt 1975). These authors discussed whether the myelin had been tagged or opsonised in some way for phagocytosis or whether the phagocytes had been rendered specific for myelin in some other way – a debate which still continues.

Similar infiltration of macrophages through the basal lamina and stripping of myelin has been described in EAN induced in chickens by immunisation with chicken nerve myelin (Ichijo et al. 1981). The only additional feature noted was

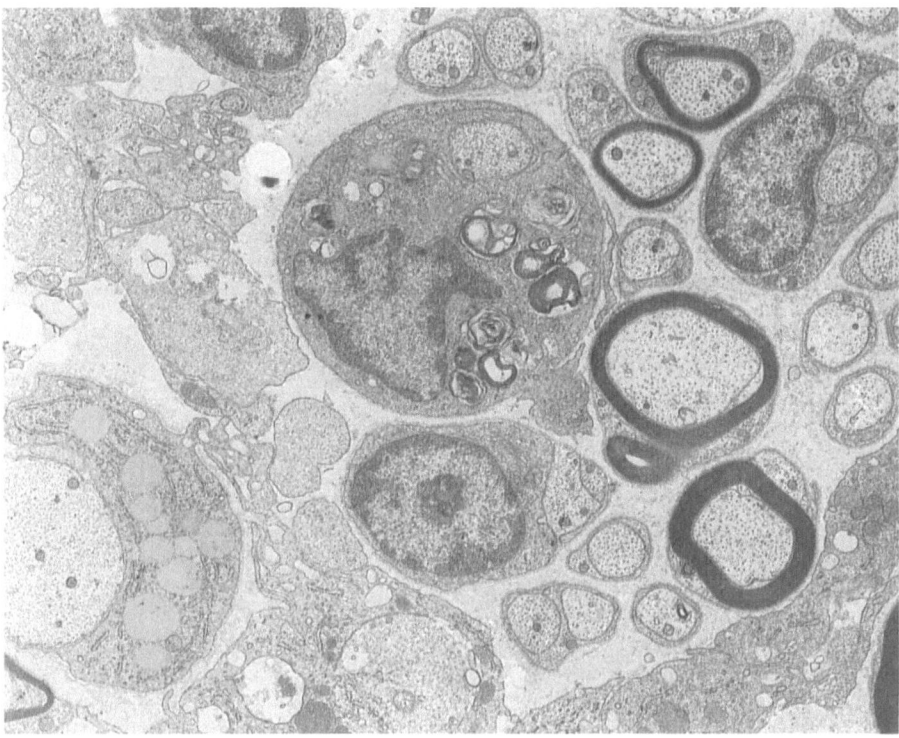

Fig. 3.6. Macrophage-mediated demyelination in the vagus nerve of a rat with EAN. A macrophage has demyelinated an axon and is leaving the basal lamina tube. A previously demyelinated axon lies nearby.

the occasional invasion of mononuclear cells into the periaxonal space, probably via the node of Ranvier, followed by stripping of the internal rather than external myelin lamellae first.

The primary target of the autoimmune reaction in EAN is the myelin sheath rather than Schwann cells. In the rat, EAN lesions occur in the vagus nerve where myelin is abundant (Fig. 3.6). The cervical sympathetic nerves, which contain unmyelinated nerve fibres, Schwann cells and almost no myelinated nerve fibres, are unaffected (Morey et al. 1985) (Fig. 3.7). The vagus and splanchnic nerves do contain myelin and become involved histologically and electrophysiologically in the majority of guinea pigs and rabbits with EAN (Tuck et al. 1981). Involvement of the autonomic nervous system is prominent in GBS and can be explained by the myelin content of human sympathetic and parasympathetic nerves. Wallerian degeneration is commonly observed in EAN and does not seem consistent with the idea that EAN is due to an immune attack on myelin. It has been proposed that the axonal degeneration in GBS is due to segmental demyelination of long lengths of axon (Asbury et al. 1969). However, two groups have noted that the axonal degeneration in the dorsal root ganglion and dorsal roots of guinea pigs with EAN occurs early, simultaneously with the demyelination, and in regions where the inflammatory process was severe (King et al. 1977; Madrid and Wisniewski 1977). Teased fibres showed either

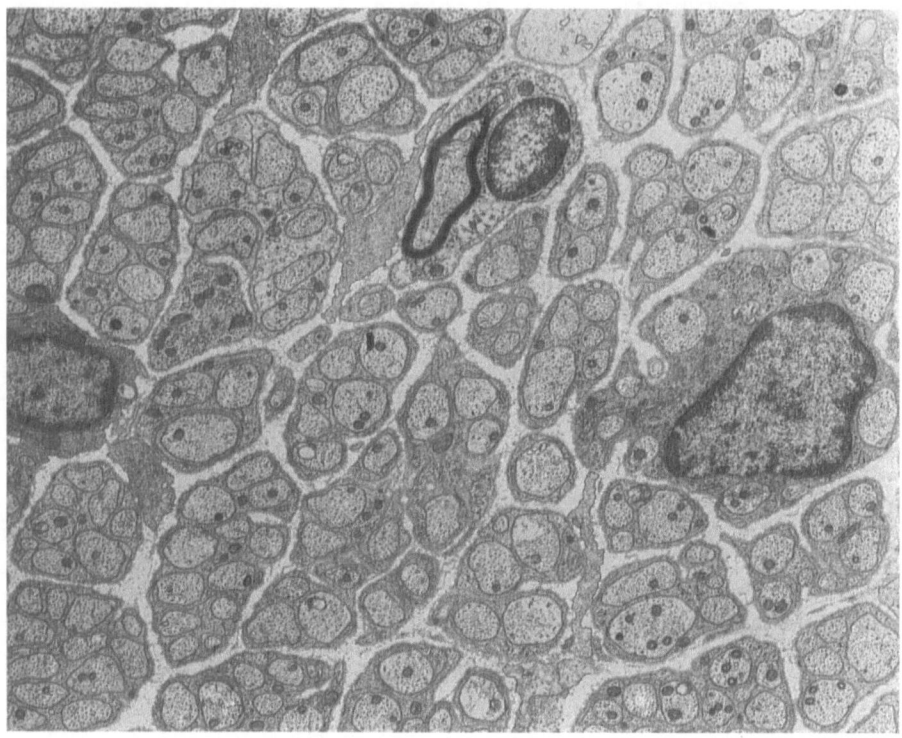

Fig. 3.7. Cervical sympathetic nerve trunk from the same rat as the vagus nerve shown in Fig. 3.6. The Schwann cells and unmyelinated axons remain normal. The myelinated nerve fibre was the only myelinated axon in the whole transverse section of this nerve. This absence of any inflammatory reaction compared with Fig. 3.6. is consistent with myelin, and not Schwann cells, being the target of the immune response in EAN.

segmental demyelination with preserved axons or axonal degeneration, never both. Wallerian degeneration might be caused by axonal damage from release of toxic products in a particularly severe inflammatory reaction directed against myelin.

The complete functional recovery made by most rats with EAN suggests that effective remyelination and restoration of conduction occur in most nerve fibres. Morphologically, remyelination and clearance of debris by macrophages are evident (Fig. 3.8).

Cells infiltrating the peripheral nerves in EAN can be identified with monoclonal antibodies and immunohistochemical techniques. The infiltrating cells are leucocytes, expressing leucocyte common antigen, and include T cells expressing the CD5 surface antigen (Olsson et al. 1984; Hughes et al. 1987; Ota et al. 1987). The T cells include both cells bearing the CD4 surface antigen and others bearing the CD8 surface antigen. The role of these two antigens is currently unclear. They do not simply reflect helper and cytotoxic/suppressor function respectively, but might relate to which class of major histocompatibility antigen is providing the context of antigen recognition by the T cell receptors. There was no evidence that one T cell type preceded the other in the early stages of the lesions. Hughes et al. (1987) noted that the expression of both MHC class

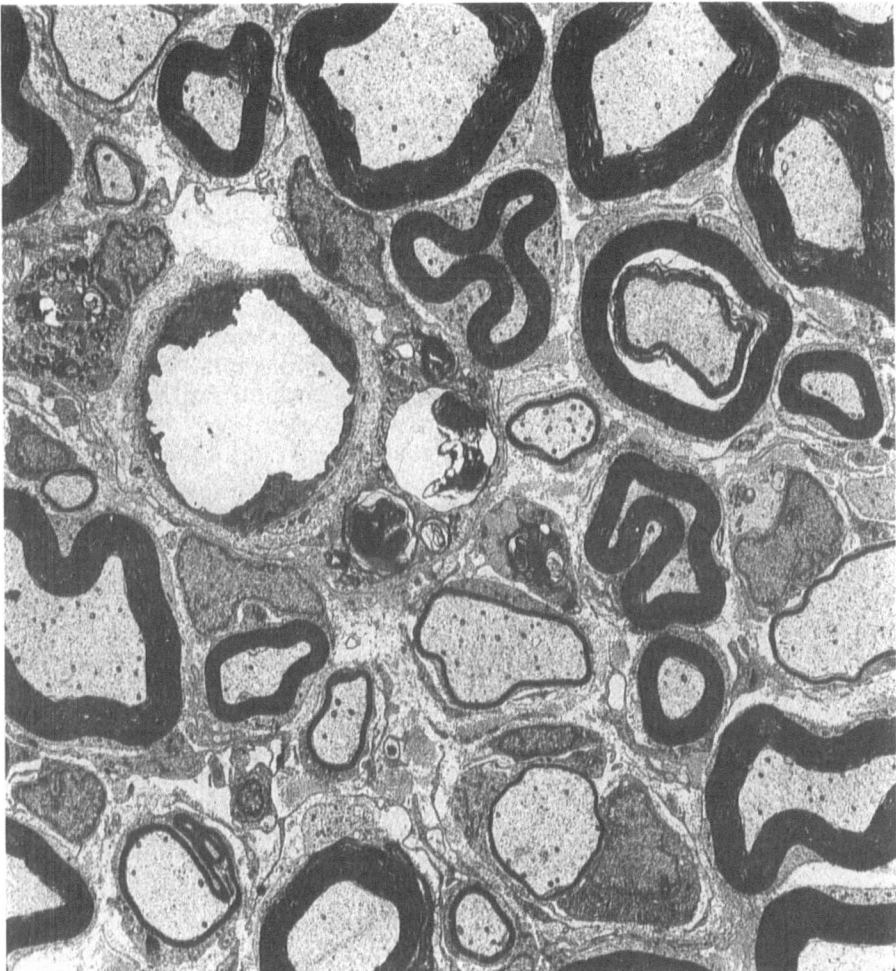

Fig. 3.8. Recovering experimental allergic neuritis. Electron micrograph from the cauda equina of a Lewis rat. Note that several axons have been remyelinated with thin myelin sheaths. Debris-laden macrophages surround a capillary. (From Leibowitz and Hughes (1983), with permission. Electron micrograph prepared by Dr C Meier.)

I and class II antigens was increased on endoneurial cells in early EAN. The endoneurial capillary endothelium never expressed MHC class II but did express class I. If endothelium can also express myelin antigens, CD8+ cells, which recognise antigen in the context of MHC class I, might be involved in initiating the breach of the blood–nerve barrier which occurs early in EAN.

Neurological Signs of EAN

The most convenient system for inducing EAN is to immunise Lewis rats with bovine spinal root myelin. The animals and antigen are readily available. The disease produced is pure EAN without contaminating EAE. The immunisation produces disease reproducibly in nearly 100% of animals (Kadlubowski and Hughes 1979, 1980; Smith et al. 1979).

EAN can also be produced in rabbits, guinea pigs, mice (Waksman and Adams 1956; Taylor and Hughes 1985), sheep (Paraf 1963), chickens (Peter and Quaglio 1967), and monkeys (Wisniewski et al. 1974). To prepare the immunising agent, the myelin or nerve is normally emulsified in oil with an emulsifying agent ("incomplete" adjuvant) to which heat-dried *Mycobacterium* bacilli are added to form "complete" adjuvant. The *Mycobacterium* in the inoculum is not necessary for EAN induction (Levine and Wenk 1963; Hughes and Kadlubowski 1980). The site of injection is important in the rat. Subcutaneous injections into the flanks and back are ineffective, but injections under the footpads of the fore or hind limbs work well. Injections into the fore limbs have the advantage that they do not interfere with assessment of hind limb function, which is the site of the most obvious neurological deficit.

About 10 days after immunisation the animals begin to lose weight, but this also happens in animals that are immunised with complete adjuvant without myelin. After 11–12 days the first definite neurological deficit, drooping of the tail, becomes apparent. The tail drags as the rat walks across the cage. If the animal is lifted up, the tip of the tail, which normally curves upwards, droops down, and the tip of the tail cannot be persuaded to curl round the examiner's finger. Within 24 hours after the tail begins to droop, weakness of one or both hind limbs becomes apparent and the rat does not turn smartly over onto its front when laid on its back. This weakness usually progresses to a stage when the rat will lie with both hind limbs sideways although if stimulated the rat is still able to get up and walk using it hind limbs albeit with a 'footdrop'. In the next stage there is complete paralysis of the hind limbs. The forelimbs sometimes show weakness of grip and difficulty supporting the body. Sometimes rats develop red or black staining round the eyes, perhaps due to inadequate grooming, with laboured breathing and have to be killed. Animals which do not proceed to this stage usually begin to improve after four or five days and proceed to recover, usually, so far as can be discerned, completely.

The disease differs from EAE in rats by the lesser severity, and lower mortality. Urinary retention and perineal soiling, which are common in Lewis rat EAE, are very rare in EAN. It has been suggested that the gait of animals with EAN is different from EAE with the presence of ataxia rather than weakness. The implication is that the animal has sensory loss rather than loss of power. This distinction is extremely difficult to determine in small animals.

Experimental Autonomic Neuropathy

A single report claimed that immunisation of rabbits with human sympathetic ganglia abolished acetylcholine-mediated reflex dilatation of the ear blood vessels (Appenzeller et al. 1965). The authors found lymphocytic infiltration of the sympathetic ganglia but did not consider that this was sufficient to explain the autonomic dysfunction observed, arguing that more significant lesions of the sympathetic efferent pathway must exist. The disease was not specific for sympathetic ganglia because sciatic nerve lesions also existed. However, similar autonomic dysfunction did not occur in rabbit EAN induced with peripheral nerve tissue. This potentially interesting model of acute acquired autonomic neuropathy has not been pursued.

Chronic Relapsing Experimental Allergic Neuritis

Although at first EAN was regarded as an acute monoplasic illness, resembling GBS in its course, this was because animals were killed soon after apparent recovery. If permitted to survive some animals develop progressive or relapsing deficits, resembling CIDP. This was first demonstrated in rhesus monkeys immunised with rabbit sciatic nerve myelin by Wisniewski et al. (1974). Some of the monkeys had recurrent episodes of neurological deficit. Whether relapses had occurred or not, histology showed a mixture of chronic demyelinated lesions and acute lesions with cell infiltrates and active phagocytosis and stripping of myelin similar to the process in acute EAN described by Lampert. In some of the chronic lesions there was accumulation of layers of Schwann cell processes around the demyelinated or remyelinating axons to form "onion bulbs". Pol-

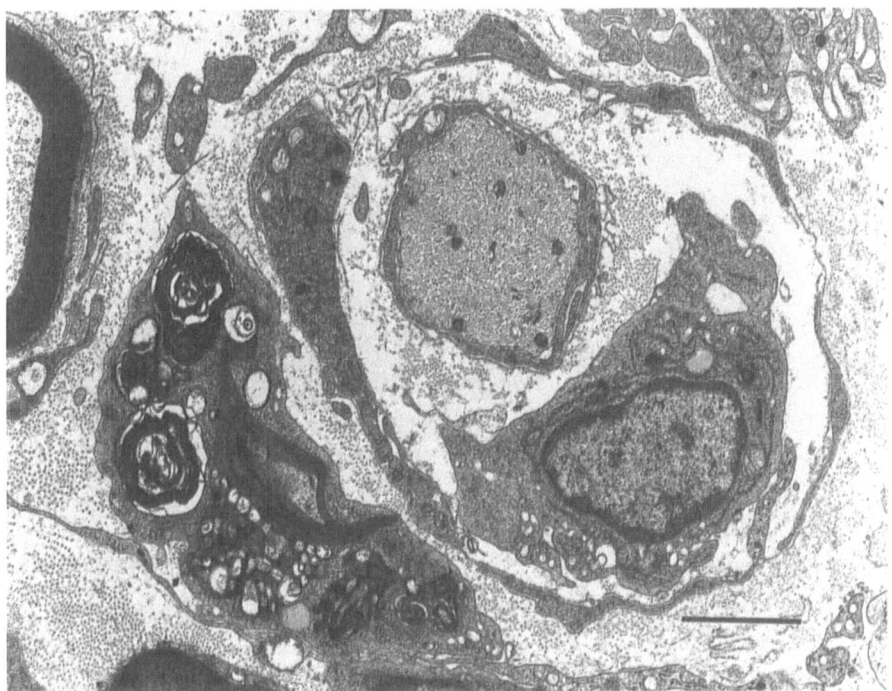

Fig. 3.9. An early stage in onion bulb formation. A demyelinated axon is surrounded by Schwann cell processes, redundant basal lamina and macrophages. Bar = 2μm. (From Adam et al. (1989), with permission.)

lard et al. (1975) found that a small proportion of guinea pigs immunised with peripheral nerve pursued a spontaneously relapsing and remitting course. Subsequent histological examination in these animals also demonstrated onion bulb formation. Similar observations were made by Madrid (1983). Craggs et al. (1986) showed that a relapsing and remitting course regularly occurred after immunisation of Lewis rats with bovine spinal root material. They were unable to influence the course by splenectomy or thymectomy. Brosnan et al. (1988) showed that a relapsing course was more common in rats immunised as weanlings rather than as adults. Adam et al. (1989) showed in the Lewis rat that onion bulb formation was common in the dorsal root ganglion and adjacent dorsal and ventral roots but rare elsewhere (Figs. 3.9–3.11). Throughout the peripheral nervous system there was persistently increased expression of major histocompatability (MHC) class I antigen by all the endoneurial cells, and there were also increased numbers of cells expressing MHC class II antigen. The Schwann cell cytoplasmic processes forming the onion bulbs expressed both MHC class I and class II antigens. The best model of CIDP so far produced is in

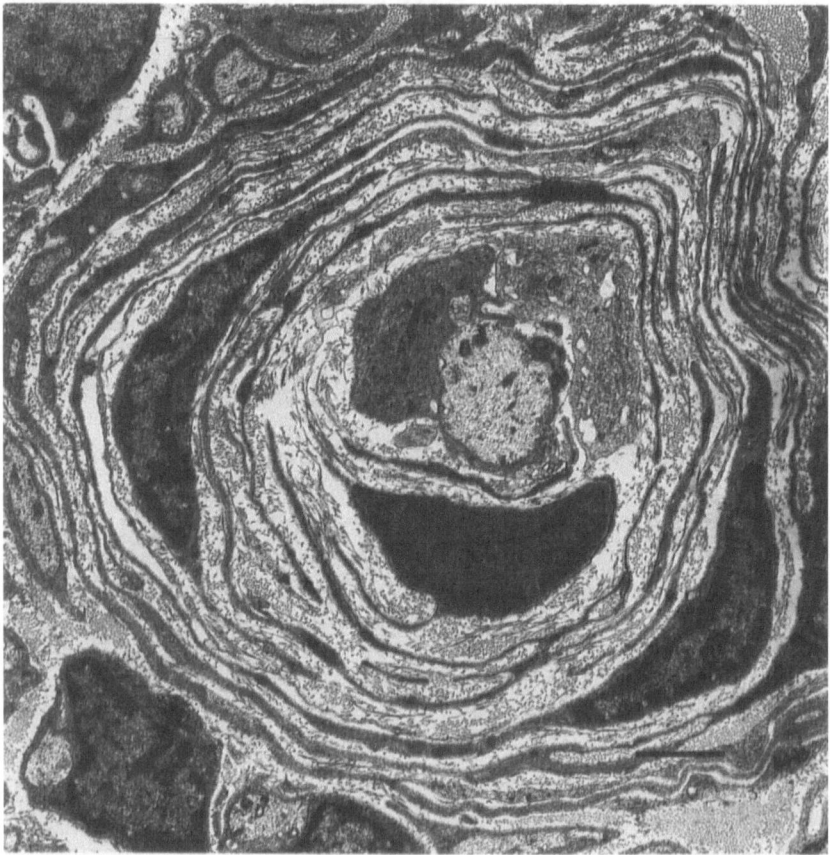

Fig. 3.10. Onion bulb. A demyelinated axon is surrounded by several wraps of Schwann cell and some fibroblast processes. Bar = 2 μm.

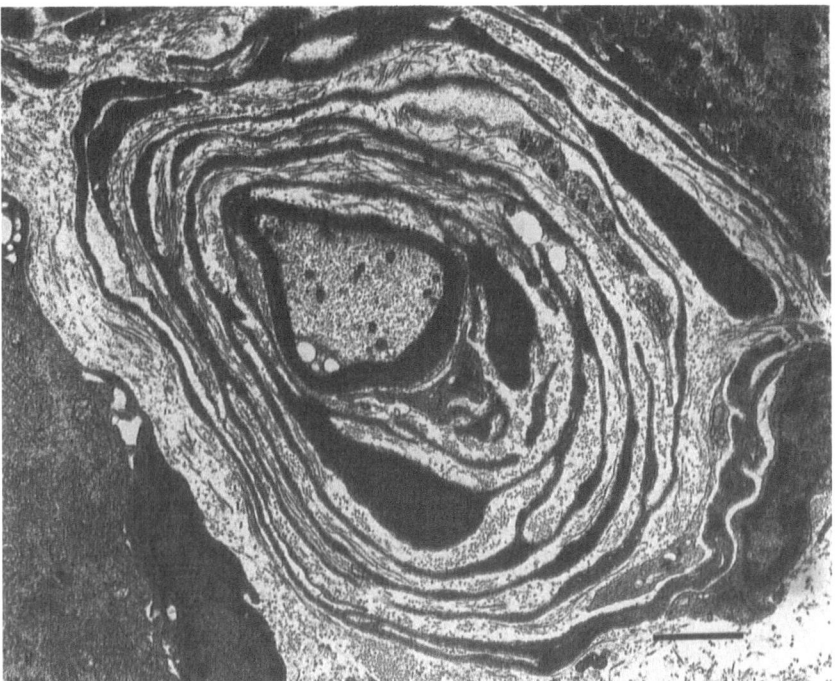

Fig. 3.11. Onion bulb. A remyelinated axon between the dorsal root ganglion cells is surrounded by layers of Schwann cell and fibroblast processes. Bar = 2 μm. (From Adam et al. (1989), with permission.)

the rabbit in which a single injection of a large dose of bovine spinal root myelin regularly produced a chronic relapsing or progressive form of EAN (Harvey et al. 1987). There was severe slowing of nerve conduction velocity and such marked dispersion of evoked muscle action potentials that the authors were unable to determine whether conduction block was present. Prominent onion bulb formation was noted in the nerve roots, and, to a lesser extent, in distal peripheral nerves.

Neurophysiology

The characteristic neurophysiological changes of a demyelinating neuropathy were first clearly demonstrated in EAN by Cragg and Thomas (1964), who immunised guinea pigs with rabbit sciatic nerve. When the animals had become paralysed, the peroneal and tibial nerves were removed for neurophysiological and then pathological study. The conduction velocity remained normal in some nerves. In others conduction was slowed with dispersion of the nerve action potential and in the most severely affected nerves conduction was blocked. They speculated that when slowing of conduction occurred it was due to loss of saltatory conduction and slower transmission of the nerve action potential across the demyelinated segments. Subsequent experiments on nerve conduction in situ have confirmed these findings (Tuck et al. 1981, 1982; Rostami et al. 1984). In addition Tuck et al. (1981) have shown that the F wave latencies following stimulation of the posterior tibial nerve were prolonged in the guinea pig and rabbit, sometimes even when the peripheral motor nerve conduction velocity remained normal. This finding implies delayed conduction due to involvement of the proximal portions of the sciatic nerve or ventral roots. These were shown by histological studies to exhibit inflammation and demyelination in some but not all the relevant animals. Meticulous recordings from the spinal nerves, dorsal root ganglia and dorsal root entry zones have shown that the dorsal root ganglion is a site of predilection for slowing of conduction and probable conduction block due to demyelination in rabbits with EAE. The existence of actual conduction block in the dorsal root ganglion was demonstrated in single-fibre studies on a few fibres (Pender and Sears 1984). It is probable that the situation with EAN in the rabbit is similar, since histologically dorsal root ganglia are particularly affected by infiltration and demyelination.

Cerebrospinal Fluid

Since an increased CSF protein concentration and normal cell count are usual, although not invariable findings in GBS, it is appropriate to consider the CSF changes in EAN. In rabbits with EAN, the cisternal CSF protein concentration was increased in most animals compared with either normal animals or animals injected with Freund's complete adjuvant without nerve tissue (Waksman and Adams 1956). There was also a modest increase in CSF cell count in some animals but much less marked than in EAE.

Identification of the Antigen Responsible for EAN

P_2 Protein

Early attempts to purify the antigen responsible for EAN from nerve tissue were frustrated by the loss of activity during chemical extraction, especially when the guinea pig was used as the experimental animal. This was disappointing because Waksman and Adams (1956) had found that the antigen was resistant to autoclaving. It was later discovered that the major antigen, P_2, is subject to critical conformational changes following delipidation. The guinea pig was not the most suitable animal because it regularly develops EAE following immunisation with minute quantities of myelin basic protein. The antigen causing EAN in the guinea pig has still not been rigorously identified.

In early experiments acid extracts of human nerve were shown to induce EAN in rabbits, rhesus monkeys and baboons (Behan et al. 1969) and guinea pigs (Sheremata and Behan 1973). However, a basic protein extracted from human peripheral nerve induced EAE and not EAN in guinea pigs (Paty 1971). The presumption that the antigen was in myelin was confirmed when unmyelinated fetal nerves failed to induce EAN in rabbits (Robinson et al. 1972) and nerves which had lost all myelin debris following Wallerian degeneration did not cause EAN in guinea pigs (McDermott and Wisniewski 1977). Curtis et al. (1979) were able to produce EAN in Lewis rats by immunisation with mammalian (human, bovine, rabbit, guinea pig and rat) peripheral nerve myelin but not with amphibian (frog) myelin.

During the 1970s there was a strong feeling that P_2 protein, which was relatively confined to the PNS rather than the CNS, ought to be the responsible antigen but proof was elusive. Brostoff et al. (1972) isolated P_2 protein from rabbit sciatic nerve and found that it produced a mixture of EAE and EAN in the monkey (whereas whole sciatic nerve produced EAN alone) and EAE in the guinea pig. Abramsky and London (1975) used the method of London (1971) to prepare a basic protein from bovine sciatic nerve which was probably the same as P_2. They reported mild neurological manifestations of EAN in guinea pigs and rabbits with mild histological lesions in the PNS and none in the CNS. This was the first demonstration that P_2 protein would induce EAN, but it is puzzling that subsequent work failed to reproduce the result in the guinea pig (Kadlubowski et al. 1980). Subsequently Abramsky et al. (1977) induced mild neurological deficit and histological signs of EAN with P_2 protein from bovine sciatic nerve. Until this time P_2 protein had been prepared following delipidation. Some of these early preparations of P_2 contained substantial amounts of histidine, which is now known to be absent from P_2 but abundant in MBP, which suggests that the preparations were contaminated by a MBP peptide (Kadlubowski et al. 1980).

Kadlubowski and Hughes (1979) purified P_2 protein from an acid extract prepared directly from bovine spinal root myelin without any preliminary delipidation and obtained a preparation which was much more active in producing EAN in Lewis rats. A sample of 5 µg was sufficient to produce EAN (Kadlubowski et al. 1980). In subsequent experiments human P_2 protein was also shown to induce EAN in the rat (Kadlubowski and Hughes 1980; Suzuki et al. 1980) and both human and bovine P_2 proteins were shown to induce EAN in the rabbit (Kadlubowski and Hughes 1980). It was not possible to induce EAN in guinea pigs with P_2. However, it remains probable that P_2 is responsible because guinea pig EAN can be suppressed by repeated injections of P_2 in saline from the day of onset of disease induced by immunisation with whole myelin (McDermott and Keith 1979). The experiment has been repeated in Lewis rats (Cunningham et al. 1983). The ability of P_2 protein to produce EAE and EAN in different species is summarised in Table 3.1.

P_2 protein not only reproduces the neurological manifestations of EAN produced by myelin but also induces the same histological changes of perivascular cellular infiltration and primary demyelination (Hughes and Powell 1984; Rostami et al. 1984). This point is stressed because in acute EAE, particularly the disease induced with MBP rather than myelin, inflammatory changes are prominent but there is little demyelination.

The ability of P_2 protein to induce EAN is thought to depend on conformational changes because of the variability of activity of different preparations. The preparation of Kadlubowski et al. (1980) produced EAN in Lewis rats at 25 µg in all and 5 µg in some animals. This preparation had been made without

Table 3.1. Disease induced by bovine and human spinal root myelin and P_2 protein in Lewis rats, guinea pigs, rabbits and mice

	Human myelin	Bovine myelin	Human P_2	Bovine P_2
SJ/L Mouse	not done	EAN	not done	EAN
Lewis rat	EAN[a]	EAN	EAN	EAN
			EAE	
Guinea pig	EAN	EAN	EAE	EAN
	EAE			**EAE**
Rabbit	EAN	EAN	not done	EAN
				EAE

Data from Kadlubowski and Hughes (1980) and Taylor and Hughes (1985).
[a] Bold type indicates that the disease in question is easily elicited and the neurological and histological manifestations are severe.

preliminary delipidation. Curtis et al. (1979) found that P_2 protein prepared from delipidated rabbit myelin was inactive, but if the pellet extracted from the myelin with acid was added back to the purified P_2 activity was restored. Ishaque et al. (1981) reported that disease-inducing activity in rats could be restored to their P_2 preparation by adding phosphatidylserine. Rostami et al. (1984) reported that disease induction by their bovine P_2 preparation was enhanced by the addition of phosphatidylserine to the emulsion and Curtis et al. (1979) were able to induce EAN with P_2 and phosphatidyl serine in Rhesus monkeys. Similarly Nagai et al. (1978) found that the disease-inducing activity in rabbits of their bovine P_2 protein was increased by addition of gangliosides. The explanation of these slightly conflicting results is probably that the disease-inducing epitope of P_2 is damaged by delipidation of myelin before extraction. However, the conformation can be altered back to one that is favourable for presenting the epitope in a form recognisable by T cells by addition to some acidic lipids or gangliosides. Kadlubowski et al. (1980) presented data that their P_2 preparation was not contaminated with lipid. Weise et al. (1980) also showed that severe EAN can be induced by purified P_2 preparations in the absence of lipid. Subsequent demonstrations that EAN can be produced with synthetic peptides clearly indicate that lipid is not required for induction of EAN.

P_2 Peptides

The peptide sequence of P_2 protein was rapidly worked out for the human, bovine and rabbit proteins (Fig. 3.12) but the identity of the sequence within the protein which was responsible for EAN remained uncertain until recently. The initial approach was to cleave the protein with cyanogen bromide which produced one large peptide, CN_1, and two smaller peptides CN_2 and CN_3. CN_2, which is at the C-terminal end, is itself composed of two peptides joined by a disulphide bridge (Weise et al. 1980) and CN_3 is at the N-terminal end. The peptides are detailed in Fig. 3.12 and illustrated diagrammatically with information about their disease-inducing activity in Fig. 3.13. In the rat, it is clear that the essential sequence is contained within the largest fragment, CN_1, re-

```
                                                    10                                              20
Human   Ac-Ser-Asn-Lys-Phe-Leu-Gly-Thr-Trp-Lys-Leu-Val-Ser-Ser-Glu-Asn-Phe-Asp-Asp-Tyr-Met-
Bovine                                                                                     Glu
Rabbit
PEPTIDE<---------------------------------------- CN3 --------------------------------->
                                                    30                                              40
Human   -Lys-Ala-Leu-Gly-Val-Gly-Leu-Ala-Thr-Arg-Lys-Leu-Gly-Asn-Leu-Ala-Lys-Pro-Thr-Val-
Bovine                                                                                     Arg
Rabbit                                                                                     Asn
PEPTIDE<---------------------------------------- CN1 ----------------------------------
                                                    50                                              60
Human   -Ile-Ile-Ser-Lys-Lys-Gly-Asp-Ile-Ile-Thr-Ile-Arg-Thr-Glu-Ser-Thr-Phe-Lys-Asn-Thr-
Bovine                                                                      Pro
Rabbit
PEPTIDE -------------------------------------- CN1 ------------------------------------
                                                    70                                              80
Human   -Glu-Ile-Ser-Phe-Lys-Leu-Gly-Gln-Glu-Phe-Glu-Gln-Thr-Thr-Ala-Asp-Asn-Arg-Lys-Thr-
Bovine
Rabbit                                                            Glu
PEPTIDE -------------------------------------- CN1 ------------------------------------
                                                    90                                             100
Human   -Lys-Ser-Ile-Val-Thr-Leu-Gln-Arg-Gly-Ser-Leu-Asn-Gln-Val-Gln-Arg-Trp-Asn-Gly-Lys-
·Bovine           Thr            Ala                                     Lys            Asn
Rabbit            Glu            Ala                                     Lys
PEPTIDE -------------------------------------- CN1 ------------------------------------
                                                   110                                             120
Human   -Glu-Thr-Thr-Ile-Lys-Arg-Lys-Leu-Val-Asp-Gly-Lys-Met-Val-Ala-Glu-Cys-Lys-Met-Lys-
Bovine                                                           Val
Rabbit
PEPTIDE ----------- CN1 ----------------------------------->< ----------- CN2 ----------
                                                   130
Human   -Gly-Val-Val-Cys-Thr-Arg-Ile-Tyr-Glu-Lys-Val-OH
Bovine     Asp
Rabbit
PEPTIDE ----------- CN2 -------------------------------->
```

Fig. 3.12. Peptide sequence of human P_2 protein (*top row*, from Suzuki et al. (1982)). The sequences of the bovine (*second row*, from Kitamura et al. (1980)) and rabbit (*third row*, from Ishaque et al. (1982)) P_2 proteins are the same except for the residues shown. The sites of cleavage into CN_1, CN_2 and CN_3 peptides are given.

presenting residues 21–129. In the rabbit and guinea pig, the situation is not yet clear, but it does seem likely that the carboxy-terminal fragment induces EAE rather than EAN. The smallest sequence responsible for EAN in the Lewis rat is not yet unequivocally identified. There are two reports that sequence 53–78 is active (Uyemura et al. 1982; Rostami et al. 1988) but this could not be confirmed by Whitaker and Seyer (1984) or Olee et al. (1989). The positive reports (Uyemura et al. 1982; Suzuki et al. 1984) stated that only histological disease was produced. The other report, which claimed more severe disease, is so far only available in abstract form (Rostami et al. 1988). On the other hand the synthetic peptides 57–81 and 60–81 produced both neurological and histological disease comparable to that produced by the whole protein (Olee et al. 1989). Olee et al. (1989) discuss the predictions that residues around 70, especially 57–76, will be the only parts of the P_2 molecule to form an

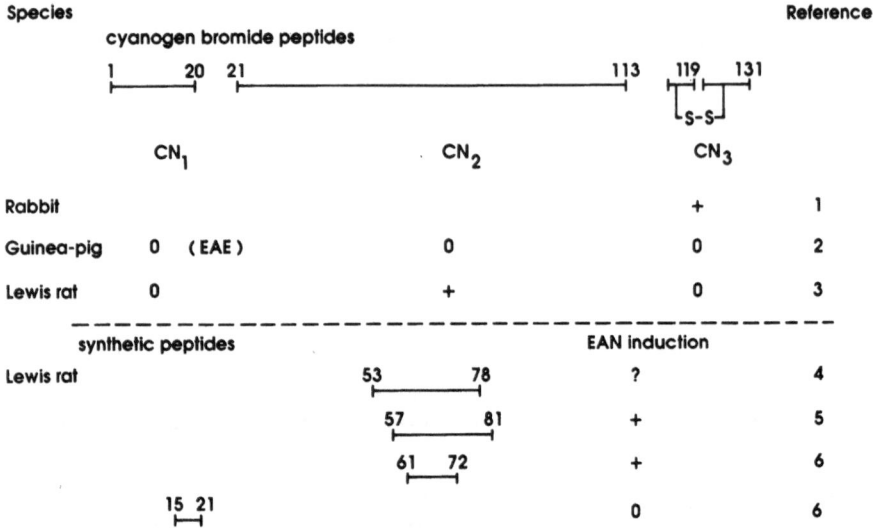

Fig. 3.13. Diagram of cyanogen bromide derived and synthetic peptides of P_2 protein and their EAN-inducing ability in the rat, rabbit and guinea pig. Note that CN_2 is composed of two sequences which are bridged by a disulphide bond. 1, Brostoff et al. (1977), Hsieh et al. (1981); 2, Brostoff et al. (1980); 3, Weise et al. (1980), Szymanska et al. (1981), Uyemura et al. (1982); 4, Uyemura et al. (1982), Rostami et al. (1988), Whitaker and Seyer (1984), Olee et al. (1988); 5, Olee et al. (1988); 6, Olee et al. (1989).

α-helical structure and that residues 61–72 will be amphipathic. This is relevant because T cell receptors tend to recognise amphipathic sections of α-helical structures (De Lisi and Berzofsky 1985). The section 61–72 is contained within both the 53–78 peptide of Uyemura et al. (1982) and the 57–81 peptide of Olee et al. (1989).

Finally Olee et al. (1989) have shown that immunisation with a synthetic peptide corresponding to residues 61–72 will induce as severe EAN as equimolar amounts of whole P_2 protein. In addition this peptide will stimulate T cell lines raised against either the whole P_2 protein or the P_2 57–81 peptide. Shortening the 61–72 peptide at the carboxy-terminal end reduces the ability of the peptide to stimulate these T cell lines whereas addition of the residues 59–60, 58–60 or 57–60 enhances this ability. Whether a peptide slightly lengthened by these residues could induce more severe EAN has not been tested. Peptide 15–21 is the other α-helical amphipathic sequence in P_2 protein but this does not induce EAN (Olee et al. 1989). The situation is summarised in Fig. 3.13.

P_0 Protein

Just as EAE has eventually been shown to be induced by the proteolipid protein (Yoshimura et al. 1985; Endoh et al. 1986), so EAN has recently been shown to be induced by P_0 protein, a glycoprotein specific for peripheral nerve myelin (Milner et al. 1987). P_0 protein was purified from the acid-insoluble residue of bovine spinal root myelin in the presence of the denaturing solvent sodium dodecylsulphate (SDS). Following dialysis of the P_0 preparation against lyso-phosphatidylcholine, a detergent required to keep the P_0 in solution, much of the SDS was removed. The absence of detectable quantities of P_2 protein was demonstrated by SDS gel electrophoresis and immunoblotting, and the P_0 protein was shown to induce inflammatory lesions in the nerve roots and peripheral nerves. Compared with P_2 the water-insoluble P_0 glycoprotein is difficult to isolate. Further experiments to investigate the morphology of the disease induced by it and of immune responses to P_0 in EAN and GBS would be worthwhile.

Genetic Factors

As well as species differences on susceptibility to EAN there are profound differences within different strains of the same species. Thus Lewis rats are particularly susceptible, Wistar and Sprague–Dawley rats relatively resistant, and Buffalo and Brown Norway rats completely resistant to induction of EAN with bovine myelin (Hoffman et al. 1980; Steinman et al. 1981). These differences correspond to the responsiveness of the lymphocytes from immunised rats of the different strains to P_2 in vitro (Steinman et al. 1981). Detailed experiments to determine whether the differences in responsiveness are determined at the level of the antigen presenting cell or T lymphocyte have not been performed. Some intriguing experiments by Linington et al. (1986) have shown that an anti-P_2 cell line derived from the Brown Norway rat strain, which is resistant to actively induced EAN, could transfer EAN to naïve Brown Norway recipients. The disease was slower in onset and less severe than in Lewis rats. A partial explanation for the difference between the two strains was that the Lewis T cell line responded to peptide 53–78, which is one of the sequences which has been reported to induce EAN, whereas the Brown Norway cell line did not. It is likely that the Brown Norway lymphocyte "sees" a different disease-inducing epitope. It remains unclear why the Brown Norway strain should be susceptible to passively induced EAN but not to active immunisation. It had been thought that the susceptibility of each strain depended on its major histocompatibility antigen and was the same as that for EAE. A recent abstract suggests that this may not be the case since a substrain of Lewis rats is resistant to EAE but susceptible to EAN (Rostami et al. 1986a).

The mouse would be a convenient subject for immunological studies of EAN but unfortunately the disease is difficult to induce and the neurological manifestations are extremely mild. However, the SJL strain is particularly susceptible and the SWR, BALB/C, C57BL6 and AKR strains are relatively or absolutely resistant (Taylor and Hughes 1985; Rostami et al. 1986b).

Antibody Production in EAN

It is easy to show that animals with EAN develop antibodies against antigens within the immunising inoculum but more difficult to show that the antibodies have anything to do with the pathogenesis of the disease. This was shown by

Caspary and Field (1965) who identified antibodies to sciatic nerve basic protein by a haemagglutination technique in rats immunised with human sciatic nerve. A curious early report from Russia was that rabbits with surgical destruction of the posterior hypothalamus had decreased ability to produce antibodies but developed more severe EAN than unoperated controls (Konovalov et al. 1971). It is difficult to draw conclusions from this experiment since the operation may also have damaged the pituitary–adrenal axis and so reduced the ability to suppress any sort of inflammation. Lewis rats immunised with bovine spinal root myelin were shown to develop antibodies to P_2 which can be detected, for instance, by radioimmunoassay (Hughes et al. 1981a). If the immunising antigen is myelin, antibodies to P_2 can only be detected in the serum a day or two after the animals have developed neurological manifestations and persist long after recovery. Furthermore, antibodies may not be detected in animals with passively transferred disease, and yet detected in others to which cells were transferred but disease did not develop (Hughes et al. 1981a). The very erratic correlation between the presence and titre of antibodies and disease manifestations makes it unlikely that circulating antibodies to P_2 have a major role in disease pathogenesis.

Nevertheless, several observations suggest that antibodies against other myelin antigens, predominantly lipids, can play a role in inducing demyelination. Complement-fixing antibodies to peripheral and central nervous system tissue have been described in the serum of rabbits with EAN (Waksman and Adams 1955). In the light of subsequent knowledge these were probably antibodies to galactocerebroside. Serum from rabbits or guinea pigs with EAN will induce demyelination of mouse dorsal root ganglion cultures (Yonezawa et al. 1968; Raine and Bornstein 1979). The effect can be abolished by removing complement or by absorbing the serum with nerve tissue, consistent with the effect being a specific complement-dependent antibody-mediated reaction. The demyelinating effect can be reproduced by antibodies to galactocerebroside (Saida et al. 1981; Raine et al. 1981) but not antibodies to P_2 (Seil et al. 1981). In the absence of complement, EAN serum, like EAE serum, produces an interesting ultrastructural change in the myelin, a widening of the lamellar spacing in the outer part of the sheath which resembles the change seen particularly with IgM paraproteinaemia and antimyelin-associated glycoprotein antibodies (Chapter 12). The demyelinating effect of acute EAN serum has been confirmed in the rat intraneural injection model. Saida et al. (1978) first showed that rabbit EAN serum induced focal demyelination following injection into the rat sciatic nerve. This effect was also shown to be complement-dependent and absorbable with peripheral nerve tissue. The demyelinating effect can be produced by antibodies to galactocerebroside or P_0 but not by antibodies to P_2, myelin basic protein or ganglioside (Saida et al. 1981; Hughes et al. 1985).

Cell-Mediated Immunity in EAN

It is also not surprising that evidence of cell-mediated immunity to myelin antigens can be detected during EAN. This was shown with skin tests to whole nerve tissue by Waksman and Adams (1955, 1956). Hughes et al. (1981a) showed that weakly positive delayed hypersensitivity-type skin tests could be elicited to P_2 during the very early stages of EAN but that the tests became negative at the height of the disease. In EAE, skin tests to myelin basic protein follow a similar pattern. This suppression of delayed hypersensitivity skin tests at the height of the disease has never been explained but it is probably non-specific since delayed hypersensitivity to tuberculin becomes depressed at the same time. In vitro tests of cell-mediated immunity have also been shown to become positive. For instance, early in the course of rat or rabbit EAN induced with myelin, lymph node and blood mononuclear cells are stimulated to transform by P_2 protein (Hughes et al. 1981a; Nomura et al. 1987; Taylor and Hughes 1988). Since these results are relevant to the interpretation of the results of similar tests in patients with GBS it is important to emphasise that the responses of the blood cells are quite low and not significantly different from controls, especially controls immunised with adjuvant, except in the early stages of the disease (Taylor and Hughes 1988). The responses of cells extracted from the cauda equina to P_2 protein are greater and persist for longer (Taylor and Hughes 1988). The response of lymph node cells from rats immunised with P_2 protein is more impressive and more prolonged than that of rats immunised with myelin (Hughes et al. 1981a): the response to the large CN_1 (disease-inducing) peptide was greater than that to the CN_2 and CN_3 (non-disease inducing) peptides (Milek et al. 1983). The evidence from skin tests and in vitro tests is consistent with a major role for cell-mediated immunity to P_2 in the pathogenesis of EAN.

Careful observations of the phenotype of circulating lymphocytes have shown a depression of the percentage of blood CD8+ cells 21 days after immunisation with Lewis rats just after the peak of the disease (Brosnan et al. 1985). Similar alterations are noted in some patients with GBS and progressive multiple sclerosis but their significance is unclear. One possibility is that CD8+ cells enter the endoneurium in increased numbers at this time. The CD8+ cells include cells serving a suppressor function as well as those having a cytotoxic function. Entry of cells with a suppressor function would be appropriate at a time just before the onset of improvement. However, studies of cells in the lesions have not shown increased proportions of CD8+ cells in the later stages (Hughes et al. 1987).

The intravenous injection of cells from one animal into another has been repeatedly shown to induce EAN whereas transfer of serum does not (Astrom and Waksman 1962; Hughes et al. 1981a). EAN lymphocytes have been shown to induce limited demyelination in trigeminal ganglion cultures (Arnason et al. 1969) and following direct injection into rat sciatic nerve (Arnason and Chelmicka-Szorc 1972; Gilbert et al. 1983). There is dispute concerning whether the small amount of demyelination seen under these circumstances is significant (Brosnan et al. 1984).

Cell transfer experiments were greatly improved by the development of techniques to maintain T cells by alternate incubations with antigen and T cell growth factor. These techniques had been developed and used to generate T cell lines against MBP which will transfer EAE. Linington et al. (1984), by an elegant series of experiments, showed that a Lewis rat T cell line which they had generated against P_2 would transfer disease to naive recipients. Signs developed about four to five days after intravenous injection, whereas active immunisation does not induce disease until after about 12 days. The disease could be transferred with as few as 50 000 cells per animal and had all the morphological and electrophysiological features of actively induced EAN (Izumo et al. 1985;

Heininger et al. 1986). This model permitted a dissection of the morphological evolution of the transferred disease which showed oedema *with* or *without* cell infiltration in the nerves and roots after four days and then extensive cell infiltration and demyelination associated with macrophage infiltration. Degranulated mast cells were noted in the early stages and occasional Wallerian degeneration was also seen, just as in actively induced EAN. Transfer of $1\,000\,000$ anti-P_2 T cell line cells induced mild disease with electrophysiological findings of slowed conduction indicating demyelination whereas twice the number of cells induced several and more persistent paralysis. In these severely affected animals the F waves became delayed and the sciatic nerves became inexcitable between four and five days after immunisation which the authors interpreted as due to acute axonal damage at root level (Heininger et al. 1986). It is notable that these transfers reproduced the range of electrophysiological findings seen both in actively induced EAN and in GBS. Rostami et al. (1985) also produced a Lewis rat anti-P_2 T cell line which transferred the neurological and histological manifestations of EAN to naive recipients. The T cell lines used in these experiments expressed the CD4 gene product which is the usual surface marker on T helper cells. Unfortunately these experiments do not resolve the question of how the demyelination is finally produced. It might be via antibody production, recruitment of macrophages or a direct myelinotoxic effect of the T-helper cells themselves. Not all anti-P_2 T cell lines induce disease. For instance, the line prepared by Taylor and Hughes (1988) did not transfer disease but did prevent the development of EAN by subsequent active immunisation. It is now necessary to overcome the problems in preparing rat T cell clones and define the peptide sequences recognised by disease-inducing and disease-suppressing clones.

Further evidence that T cells are essential for the induction of EAN comes from an experiment in which Lewis rats were depleted of T cells by adult thymectomy, lethal irradiation and reconstitution with bone marrow cells. Only a small proportion of rats so treated developed EAN and in those rats which did develop EAN the T cell depletion was found to have been incomplete (Brosnan et al. 1987). Similarly, it has been shown that treatment with cyclosporin A, which predominantly affects T cell function, will suppress EAN (King et al. 1983). Although these experiments establish the need for T cells it still leaves open the question of the mechanism.

Galactocerebroside Neuritis

The glycolipid, galactocerebroside, produces a demyelinating neuropathy which is different from the type of EAN that has been discussed so far but which is a very interesting model in its own right. The disease has already been discussed briefly in Chapter 2. It was first induced by T. Saida et al. (1979b) with repeated injections of galactocerebroside into rabbits, which is the only species discovered to be susceptible. Weight loss and progressive limb weakness, sensory impairment, tremor and unsteadiness began two to four months after immunisation. These features evolved over two or three weeks and sometimes caused tetraplegia and respiratory embarrassment (Saida et al. 1981). The first electrophysiological abnormality was prolongation of somatosensory-evoked responses at a stage when peripheral nerve conduction remained normal. After several months conduction velocity became slowed and F wave and distal motor latencies were prolonged in a manner suggesting demyelination. There was no marked decrease of proximally compared with distally evoked muscle action potentials which would have indicated conduction block (Stoll et al. 1986). Pathologically the prominent finding has been perivascular demyelination especially in nerve roots with macrophage-mediated stripping and phagocytosis of myelin debris followed by remyelination. The lesions differ from those of P_2-induced demyelination because there is an absence of perivascular lymphocytic infiltration (Saida et al. 1981; Stoll et al. 1986). Since antigalactocerebroside serum will produce demyelination in tissue culture (T. Saida et al. 1979a; Raine et al. 1981) and conduction block and demyelination following intraneural injection (K. Saida et al. 1979; Sumner et al. 1982; Hughes et al. 1985), it is tempting to assume that antigalactocerebroside antibodies are responsible for the demyelination in this model. This may be true but it is puzzling that antibodies persist in rabbits while the disease is recovering and remyelination is proceeding. Although some of its morphological and electrophysiological characteristics would make this an attractive model for CIDP, and possibly GBS, significant amounts of antibody to galactocerebroside have not been identified in any human demyelinating disease.

Treatment of EAN

Several experiments have been published demonstrating prevention or treatment of EAN with immunomodulatory experiments directed towards investigating the immune mechanisms of the disease or identifying treatments which would be worth evaluating in human GBS. Steroids have a powerful anti-inflammatory effect and non-specifically suppress all components of the immune response. An initial study of the effect of relatively small doses of hydrocortisone (about 5 mg/kg), failed to alter the course of EAN in five rabbits: one died of bacteraemia, a reminder of the potential hazards of such a regime (Heitman and Mannweiler 1956). In rats, prednisolone 1 mg/kg was shown to reduce the severity of disease when started just before (Hughes et al. 1981b) or after the appearance of the first sign (King et al. 1985). Furthermore, even larger doses of methylprednisolone given after the onset of signs dramatically reduced the neurological deficit and the severity of cell infiltration and demyelination (Watts et al. 1989). Methylprednisolone 5 mg/kg partially suppressed or prevented EAN in rabbits (Ohno et al. 1988). Dexamethasone 4 mg/kg has been shown to inhibit markedly the transfer of EAN induced by a P_2 cell line (Heininger et al. 1988). The successful results of treatment of EAN with large doses of steroids soon after the onset of the disease have been one of the factors leading to a re-evaluation of the usefulness of steroids in GBS (Chapter 8).

Treatments aimed at both the humoral and cell-mediated arms of the immune response have been successful in suppressing or treating EAN. Complement depletion with cobra venom factor just before the expected onset of EAN has been shown to delay the onset of EAN for as long as complement levels in the blood stay low and to reduce the amount of demyelination in the spinal roots

(Feasby et al. 1987). This suggests a significant role for complement and probably complement-fixing antibodies, in the early stages of EAN. However, complement may contribute to other components of the inflammation in EAN. It has been shown that P_0 will activate complement via the alternative complement pathway (Chapter 2) which might contribute to the inflammation induced by any form of myelin damage. Plasma exchange performed during the incubation period (Antony et al. 1981) or following the onset of neurological deficit (Gross et al. 1983) reduced the severity of neurological and histological manifestations of EAN in rabbits. The presumption is that the main effect of plasma exchange is removal of antibody. The possibility of other effects including the removal of lymphokines involved in T cell activation cannot be excluded.

Treating rats with monoclonal antibodies to T cells or T cell subsets usually suppresses EAN. Antibodies to CD4, CD8 and MHC class II antigens all reduced the severity of the disease when given on day 9 after immunisation, just before the onset of neurological signs (Strigard et al. 1988). Similarly another monoclonal antibody (called W3/13) which labels T cells and polymorphonuclear leucocytes was followed by less severe disease (Strigard et al. 1988, 1989). On the other hand a monoclonal antibody to CD5 lymphocytes (OX19) was followed by *worse* disease. The authors suggest that OX19 merely modulates CD5 expression and does not eliminate T cells and that CD5 expression may play a role in suppressing immune responses. This is an interesting idea and it will be important to confirm the experimental observation on which it is based. A monoclonal antibody directed against the interleukin-2 receptor, which is present on activated T cells, inhibited the development of passively transferred EAN (Hartung et al. 1987).

Prophylaxis or treatment with cyclosporin A of guinea pigs or rats has been shown to suppress or reduce the neurological and histological manifestations of EAN (King et al. 1983). Cyclosporin A also suppressed EAN induction in the recipients of anti-P_2 T cell line cells (Hartung et al. 1987). Since the predominant effect of cyclosporin A is to interfere with IL-2 production and proliferation of T cells this result provides further support for the notion that T cell-mediated mechanisms are important in EAN.

Previous administration of nerve antigens in saline reduces the incidence and severity of EAN following subsequent standard immunisation with nerve in complete Freund's adjuvant (Lehrich and Arnason 1971; Brosnan et al. 1988). Similar experiments have been undertaken in EAE in which it has been suggested that antibodies or T lymphocytes generated by the previous immunisation exert a suppressor function. In modern immunological terms these could be regarded as anti-idiotypic antibodies or T lymphocytes. McDermott and Keith (1979) used a similar technique to demonstrate that bovine P_2, but not MBP, would inhibit the induction of EAN in guinea pigs by whole myelin. Brosnan and Tansey (1984) performed an imaginative experiment in which the onset of EAN was delayed by depleting rats of serotonin with an injection of reserpine just before the expected disease onset. Serotonin is the most vasoactive amine known, being 100 times more powerful than histamine. It was argued that if mast cells mediated the increased vascular permeability and contributed to inflammation in delayed hypersensitivity reactions, serotonin depletion should suppress the development of the disease. Their prediction was fulfilled in that the disease was delayed though the treated animals did still develop disease, albeit two or three days later than usual. However, the result does not clinch a role for mast cells in EAN because serotonin might be derived from other cells, and its release might be mediated by antibody rather than delayed hypersensitivity reactions.

The same authors performed another imaginative experiment on the importance of macrophages in EAN (Tansey and Brosnan 1982). Injections of silica quartz dust had been shown to inhibit macrophage function and so they gave intraperitoneal injections of 200 mg silica dust to Lewis rats "incubating" EAN. Injections 8 days, or 11 days, or 8 and 11 days after immunisation reduced the incidence and severity of neurological manifestations significantly. There was also a reduction in the histological manifestations of the disease. The effect of this treatment was quite dramatic and supports a role for macrophages in EAN which might operate at several levels from antigen presentation to T cells to the final common pathway of macrophage stripping of myelin. Since silica has been shown to inhibit markedly the development of EAN in rats injected with a P_2 cell line it is likely that silica inhibits the effector pathway, i.e. macrophage stripping (Heininger et al. 1988). Macrophages generate the inflammatory mediators prostaglandins via the cyclo-oxygenase pathway and leucotrienes via the lipoxygenase pathway. Inhibiting prostaglandin synthesis with indomethacin attenuated the disease, whereas a selective lipoxygenase blocker, nafazatrom, had little effect (Hartung et al. 1988). Silica and dexamethasone, but not indomethacin or nafazatrom, markedly inhibited the EAN induced by transferring a P_2-specific T cell line (Heininger et al. 1988). This rather unexpected result suggests that the prostaglandin pathway plays little part in the pathogenesis of T cell-line induced EAN. The inhibition of actively induced EAN by indomethacin might be due to interference with an early phase in disease induction, or possibly to inhibition of antibody and complement-mediated components of the inflammation.

Taylor et al. (1988) fed Lewis rats on fish oil which is rich in eicosapentanoic acid. This long-chain fatty acid is metabolised to the leukotriene, LTB_5, a less-inflammatory compound than LTB_4 which is the usual leukotriene produced from arachidonic acid in the normal diet. The fish oil diet successfully induced synthesis of LTB_5 by leucocytes but did not affect the neurological or histological manifestations of EAN any more than a diet of beef tallow. This may have been because LTB_4 synthesis was not suppressed.

Summary

EAN was first described by Waksman and Adams (1955) and can be induced in many animal species by immunisation with peripheral nerve myelin in adjuvant. The best-studied disease is induced in the Lewis rat by immunisation with bovine myelin. A basic protein P_2, molecular weight 15000, in peripheral nerve myelin is the major responsible antigen. A glycoprotein P_0, molecular weight 28000, has also been reported to induce disease. The important epitope in P_2 protein is probably a short amphipathic sequence, one of the α-helical parts of the molecule, residues 61–72. The neurological manifestations are rapidly progressive weakness followed by improvement over about 10 days and then persistent minor deficit with a spontaneously relapsing and remitting course.

Electrophysiologically there is slowing of conduction followed by conduction block. Pathologically leakage of immunoglobulin due to impairment of the normally restrictive blood–nerve barrier, endoneurial oedema and cellular infiltration develop simultaneously. The cell population consists of T lymphocytes and macrophages. Loss of myelin is largely due to primary demyelination, is always cell-associated, and proceeds predominantly by macrophage stripping and phagocytosis. Several affected animals also have significant amounts of axonal degeneration. The precise mechanism whereby the macrophages are triggered to invade the Schwann cell basal lamina and mesaxon remains unclear. However, the disease can be transferred to naive recipients with a T-helper cell line and prevented or suppressed by T cell depletion. These experiments indicate that T cells are important but do not rule out an additional role for antibody. EAN has many neurophysiological and pathological similarities with GBS and also CIDP.

A separate model of macrophage-mediated demyelinating polyradiculoneuropathy has been induced in rabbits by immunisation with galactocerebroside in adjuvant. Some, but not all, of the evidence supports the idea that antibodies to galactocerebroside are important in the pathogenesis of this "galactocerebroside neuritis", which has neurophysiological and pathological similarities to CIDP.

References

Abramsky O, London Y (1975) Purification and partial characterization of two basic proteins from human peripheral nerve. Biochim Biophys Acta 393:556–562

Abramsky O, Teitelbaum D, Arnon R (1977) Experimental allergic neuritis induced by a basic neuritogenic protein (P1L) of human peripheral nerve origin. Eur J Immunol 7:213–217

Acha-Orbea H, Mitchell DJ, Timmermann L, et al. (1988) Limited heterogeneity of T cell receptors from lymphocytes mediating autoimmune encephalomyelitis allows specific immune intervention. Cell 54:263–273

Adam AM, Atkinson PF, Hall SM, Hughes RAC, Taylor WA (1989) Chronic experimental allergic neuritis in Lewis rat. Neuropathol Appl Neurobiol 15:249–264

Allt G (1975) The node of Ranvier in experimental allergic neuritis. An electron microscopic study. J Neurocytol 4:63–76

Antony JH, Pollard JD, McLeod JG (1981) Effects of plasmapheresis on the course of experimental allergic neuritis in rabbits. J Neurol Neurosurg Psychiatry 44:1124–1128

Appenzeller O, Arnason BGW, Adams RD (1965) Experimental autonomic neuropathy: an immunologically induced disorder of reflex vasomotor function. J Neurol Neurosurg Psychiatry 28:510–515

Arnason BGW, Chelmicka-Szorc E (1972) Passive transfer of experimental allergic neuritis in Lewis rats by direct injection of sensitised lymphocytes into the sciatic nerve. Acta Neuropathol 22:1–6

Arnason BGW, Winkler GF, Hadler NM (1969) Cell-mediated demyelination of peripheral nerve in tissue culture. Lab Invest 21:1–10

Asbury AK, Arnason BG, Adams RD (1969) The inflammatory lesion in idiopathic polyneuritis. Its role in pathogenesis. Medicine 48:173–215

Askenase PW, Bursztajn S, Gershon NID, Gershon RK (1980) T cell dependent mast cell degranulation and release of serotonin in immune delayed-type hypersensitivity. J Exp Med 152:1358–1374

Astrom KE, Waksman BH (1962) The passive transfer of experimental allergic encephalomyelitis and neuritis with living lymphoid cells. J Pathol Bacteriol 83:89–100

Ballin RHM, Thomas PK (1968) Electron microscopic observations on demyelination and remyelination in experimental allergic neuritis. Part 1 (Demyelination). J Neurol Sci 8:1–18

Behan PO, Lamarche JB, Behan WB, Feldman RG (1969) Immunopathological mechanisms of allergic neuritis in animals, primates and man. Trans Am Neurol Assoc 94:219–222

Brosnan CF, Lyman WD, Tansey FA, Carter TH (1985) Quantitation of mast cells in experimental allergic neuritis. J Neuropathol Exp Neurol 44:196–203

Brosnan JV, Craggs RI, King RHM, Thomas PK (1984) Attempts to transfer experimental allergic neuritis with lymphocytes. J. Neuroimmunol 6:373–385

Brosnan JV, Fellowes R, Craggs RI, King RHM, Bowley TJ, Thomas PK (1985) Changes in lymphocyte subsets during the course of experimental allergic neuritis. Brain 108:315–334

Brosnan JV, Craggs RI, King RHM, Thomas PK (1987) Reduced susceptibility of T cell deficient rats to induction of experimental allergic neuritis. J Neuroimmunol 14:267–282

Brosnan JV, King RHM, Thomas PK, Craggs RI (1988) Disease patterns in EAN in the Lewis rat. J Neurol Sci 88:261–276

Brosnan SF, Tansey FA (1984) Delayed onset of experimental allergic neuritis in rats treated with reserpine. J. Neuropathol Exp Neurol 43:84–93

Brostoff S, Burnett S, Lampert P, Eylar E (1972) Isolation and characterisation of a protein from sciatic nerve myelin responsible for experimental allergic neuritis. Nature New Biol 235:210–212

Brostoff SW, Levit S, Powers JM (1977) Induction of experimental allergic neuritis with a peptide from myelin P_2 basic protein. Nature 268:752–753

Brostoff SW, Powers JM, Weise MJ (1980) Allergic encephalomyelitis induced in guinea pigs by a peptide from the NH_2 terminus of bovine P_2 protein. Nature 285:103–104

Caspary EA, Field FJ (1965) Antibody response to central and peripheral nerve antigens in rat and guinea pig. J Neurol Neurosurg Psychiatry 28:179–182

Cohen IR, Weiner HL (1988) T cell vaccination. Immunol Today 9:332–334

Cragg BG, Thomas PK (1964) Changes in nerve conduction in experimental allergic neuritis. J Neurol Neurosurg Psychiatry 27:106–115

Craggs R, Brosnan JV, King RHM, Thomas PK (1986) Chronic relapsing experimental allergic neuritis in Lewis rats: effects of thymectomy and splenectomy. Acta Neuropathol 70:22–29

Cunningham JM, Powers JM, Brostoff SW (1983) Prevention of experimental allergic neuritis in the Lewis rat with bovine P_2 protein. Brain Res 258:285–289

Curtis BM, Forno LS, Smith PE (1979) Reactivation of neuritogenic activity of P_2 protein from peripheral nervous system myelin. Brain Res 175:387–391

De Lisi G, Berzofsky JA (1985) T cell antigenic sites tend to be amphipathic structures. Proc Natl Acad Sci USA 82:7048–7052

Ellerman KE, Powers JM, Brostoff SW (1988) A suppressor T-lymphocyte cell line for autoimmune encephalomyelitis. Nature 331:256–267

Endoh M, Tabira T, Kunishita T, Sakai K, Yamamura T, Taketomi T (1986) DM-20, a proteolipid apoprotein, is an encephalitogen of acute and relapsing autoimmune encephalomyelitis in mice. J Immunol 137:3832–3835

Feasby TE. Gilbert JJ, Hahn AF, Neilson M (1987) Complement depletion suppresses Lewis rat experimental allergic neuritis. Brain Res 419:97–103

Gilbert JJ, Feasby TE, Hahn AF (1983) Intraneural injection of lymphocytes in experimental allergic neuritis. Acta Neuropathol 61:61–64

Gross MLP, Craggs RI, King RHM, Thomas PK (1983) The treatment of experimental allergic neuritis by plasma exchange. J Neurol Sci 61:149–160

Hahn AF, Feasby TE, Gilbert JJ (1985) Blood–nerve barrier studies in experimental allergic neuritis. Acta Neuropathol 68:101–109

Happ MP, Kiraly AS, Offner H, Vandenbark A, Heber-Katz E (1988) The autoreactive T cell population in experimental allergic encephalomyelitis: T cell receptor B-chain rearrangements. J Neuroimmunol 19:191–204

Hartung HP, Schafer B, Heininger K, Stoll G, Toyka KV (1988) The role of macrophages and line-mediated experimental autoimmune neuritis in the rat. Neurosci Lett 83:195–200

Hartung H-P, Schafer B, Heininger K, Stoll G, Toyka KV (1988) The role of macrophages and eicosanoids in the pathogenesis of experimental allergic neuritis. Brain 111:1039–1059

Harvey GK, Pollard JD, Schindhelm K, Antony J (1987) Chronic experimental allergic neuritis. An electrophysiological and histological study in the rabbit. J Neurol Sci 81:215–226

Hashim GA (1980) T cell activation and suppression. In: Davison AN, Cuzner ML (eds) Experimental allergic encephalomyelitis and multiple sclerosis. Academic Press, New York, pp 79–104

Heber-Katz E, Acha-Orbea H (1989) The V-region hypothesis: evidence from autoimmune encephalomyelitis. Immunol Today 10:164–169

Heininger K, Stoll G, Linington C, Toyka KV, Wekerle H (1986) Conduction failure and nerve conduction slowing in experimental allergic neuritis induced by P_2-specific T cell lines. Ann

Neurol 19:44–49

Heininger K, Schafer B, Hartung HP, Fierz W, Linington C, Toyka KV (1988) The role of macrophages in experimental allergic neuritis induced by a P_2 specific T cell line. Ann Neurol 23: 326–331

Heitman NR, Mannweiler KL (1956) Experimental animal studies on allergic polyneuroses. Dtsch Z Nervenheilk 177:28–47

Hoffman PM, Powers JM, Weise MJ, Brostoff SW (1980) Experimental allergic neuritis I. Rat strain differences in the response to bovine myelin antigens. Brain Res 195:355–362

Hsieh DL, Weise MJ, Levit S, Powers JM, Brostoff SW (1981) Structure of bovine P_2 basic protein: sequence of a carboxyterminal segment that is a neuritogen in rabbits. J Neurochem 36:913–916

Hughes RAC (1974) Protection of rats from experimental allergic encephalomyelitis with antiserum to guinea pig spinal cord. Immunology 26:703–711

Hughes RAC, Kadlubowski M (1980) Experimental allergic neuritis in the rat. In: Rose FC, Behan PO (eds) Animal models of neurological disease. Pitman Medical, Tunbridge Wells, pp 95–103

Hughes RAC, Powell HC (1984) Experimental allergic neuritis: demyelination induced by P_2 alone and non-specific enhancement by cerebroside. J Neuropathol Exp Neurol 43:154–161

Hughes RAC, Kadlubowski M, Gray IA, Leibowitz S (1981a) Immune responses in experimental allergic neuritis. J Neurol Neurosurg Psychiatry 44:565–569

Hughes RAC, Kadlubowski M, Hufschmidt A (1981b) Treatment of acute inflammatory polyneuropathy. Ann Neurol 9 Suppl:125–133

Hughes RAC, Powell HC, Braheny SL, Brostoff SW (1985) Endoneurial injection of antisera to myelin antigens. Muscle Nerve 8:516–522

Hughes RAC, Atkinson PF, Gray IA, Taylor WA (1987) Major histocompatibility antigens and lymphocyte subsets during experimental allergic neuritis in the Lewis rat. J Neurol 234:390–395

Ichijo K, Fujimoto Y, Okada K (1981) Ultrastructural study of experimental allergic neuritis in the chicken. I. Cell migration, granuloma formation and demyelination. Zentralbl Veterinarmed [B] 28:210–225

Ishaque A, Szymanska I, Ramwani J, Eylar EH (1981) Allergic neuritis: phospholipid requirement for the disease inducing conformation of the P_2 protein. Biochim Biophys Acta 669:28–32

Ishaque A, Hofmann T, Eylar EH (1982) The complete amino acid sequence of the rabbit P_2 protein. JAMA 257:592–595

Izumo S, Linington C, Wekerle H, Meyermann R (1985) Morphological study on EAN mediated by T cell line specific for bovine P_2 protein in Lewis rats. Lab Invest 53:209–218

Kadlubowski M, Hughes RAC (1979) Identification of the neuritogen responsible for experimental allergic neuritis. Nature 277:140–141

Kadlubowski M, Hughes RAC (1980) The neuritogenicity and encephalitogenicity of P_2 in the rat, guinea pig and rabbit. J Neurol Sci 48:171–178

Kadlubowski M, Hughes RAC, Gregson NA (1980) Experimental allergic neuritis in the Lewis rat: characterisation of the activity of peripheral myelin and its major basic protein P_2. Brain Res 184: 439–454

King RHM, Thomas PK, Pollard JD (1977) Axonal and dorsal root ganglion cell changes in experimental allergic neuritis. Neuropathol Appl Neurobiol 3:471–486

King RHM, Craggs RI, Gross MLP, Tompkins C, Thomas PK (1983) Suppression of experimental allergic neuritis by cyclosporin A. Acta Neuropathol 59:262–268

King RHM, Craggs RI, Gross MLP, Thomas PK (1985) Effects of glucocorticoids on experimental allergic neuritis. Exp Neurol 87:9–19

Kitamura K, Suzuki M, Suzuki A, Uyemura K (1980) The complete amino acid sequence of the P_2 protein in bovine peripheral nerve myelin. FEBS Lett 115:27–30

Konovalov G, Korneva E, Khai L (1971) Effect of destruction of the posterior hypothalamic area on experimental allergic neuritis. Brain Res 29:383–386

Lampert PW (1967) Electron microscopic studies on ordinary and hyperacute experimental allergic encephalomyelitis. Acta Neuropathol 9:99–126

Lampert PW (1969) Mechanism of demyelination in experimental allergic neuritis. Electron microscopic studies. Lab Invest 20:127–138

Lassmann H (1983) Comparative neuropathology of chronic experimental allergic encephalomyelitis and multiple sclerosis. Springer-Verlag, Berlin

Lassmann H, Brunner C, Bradl M, Linington (1988) Experimental allergic encephalomyelitis: the balance between encephalitogenic T lymphocytes and demyelinating antibodies determines the size and structure of demyelinated lesions. Acta Neuropathol 75:566–576

Lehrich JR, Arnason BG (1971) Suppression of experimental allergic neuritis in rats by prior immunization with nerve in saline. Acta Neuropathol 18:144–155

Leibowitz S, Hughes RAC (1983) Immunology of the nervous system. Edward Arnold, London, pp 1–304

Levine S, Wenk EJ (1963) Allergic neuritis induced in rats without the use of mycobacteria. Proc Soc Exp Biol Med 113:898–900

Linington C, Izumo S, Suzuki M, Uyemura M, Meyermann R, Wekerle H (1984) A permanent rat T cell line that mediates experimental allergic neuritis in the rat in vitro. J Immunol 133: 1946–1950

Linington C, Mann A, Izumo S et al. (1986) Induction of experimental allergic neuritis in the BN rat: P$_2$ protein-specific T cells overcome resistance to actively induced disease. J Immunol 137: 3826–3831

Linington C, Bradl M, Lassmann H, Brunner C, Vass K (1988) Augmentation of demyelination in rats: acute allergic encephalomyelitis directed against a myelin/oligodendrocyte glycoprotein. Am J Pathol 130:443–454

London Y (1971) Ox peripheral nerve myelin membrane. Purification and partial characterization of two basic proteins. Biochim Biophys Acta 249:188–196

Madrid RE (1983) Chronic progressive and relapsing EAN in guinea pigs. In: Battistin L, Hashim GA, Lajtha A (eds) Clinical and biological aspects of peripheral nerve diseases. Alan R. Liss, New York, pp 265–276

Madrid RE, Wisniewski HM (1977) Axonal degeneration in demyelinating disorders. J Neurocytol 6:103–117

McDermott JR, Keith AB (1979) Suppression of experimental allergic neuritis with P$_2$ protein of peripheral nervous system myelin. Z Naturforsch [C] 34:641–643

McDermott JR, Wisniewski HM (1977) Studies on the myelin protein changes and antigenic properties of rabbit sciatic nerves undergoing Wallerian degeneration. J Neurol Sci 33:81–94

Milek DJ, Cunningham JM, Powers JM, Brostoff SW (1983) Experimental allergic neuritis: Humoral and cellular responses to the cyanogen bromide peptides of the P$_2$ protein. J Neuroimmunol 4:105–116

Milner P, Lovelidge GA, Taylor WA, Hughes RAC (1987) P$_0$ myelin protein produces experimental allergic neuritis in Lewis rats. J Neurol Sci 79:275–285

Morey MK, Hughes RAC, Powell HC (1985) Are Schwann cells involved in experimental allergic neuritis? Acta Neuropathol 67:75–80

Nagai Y, Uchida T, Takeda S, Ikuta F (1978) Restoration of activity for induction of experimental allergic peripheral neuritis by a combination of myelin basic protein P$_2$ and gangliosides from peripheral nerve. Neurosci Lett 8:247–254

Nomura K, Hamaguchi K, Ohno R et al. (1987) Cell-mediated immunity to bovine P$_2$ protein and neuritogenic synthetic peptide in experimental allergic neuritis. J Neuroimmunol 15:25–35

Ohno R, Hamaguchi K, Nomura K et al. (1988) Immune responses in experimental allergic neuritis treated with corticosteroids. Acta Neurol Scand 77:468–473

Olee T, Powers JM, Brostoff SW (1988) A T cell epitope for experimental allergic neuritis. J Neuroimmunol 19:167–173

Olee T, Weise M, Powers J, Brostoff SW (1989) A T cell epitope for experimental allergic neuritis is an amphipathic α-helical structure. J Neuroimmunol 21:235–240

Olsson T, Holdmahl R, Klareskog L, Forsum U, Kristinsson K (1984) Dynamics of Ia expressing cells and T lymphocytes of different subsets during experimental allergic neuritis in Lewis rats. J Neurol Sci 66:141–149

Ota K, Irie H, Takahashi K (1987) T cell subsets and Ia-positive cells in the sciatic nerve during the course of experimental allergic neuritis. J Neuroimmunol 13:283–292

Paraf A (1963) Polyradiculonévrite experimentale chez le mouton, ses rapports avec la syndrome de Guillain–Barré chez l'homme. Ann Inst Pasteur 104:208–218

Paterson PY, Harwin SM (1963) Suppression of allergic encephalomyelitis in rats by means of antibrain serum. J Exp Med 117:755–762

Paty DW (1971) An encephalitogenic basic protein from human peripheral nerve. Eur Neurol 5: 281–287

Pender MP, Sears TA (1984) The pathophysiology of acute experimental allergic encephalomyelitis in the rabbit. Brain 107:699–726

Peter M, Quaglio GL (1967) Experimental allergic neuritis in the chicken. Pathol Vet 4:464–476

Pollard JD, King RHM, Thomas PK (1975) Recurrent experimental allergic neuritis: an electron microscope study. J Neurol Sci 24:365–383

Powell HC, Braheny SL, Myers RR, Rodriguez M, Lampert PW (1983) Early changes in experimental allergic neuritis. Lab Invest 83:332–338

Raine CS, Bornstein MB (1979) Experimental allergic neuritis – ultrastructure of serum induced myelin aberrations in peripheral nervous system cultures. Lab Invest 40:423–432

Raine CS, Johnson AB, Marcus DM, Suzuki A, Bornstein MB (1981) Demyelination in vitro – absorption studies demonstrate that galactocerebroside is a major target. J Neurol Sci 52:117–131

Rivers TM, Schwentker FF (1935) Encephalomyelitis accompanied by myelin destruction experimentally produced in monkeys. J Exp Med 61:689–703

Robinson HC, Allt G, Evans DHL (1972) A study of the capacity of myelinated and unmyelinated nerves to induce experimental allergic neuritis. Acta Neuropathol 21:99–108

Rostami A, Brown MJ, Lisak RP, Sumner AJ, Zweiman B, Pleasure DE (1984) The role of myelin P_2 protein in the production of experimental allergic neuritis. Ann Neurol 16:680–685

Rostami A, Burns JB, Brown MJ et al. (1985) Transfer of experimental allergic neuritis with P_2-reactive T cell lines. Cell Immunol 91:354–361

Rostami A, Rosen JL, Hickey WF, Brown MJ (1986a) Experimental allergic neuritis can occur in experimental allergic encephalomyelitis-resistant Lewis rats. Neurology 36:312

Rostami A, Rosen JL, Cancro MP, Brown MJ, Pleasure DE (1986b) The variable susceptibility of inbred mice to experimental allergic neuritis. Neurology 36:304–305

Rostami AM, Ventura E, Kimura H, Brown MJ, Pleasure DE (1988) Induction of severe experimental allergic neuritis with a synthetic peptide corresponding to the 53–78 amino acid sequence of the myelin P_2 protein. Neurology 38:375

Saida K, Saida T, Brown MJ (1979) In vitro demyelination induced by intraneural injection of anti-galactocerebroside serum: a morphological study. Am J Pathol 95:99–116

Saida T, Saida K, Silberberg DH, Brown MJ (1978) Transfer of demyelination by injection of experimental allergic neuritis serum. Nature 272:639–641

Saida T, Saida K, Dorfman SH (1979a) Experimental allergic neuritis induced by sensitisation with galactocerebroside. Science 204:1103–1106

Saida T, Saida K, Silberberg DH (1979b) Demyelination produced by experimental allergic neuritis serum and anti-galactocerebroside antiserum in CNS cultures. An ultrastructural study. Acta Neuropathol 48:18–25

Saida T, Saida K, Silberberg DH, Brown MK (1981) Experimental allergic neuritis induced by galactocerebroside. Ann Neurol 9 suppl:87–101

Seil FJ, Kies MW, Bacon ML (1981) A comparison of demyelinating and myelination inhibiting factor induction by whole peripheral nerve tissue and P_2 protein. Brain Res 210:441–448

Sheremata WA, Behan PO (1973) Experimental allergic neuritis: a new experimental approach. J Neurol Neurosurg Psychiatry 36:139–145

Smith ME, Forno LS, Hoffman WW (1979) Experimental allergic neuritis in the Lewis rat. J Neuropathol Exp Neurol 38:377–391

Steinman L, Smith ME, Forno LS (1981) Genetic control of susceptibility to experimental allergic neuritis and the immune response to P_2 protein. Neurology 31:950–954

Stoll G, Schwendemann G, Heininger K et al. (1986) Relation of clinical, serological, morphological and electrophysiological findings in galactocerebroside induced experimental allergic neuritis. J Neurol Neurosurg Psychiatry 49:258–264

Strigard K, Olsson T, Larsson P, Holmdahl R, Klareskog L (1988) Modulation of experimental allergic neuritis in rats by in vivo treatment with monoclonal anti T cell antibodies. J Neurol Sci 83:283–291

Strigard K, Larsson P, Holmdahl R, Klareskog L. Olsson T (1989) In vivo monoclonal antibody treatment with Ox19 (anti-rat CD5) causes disease relapse and terminates P_2-induced immunospecific tolerance on experimental allergic neuritis. J Neuroimmunol 23:11–18

Sumner AJ, Saida K, Saida T, Silberberg DH, Asbury AK (1982) Acute conduction block associated with experimental antiserum mediated demyelination of peripheral nerve. Ann Neurol 11:469–477

Suzuki M, Kitamura K, Uyemura K, Ogawa Y, Ishihara Y, Matsuyama H (1980) Neuritogenic activity of peripheral nerve proteins in Lewis rats. Neurosci Lett 19:353–358

Suzuki M, Kitamura K, Sakamoto T, Uyemura K (1982) The complete amino acid sequence of human P_2 protein. J Neurochem 39:1759–1762

Suzuki M, Kitamura K, Uyemura K, Ogawa Y, Nozaki S, Marumatsu I (1984) Synthesis and neuritogenic activity of the peptides related to P_2 protein. In: Alvord EC, Kies MW, Suckling AJ (ed) Experimental allergic encephalomyelitis – a useful model for multiple sclerosis. Alan R Liss, New York, pp 478–492

Szymanska I, Ishaque A, Ramwani J'AI, Eylar EH (1981) Allergic neuritis: a neuritogenic peptide from the P_2 protein that induces disease in rats. J Immunol 126:1203–1206

Tansey FA, Brosnan CF (1982) Protection against experimental allergic neuritis with silica quartz dust. J Neuroimmunol 3:169–179

Taylor WA, Hughes RAC (1985) Experimental allergic neuritis induced in SJL mice by bovine P_2. J Neuroimmunol 8:153–157

Taylor WA, Hughes RAC (1988) Responsiveness to P_2 of blood and cauda equina derived lymphocytes in experimental allergic neuritis: preliminary characterisation of a cauda equina derived P_2 specific T cell line. J Neuroimmunol 19:279–289

Taylor WA, Hughes RAC, Lee T (1988) Lack of effect of fish oil enriched diet on experimental allergic neuritis in Lewis rats. J Neuroimmunol 17:193–197

Tuck RR, Pollard JD, McLeod JG (1981) Autonomic neuropathy: experimental allergic neuritis: an electrophysiological and histological study. Brain 104:187–208

Tuck RR, Antony JH, McLeod JG (1982) F-wave in experimental allergic neuritis. J Neurol Sci 56:173–184

Uyemura K, Suzuki M, Kitamura K et al. (1982) Neuritogenic determinant of bovine P_2 protein in peripheral nerve myelin. J Neurochem 39:895–898

Waksman BH, Adams RD (1955) Allergic neuritis: experimental disease of rabbits induced by the injection of peripheral nervous tissue and adjuvants. J Exp Med 102:213–225

Waksman BH, Adams RD (1956) A comparative study of experimental allergic neuritis in the rabbit, guinea pig and mouse. J Neuropathol Exp Neurol 15:293–310

Watts PM, Taylor WA, Hughes RAC (1989) High-dose methylprednisolone suppresses experimental allergic neuritis in the Lewis rat. Exp Neurol 103:101–104

Weise MJ, Hsieh D, Hoffman PM, Powers JM, Brostoff SW (1980) Bovine peripheral nervous system myelin P_2 protein: chemical and immunological characterization of the cyanogen bromide peptides. J Neurochem 35:393–400

Whitaker JN, Seyer JM (1984) Degradation of bovine P_2 protein by bovine brain cathepsin D. Neurochem Res 9:1431–1443

Wisniewski H, Prineas J, Raine CS (1969) An ultrastructural study of experimental demyelination and remyelination. 1 Acute experimental allergic encephalomyelitis in the peripheral nervous system. Lab Invest 21:105–118

Wisniewski HM, Brostoff SW, Carter H, Eylar EH (1974) Recurrent experimental allergic polyganglioradiculoneuritis. Arch Neurol 30:347–358

Yonezawa T, Ishihara Y, Matsuyama H (1968) Studies on experimental allergic neuritis. I. Demyelinating patterns studied in vitro. J Neuropathol Exp Neurol 27:453–463

Yoshimura T, Kunishita T, Sakai K, Endoh M, Namikawa T, Tabira T (1985) Chronic experimental allergic encephalomyelitis in guinea pigs induced by proteolipid protein. J Neurol Sci 69:47–58

Zamvil SS, Mitchell DJ, Moore AC, Kitamura K, Steinman L, Rothbard JB (1986) T cell epitope of the autoantigen myelin basic protein that induces encephalomyelitis. Nature 324:258–260

Chapter 4

Pathology of Guillain–Barré Syndrome

Introduction

Descriptions of myelin breakdown and inflammatory changes in peripheral nerves of patients with acute "multiple neuritis" or Landry's ascending paralysis began to appear during the late nineteenth century, long before Guillain, Barré and Strohl's paper on acute radiculoneuritis with albumino-cytological dissociation was published in 1916. However, the inflammatory changes were regarded as inconstant, late and secondary in an influential review and study of 50 personal cases by Haymaker and Kernohan in 1949. Twenty years later the modern history of Guillain–Barré syndrome began with the landmark study of Asbury et al. (1969) who emphasised the early appearance of lymphocytic infiltration and drew the important analogy with EAN. Subsequent electron microscopic studies, notably by Prineas (1981), have indicated a consistent association between macrophage infiltration and demyelination, similar to that noted in EAN (see Chapter 3). This chapter reviews these landmark studies in the light of our present understanding of ways in which demyelination may be induced by immunological means (see Chapter 2). In this review it is important to remember the alternative possibilities that demyelination neuropathy may be induced either by T cell-mediated responses to P_2 protein or by antibody-mediated responses to galactocerebroside. Another strand running through the interpretation of the pathology of GBS is that its clinical course is heterogeneous. While most patients make satisfactory recoveries, a few, often those with an explosive onset, develop wasting, persistent weakness and disability, and have electrophysiological evidence of axonal degeneration rather than demyelination.

In a very lengthy paper Haymaker and Kernohan (1949) reviewed the early literature about Landry's ascending paralysis and Guillain and Barré's radiculoneuritis with albumino-cytological dissociation. They summarised previous work as having shown no changes in the CNS except for the degenerative changes in the anterior horn cells and cranial nerve nuclei anticipated from peripheral axonal damage. Some cases had shown minor hyperaemia and thickening of the spinal arachnoid trabeculae and lymphocytic infiltration of the spinal cord meninges adjacent to the roots. There was also a variable degree of inflammatory cell infiltration in the spinal roots, dorsal root ganglia and peripheral nerves. However, Haymaker and Kernohan concluded that the "infrequent occurrence of inflammatory cells in cases of brief duration has been taken as evidence that the presence of such cells in cases of longer duration constitutes a response to neuronal damage not an integral part of the initial process".

With this impression of the literature in mind Haymaker and Kernohan (1949) presented a personal series of 50 autopsied cases subjected to post-mortem examination. Material was available from all stages of the disease from 2 to 46 days. The clinical information is so scanty that it is necessary to take the diagnosis on trust. A further difficulty in interpreting the paper is that no information is given about the amount of pathological material available in each case. They summarised their pathological findings as follows. The CNS showed only sparse perivascular collections of lymphocytes in the cerebral white matter and subependymal region in 20% of the cases, proliferation of meningeal cells in 30% and minor meningeal infiltration in 10%. In the peripheral nerves the earliest change was a bland oedema. In none of the nine early cases which died between the second and fourth day of the disease "was there any local cellular reaction or inflammatory exudate". After five days, fragmentation of myelin and swelling and beading of axis cylinders were noted, but it was not until nine days that "scattered small groups of lymphocytes were occasionally observed". At later stages until their last observation at 46 days after death myelin and axis cylinders disappeared and there was infiltration by lymphocytes and macrophages, which they regarded as having a reparative function. The cranial nerves were reported to show similar changes. There was oedema and lymphocytic infiltration of one superior cervical sympathetic ganglion in the limited material available from the autonomic system. Other relevant abnormalities were minor skeletal muscle changes in half the cases in which this was examined and in 14% mild focal myocarditis, i.e. perivascular collections of lymphocytes and macrophages in the myocardium and lymphocytes in the pericardium. Not surprisingly 33 of the 50 had bronchopneumonia.

Autopsy Observations of Guillain–Barré Syndrome

Krücke (1955) drew attention to the early appearance of intense lymphocytic infiltration in the PNS, especially the spinal roots where they join the dorsal root ganglion which had already been reported in the French and German literature and which was evident from his own material. The primacy of the lymphocyte in the pathological process was stressed by Asbury et al. (1969) who had access not only to the autopsy material but also detailed neurological records of 19 cases. These represented all the cases of GBS which had come to autopsy at the Massachusetts General Hospital in Boston during a 14-year period. They came to this study with a knowledge of the lymphocytic infiltration and segmental demyelination in EAN and a recognition of the importance of lymphocytes "as initiators of cytodestructive events in tissues rather than as ill-defined vehicles of repair". Seventeen of their 19 cases fulfilled the criterion of reaching their nadir (apart from death) within four weeks from the onset of neuropathic symptoms and so qualify as cases of GBS according to currently accepted criteria (see Chapter 1). The cases included four very early cases in which death occurred between 1 and 6 days after onset, several who died between the second week and 3 months, and three who were thought to have recovered when autopsy was performed for unrelated reasons months or years later. Lymphocytic infiltration of the PNS was found in all cases. In particular it was severe in cases examined on days 1, 3, 4 and 6 after onset and there were occasional associated polymorphonuclear leucocytes. During the next six weeks severe lymphocytic infiltration persisted in all the cases. It was perivascular, surrounding predominantly endoneurial but also epineurial vessels, and multifocal, sparing one nerve or nerve root while affecting adjacent tissue severely. In addition to typical lymphocytes there were somewhat larger cells, interpreted as lymphoblasts, resembling cells seen in EAN and, particularly where myelin destruction was severe, macrophages. The whole of the PNS was affected with an approximate correlation between the anatomical sites of pathological lesions corresponding to the clinically observed deficits. In cases with ophthalmoplegia, severe inflammatory demyelination or destructive lesions were found in cranial nerves 3, 4 and 6. In two cases with purely motor involvement the anterior roots were predominantly affected while in cases with prominent sensory involvement the posterior roots were severely involved. The multifocal nature of the involvement of the PNS was emphasised by a case in which the spinal roots were completely spared while the peripheral nerves were severely affected more distally. Even in cases which were subjected to autopsy five or six years after recovery from GBS there were still sparse perivascular lymphocytes in the PNS. The myelin damage was most marked in close relation to the inflammatory cell infiltrates and was largely due to segmental demyelination with phagocytosis of myelin debris. However, axonal degeneration was commonly seen especially in the vicinity of the most severe inflammatory changes. Apart from anterior horn cell chromatolysis and degeneration of the posterior white columns in some severely affected late cases, the CNS was almost normal. Only rare perivascular clusters of inflammatory cells were found.

Asbury et al. (1969) argued from their material that the lymphocytic infiltration was the prime mover, and that a delayed hypersensitivity mechanism, similar to that thought responsible for EAN, might be triggered by one of the several viral infections and other insults which were already known to be followed by GBS. The precise mechanism whereby the demyelination occurred could not be resolved at the light microscope level.

The paper by Asbury et al. (1969) has dominated thinking about GBS to the extent that it has been accepted that the lymphocyte is an essential early feature in every GBS lesion. However, the evidence from biopsies of GBS patients is that lymphocytes are rarely found and so it is pertinent to question whether their absence is merely a sampling artefact or whether they are indeed a less constant and prominent feature than has been suggested. Review of the eleven autopsy reports published since 1969 (Table 4.1) has shown that lymphocytic infiltration is variable. The extent of lymphocytic infiltration in different parts of the PNS was usually described as "mild" or "sparse" except in the case of Van Zandycke et al. (1982) in which there were multiple infiltrates in the dorsal root ganglia and in the case of Carpenter (1972) in which there was "prominent" lymphocytic infiltration. The intervals from onset to death ranged from 11 to 50 days, similar to the intervals in the series in which Asbury et al. (1969) found severe lymphocytic infiltration, albeit multifocal, in every case. Similar conclusions were reached in another review of the autopsy findings of 24 cases of fatal neuromuscular paralysis from Mexico: the authors emphasised that inflammatory lesions, though eventually found in every case, were sparse (Ramos-Alvarez et al. 1969). Cross-sections of ventral and dorsal roots generally showed no infiltration. Careful search of multiple longitudinal sections of cauda equina and lower limb nerves eventually revealed infiltration, "often of minimal de-

Table 4.1. Autopsy studies of Guillain–Barré syndrome published since 1969

Reference	Duration (days)	Age	Sex	Steroid treatment	Cause of death	Mononuclear cell infiltration			Demyelination			Axonal degeneration			CNS	Notes
						Anterior roots	Posterior roots	Peripheral nerve	Anterior roots	Posterior roots	Peripheral nerves	Anterior roots	Posterior roots	Peripheral nerves		
Kanda et al. (1989)	8	47	M	PE[a]	Sudden bradycardia	+++	+++	+++	++	++	++	0	0	0	Brainstem motor nuclei chromatolysis	Small lymphocytes few
Arstila et al. (1971)	11	66	M	ACTH	Pulmonary embolus	+	+		+	+					Normal	
Wisniewski et al. (1969)	14	21	F	0	Cardiac arrest	+	+	+	+	+	+	0	0	0	Normal	
van Zandycke et al. (1982)	17	55	F	0	Sudden	+	++				+ Facial			+ Facial	Lymphocytic infiltration of spinal cord	
Mei Liu (1970)	17	21	F	0	Tracheostomy haemorrhage											Limited e.m. study only
Carpenter (1972)	18	47	M	+	Pneumonia	++	+	+	+	+	+	0	0	0	Normal	
Feasby et al. (1986)	28	64	F	0	Septicaemia	+	+	+	0	0	0	+++	+++	+++	Anterior horn cell chromatolysis	
Phillips et al. (1984)	30	67	F	0	Sudden	+	+	+	++	++	++	0	0	0	Normal	Miller Fisher syndrome
de Haene et al. (1986)	40	52	F	0	Septicaemic shock		+	+	++	++	++	0	0	0	Rare anterior horn cell and oculomotor chromatolysis	
Prineas (1972)	50			+	Cardiac arrest	0	0		++	++		+	+	0		Remyelination prominent
Behar et al. (1986)	54	73	M	+	Sudden	++	++	0	+	+	0	0	0	0	Normal	
Best (1985)	63	72	F	0	General deterioration	++	++		++	++	++	0	0	0		

Where no information is entered, no report was provided.
Grades : 0 = none, + = mild, ++ = moderate, +++ = severe.
[a] PE = plasma exchange.
From Honavar et al. (1990), with permission.

gree". The children had died between 4 and 30 days after onset of neuropathy and no mention was made of steroids being given.

In a series of nine autopsies from GBS patients (Honavar et al. 1990), lymphocytic infiltration has been a constant but variable feature (Tables 4.2 and 4.3). Six of the cases were examined between 10 and 32 days after the onset of neuropathy and in three of these endoneurial infiltration was only considered mild. In three cases moderate mononuclear cell infiltration was found round blood vessels in some nerves (Fig. 4.1). The infiltrating cells consisted of both lymphocytes and macrophages. The frequency and degree of infiltration was less than had been described by Asbury et al. (1969) who had found infiltration at this stage to be constant and severe. The absence of more florid lymphocyte infiltration could not be accounted for by the use of steroids (Asbury and Johnson 1978) which were only used in one of our cases.

One of the recently published reports documented severe macrophage-mediated demyelination with very little lymphocytic infiltration (Kanda et al. 1989). The patient died 8 days after the onset of an illness with the typical course of GBS. Post-mortem examination demonstrated multifocal demyelination throughout the peripheral nervous system, especially in the nerve roots, in the absence of significant cellular infiltration. The autopsy was performed only three hours after death and the excellent fixation permitted evaluation of the size of the demyelinating fibres which were predominantly the medium sized myelinated the fibres. The endoneurium was oedematous and infiltrated with macrophages,

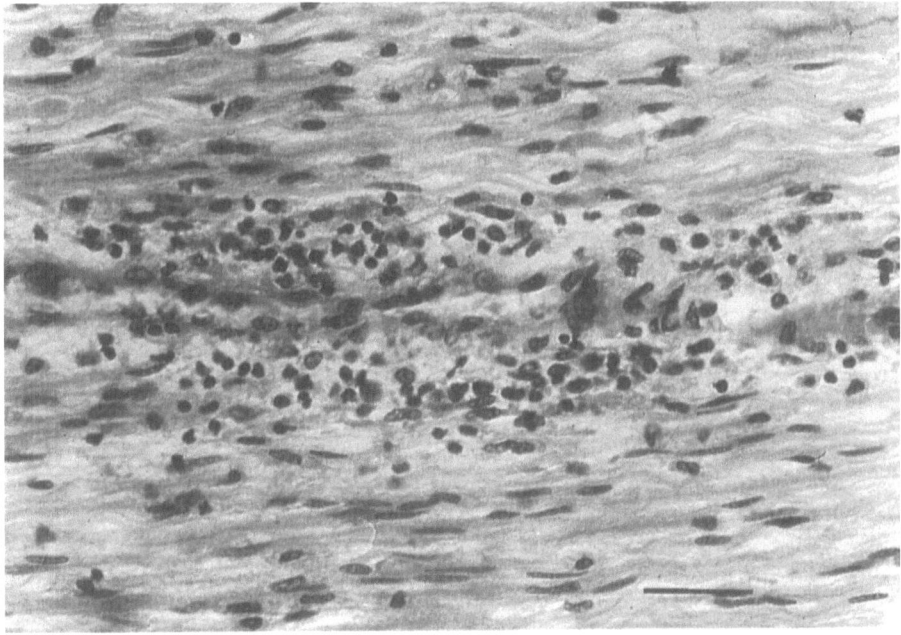

Fig. 4.1. Posterior tibial nerve removed at post-mortem examination 30 days after the onset of GBS. Note moderately severe perivascular cuff of lymphocytes and macrophages. Haematoxylin and eosin. Bar = 50 μm.

Table 4.2. Clinical features of nine fatal cases of GBS studied by the author

Case	Duration	Age	Sex	Steroid treatment	Antecedent event	Duration to being bed bound (days)	Cause of death	Notes
1	10 days	72	M	0	Chest infection	9	Respiratory arrest	Non-insulin-dependent diabetes mellitus
2	21 days	13	F	0 PE[a]	Tetanus toxoid	7	Cardiorespiratory	Intraventricular pressure monitor inserted 7 days before death
3[b]	29 days	72	M	0	Chest infection	2	Septicaemia	
4[b]	30 days	74	M	0	Influenza vaccine	2	Asystole	Paroxysmal atrial fibrillation, fluctuating blood pressure
5	31 days	63	M	0	Chest infection	2	Pneumothorax	
6	32 days	44	M	0 PE	Influenza-like illness	14	Pulmonary embolus	
7	4 mths	72	M	0	Influenza-like illness	4	Asystole	Episodes of tachycardia, hypertension and sweating
8	8 mths	1.2	M	0	Chicken-pox	–	Cerebral anoxia	Two attacks
9[b]	1 year	60	F	0	Gastroenteritis	2	Asystole	

From Honavar et al. (1990), with permission.
[a] PE = plasma exchange.
[b] In series of Winer et al. (1988).

Table 4.3. Pathological features of nine fatal cases of GBS studied by the author

Case	Mononuclear cell infiltration				Myelin loss				Axonal degeneration				CNS
	Ant. root	Post. root	Peripheral nerve	Cranial nerve	Ant. root	Post. root	Peripheral nerve	Cranial nerve	Ant. root	Post. root	Peripheral nerve	Cranial nerve	
1	0	+	+-++	NE[a]	0	+	+-+++	NE	0	0	+	NE	Mild PCD[b]
2	0	0	++	+-++	+	+	+++	++	+	+	0-++	0	Meningitis
3	0	0	+	0	0-+	0	+++	0-+	0	0	++	0	Normal
4	0	NE	+-++	NE	+	NE	+-+++	NE	0	NE	+	NE	NE
5	0	0	+	+	+	+	+++	+-+++	0	0	++	+	Normal
6	+	+	+	0	++	++	0	0	0	0	0	0	PCD
7	NE	NE	+	0	NE	NE	+	-	NE	NE	0	0	NE
8	0	0-++	+-++	NE	++	++	++-+++	NE	++	++	+++	NE	AHC[c], PCD
9	0	0	NE	NE	NE	NE	NE	NE	NE	NE	NE	NE	PCD

[a] NE, not examined.
[b] PCD, posterior column degeneration.
[c] AHC, anterior horn cell loss
From Honavar et al. (1990), with permission.

many containing myelin debris, but not lymphocytes. The patient had been treated with plasma exchange but not steroids (Kanda et al. 1989).

Feasby et al. (1986) have drawn attention to rare patients with polyneuropathy of explosively rapid onset which is clinically compatible with GBS. These patients are distinguished electrophysiologically by having inexcitable motor and sensory nerves. An autopsy on one such patient showed severe axonal degeneration in peripheral nerves and spinal roots and abundant macrophages with myelin debris but only a few scattered lymphocytes. Teased fibres from peripheral nerves showed axonal degeneration in 31%–39% of fibres but demyelination in only 5%. Electron microscopy also revealed loss of unmyelinated nerve fibres. This case typifies an illness with clinical features which are difficult or impossible to distinguish from GBS but which has as its pathological substrate an acute axonal neuropathy without inflammatory changes. This is different from the usual pathological picture of demyelinating polyradiculoneuropathy with relative preservation of axons. GBS is defined in clinical terms and its pathological substrate is not homogeneous.

Most post-mortem electron microscopic studies of GBS have been difficult to interpret because of artefactual vesiculation of the myelin. At first it was thought that this might be a significant mechanism of myelin breakdown (Wisniewski et al. 1969). Vesicular dissolution of myelin has also been discovered in every case in which electron microscopic studies have been undertaken (Fig. 4.2). However, similar ultrastructural changes are present in nerves and nerve roots from patients who have died without neurological disease and are due to post-mortem artefact (Fig. 4.3). The most helpful information concerning the ultrastructural changes of nerves in GBS has come from biopsy studies.

Biopsy Studies in Guillain–Barré Syndrome

In the first ultrastructural studies of sural nerve biopsies from patients with GBS the authors concluded that the primary change was degeneration of the whole nerve fibre and not primary demyelination (Finean and Woolf 1962; Miyakawa et al. 1971) but the preparations were marred by artefact and these conclusions have been contradicted. Subsequent ultrastructural studies of nerve biopsies in GBS have revealed that the mechanism of demyelination is a process of macrophage stripping of myelin lamellae resembling that in EAN. The situation was clarified by Prineas (1972) who described nine sural nerve biopsies and one early autopsy between two and 43 days after the onset of GBS. Lymphocytic infiltration was detected in paraffin sections in only three but six had received steroids which may have influenced this finding. We have found lymphocytic infiltration in three of ten biopsies (Fig. 4.4). Endoneurial oedema, reduction of myelinated nerve fibres, macrophage-associated demyelination and ongoing axonal degeneration were more common findings (Figs. 4.5–4.6). In six of Prineas' nine cases there was active demyelination in which mononuclear cells penetrated the Schwann cell basement membrane, and displaced the Schwann cell from the myelin sheath (Fig. 4.7). Macrophage processes then burrowed into the myelin lamellae along the minor dense lines (which represent the apposed extracellular surfaces of the Schwann cell) and phagocytosed myelin

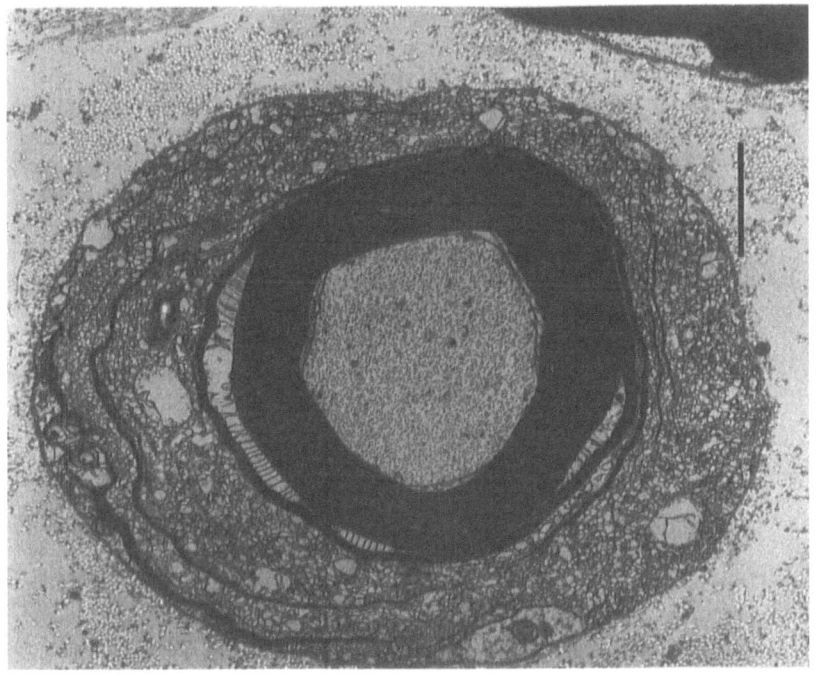

Fig. 4.3. Transverse section of a spinal root from a patient who had had no neurological disease. The root was removed at post-mortem examination 60 hours after death. Electron micrograph. Bar = 2 μm. (From Honavar et al. (1990), with permission.)

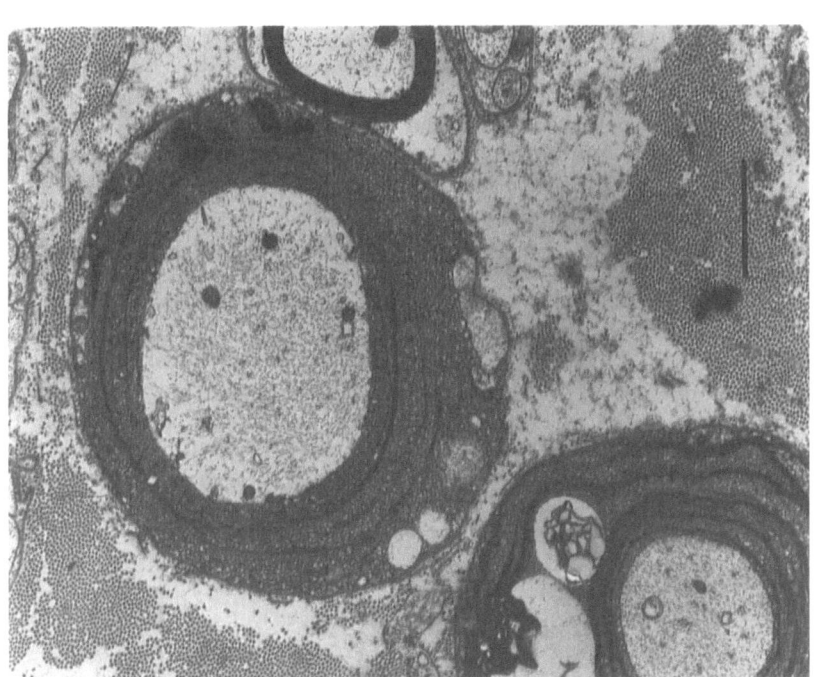

Fig. 4.2. Transverse section of a spinal root from a patient who died 10 days after the onset of GBS. The root was removed at post-mortem examination 24 hours after death. Electron micrograph. Bar = 2 μm. (From Honavar et al. (1990), with permission.)

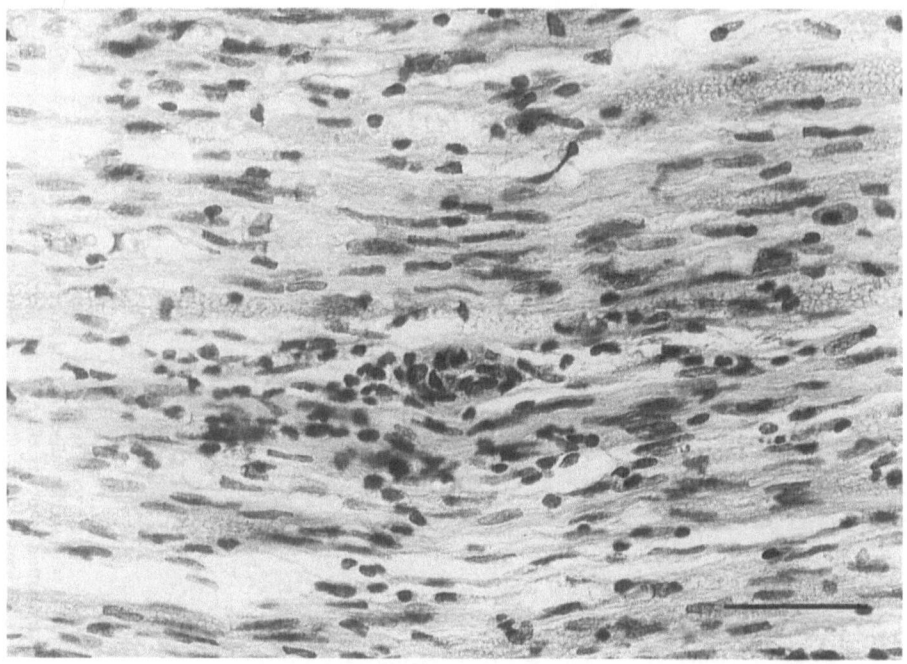

Fig. 4.4. Sural nerve biopsy 30 days after the onset of GBS showing a small perivascular collection of inflammatory cells. Haematoxylin and eosin. Bar = 50 μm.

fragments which were engulfed and digested in secondary lysosomes. Myelin vesicular dissolution was discovered in only two cases and was always associated with cellular infiltration. The process of macrophage invasion of the basement membrane and subsequent stripping of the myelin sheath has been amply confirmed by subsequent studies (Julien et al. 1980; Prineas 1981; Cornblath et al. 1987a). Such macrophage-mediated demyelination was identified in seven of ten biopsies from patients with GBS (Figs. 4.7–4.9).

Brechenmacher et al. (1987) studied 65 superficial peroneal nerve biopsies and found macrophage invasion of the Schwann cell basal lamina in 48 cases and macrophage stripping in 32 cases. Sometimes the macrophages were located between the axon and the myelin sheath, suggesting that they had penetrated from the paranodal region and in five cases some were actually within the axon (Brechenmacher et al. 1981). Myelin vesicular disruption was rarely seen and when present was related to macrophage invasion. Of the 17 cases without macrophage invasion 15 showed demyelinated or remyelinating axons. A noteworthy feature was that mononuclear cell infiltration was only observed in five out of the 57 biopsies from which paraffin-embedded material was available. Some mononuclear cell infiltration was reported in all six sural nerve biopsies in another series but the details and presence of lymphocytes were not described (Hausmanowa-Petrusewicz et al. 1979).

The variability of lymphocytic infiltration in biopsies and autopsy material from patients with GBS cannot be accounted for by sampling errors. The underlying pathological process is likely to be heterogeneous. In cases in which

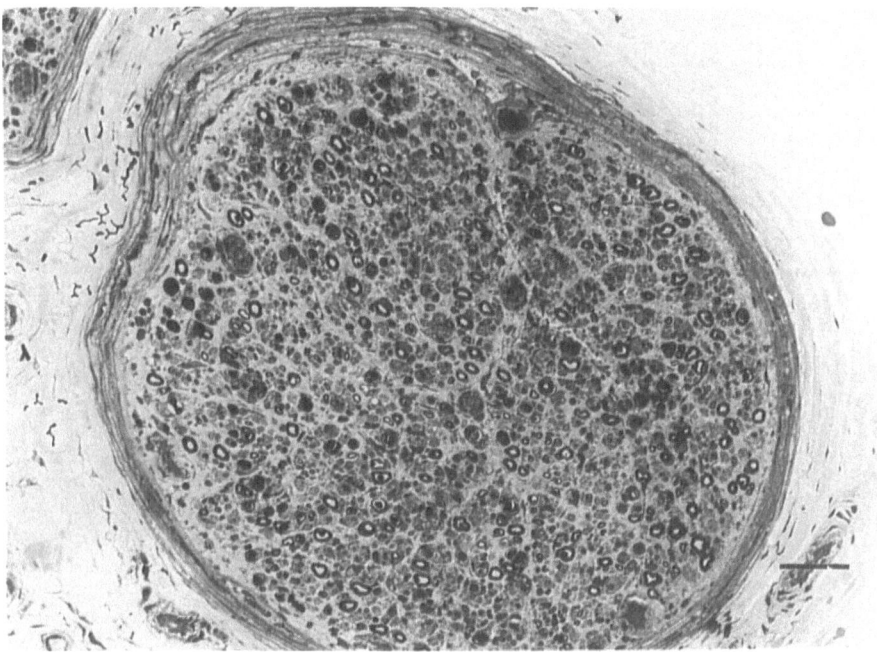

Fig. 4.5. Transverse section of another fascicle from the nerve biopsied in Fig. 4.4 showing endoneurial oedema and a marked reduction of myelinated nerve fibres. Thionin and acridine orange. Bar = 50 μm.

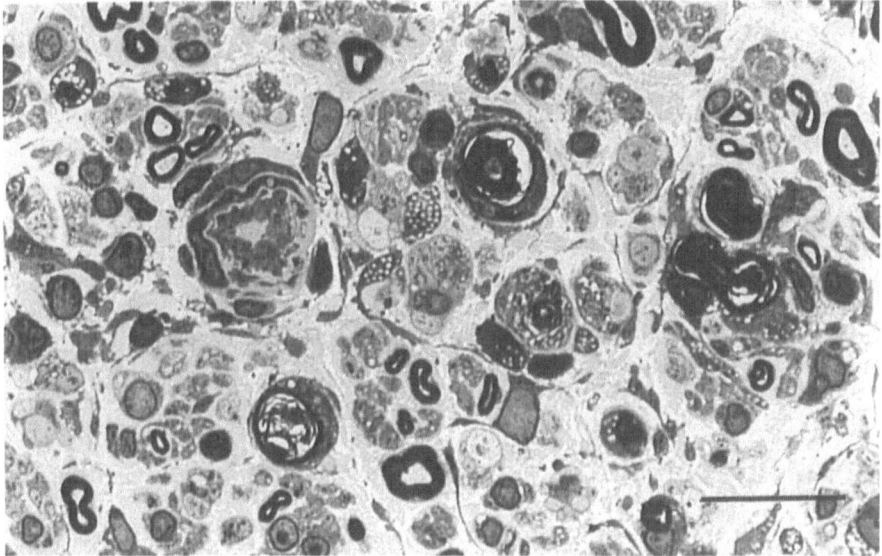

Fig. 4.6. Higher magnification of a portion of the fascicle in Fig. 4.4 showing many completely demyelinated axons and some fibres undergoing Wallerian degeneration. Thionin and acridine orange. Bar = 2 μm.

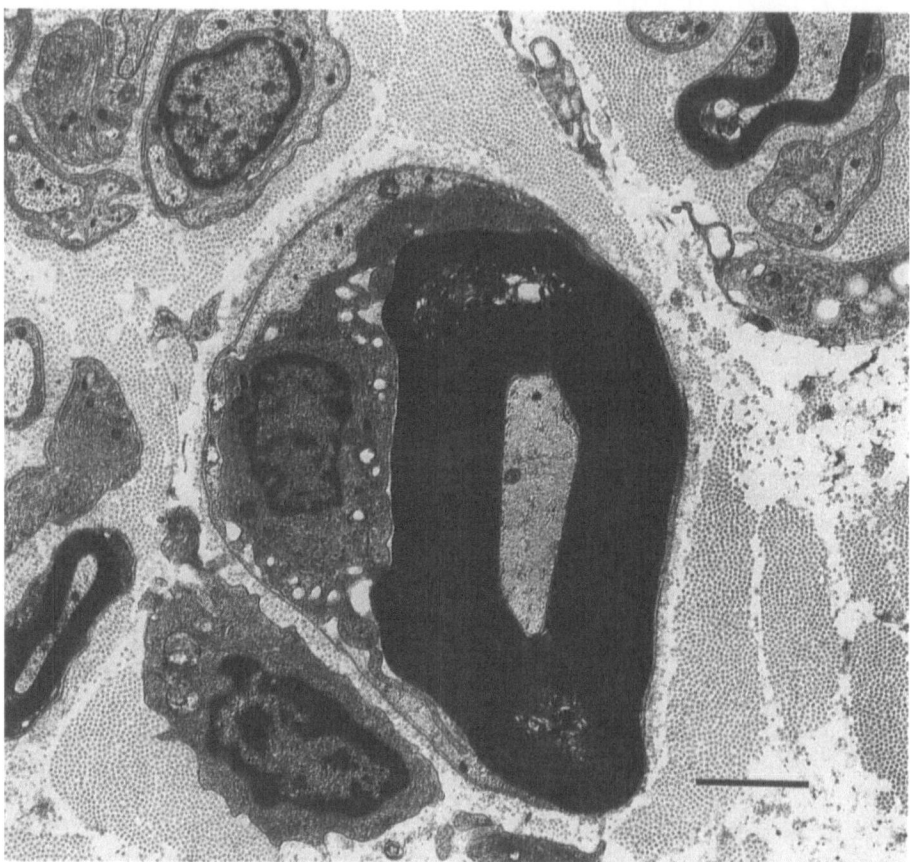

Fig. 4.7. Transverse section of a myelinated nerve whose basal lamina has been invaded by a macrophage although the myelin remains intact. Bar = 2 μm. Electron micrograph provided by Dr S.M. Hall. (From Hughes RAC, Atkinson PF, Hall SM, Leibowitz S (in preparation).)

lymphocytes are prominent at an early stage a T cell-mediated disease similar to that induced in Lewis rats with EAN transferred by T-helper cell lines directed against P_2 protein is a possibility. In cases in which macrophage-mediated demyelination is prominent in the absence of lymphocytic infiltration, antibody-mediated demyelination such as can be induced by injecting antibody to galactocerebroside is more likely.

Immunohistochemical Studies of Nerves in GBS

Immunohistochemical techniques have shown that there is simultaneous loss of the myelin proteins P_0, MBP, P_2 and MAG in the lesions of GBS (Schober et al. 1981). There is no early or preferential loss of any one of these constituents. This is consistent with the hypothesis that the pathogenetic mechanism involves the

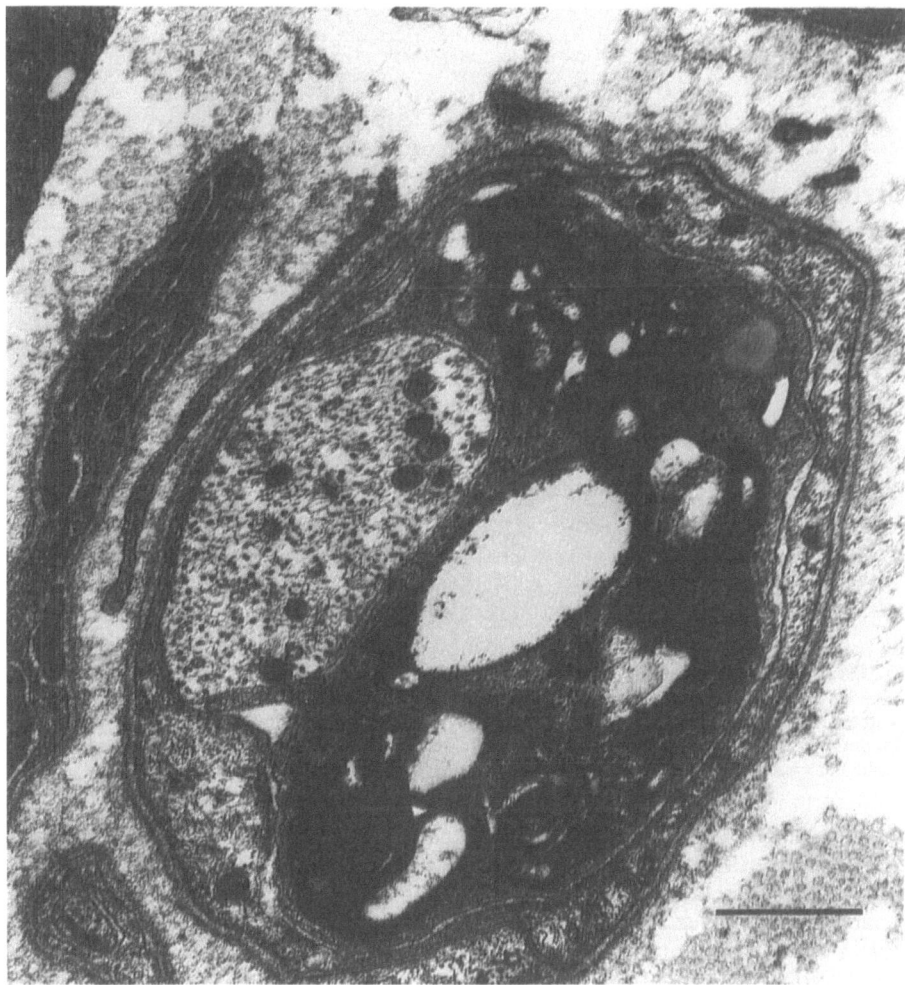

Fig. 4.8. Transverse section of a demyelinated nerve fibre. A macrophage containing debris remains within the basal lamina tube. Electron micrograph. Bar = 1 μm.

myelin sheath and not initial Schwann cell damage. Unfortunately an attack on either a myelin protein or a Schwann cell cytoplasmic component would be likely to be followed so rapidly by myelin breakdown of the type observed that these observations, while consistent with the hypothesis of primarily myelin-directed damage, do not rigorously exclude other possibilities.

According to one report complement components and IgM, but not IgG or IgA, can be detected on myelin sheaths by direct immunofluorescent labelling of frozen nerve biopsy sections from patients with GBS but not from controls (Luitjen and Faille-Kuyper 1972). We have not been able to confirm this finding in a series of 10 biopsies from patients with GBS. Brechenmacher et al. (1987) also failed to find immunoglobulin deposited on myelin in 17 biopsies, although in four of these cases various patterns of perivascular IgG, IgM or C3 deposition

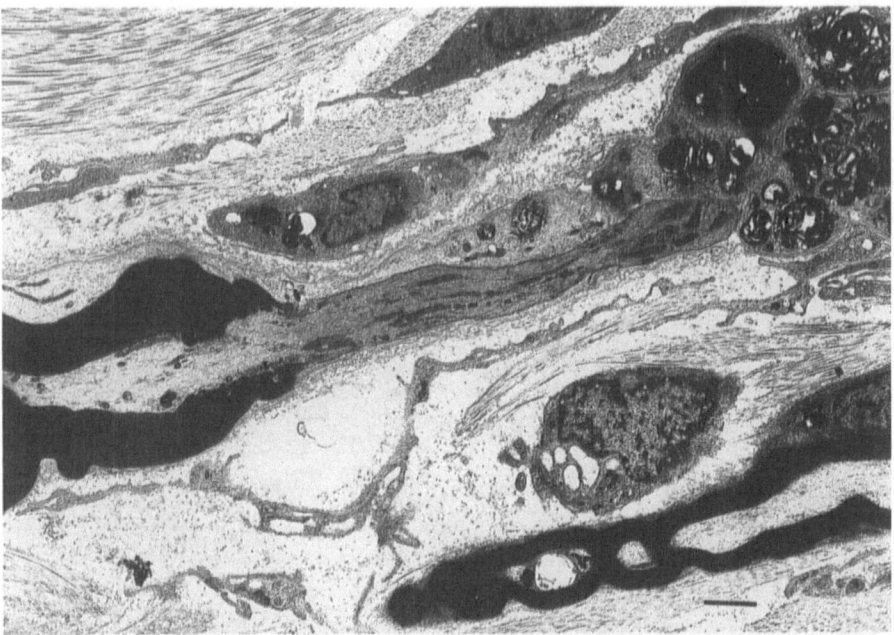

Fig. 4.9.   Sural nerve biopsy 23 days after the onset of GBS. Longitudinal section of a myelinated nerve fibre showing one heminode stripped of myelin while the other remains normal. A macrophage containing myelin debris lies within the basal lamina of the demyelinated heminode. Mononuclear cells, probably macrophages, lie in the endoneurial space nearby. Electron micrograph. Bar = 2 μm.

were identified. Nyland et al. (1981) reported finding deposits of IgG but not IgM along the myelin sheaths of one of five GBS biopsies. They also found deposits of C3 in two of the five cases. Tsukada et al. (1987) reported finding hepatitis type B virus surface antigen, C3 and IgG in the endoneurium of all of four biopsies from patients with GBS associated with that virus. They did not report finding C3, IgG or IgM actually on the myelinated fibres although they would presumably have seen it if it was there. They interpreted their result as indicating the deposition of immune complexes of viral antigen, C3 and IgG in the endoneurium. The C3d fragment released from C3 following activation via either the classical or alternative pathway was present on the myelin sheaths in both of two cases in one series (Hays et al. 1988).

In a study of nerves obtained at post-mortem deposits of immunoglobulins and C3 were sought by direct immunofluorescence and not found (Ammoumi et al. 1980). Koski et al. (1987) described the deposition of C5–9 membrane attack complex on the myelin sheaths in paraffin-embedded sections of one autopsy case. The weight of evidence is that immunoglobulins cannot be detected on myelin sheaths of GBS patients with current techniques. The complement membrane attack complex may be present on myelin sheaths but this finding needs to be confirmed in other patients.

It is possible to identify the cells infiltrating nerves with monoclonal antibodies to lymphocytes and macrophages. Such studies are still at an early stage. Pollard et al. (1987) first reported that endoneurial cells in the nerves of patients with GBS express increased amounts of MHC class II antigen. It has subsequently been shown that such increased expression is not specific to GBS but is also a feature of diabetic and hereditary neuropathies (Mancardi et al. 1988). Cornblath et al. (1987b) have found increased numbers of T cells identified by antibody to CD3 antigen in biopsies of two patients with GBS, one with and one without HIV infection. The cells consisted predominantly of CD8+ cells in the one patient who had HIV infection and a low blood CD4+ cell count. In the other patient there were more CD4+ than CD8+ cells. In our own immunohistochemical studies of nerve biopsy T lymphocytes were only found in a small proportion but "activated" macrophages have been identified in most (Fig. 4.10). It is intended to pursue these studies with immunoelectronmicroscopy to identify the cells penetrating the basement membrane, which are probably macrophages. It should also be possible to use in situ hybridisation techniques to discover whether these cells are expressing MHC class II antigen and secreting mediators such as tumour necrosis factor which might play a role in causing demyelination.

Remyelination

Even before the inflammation has subsided the process of remyelination begins. The demyelinated axons are stripped bare and the myelin debris is ingested and

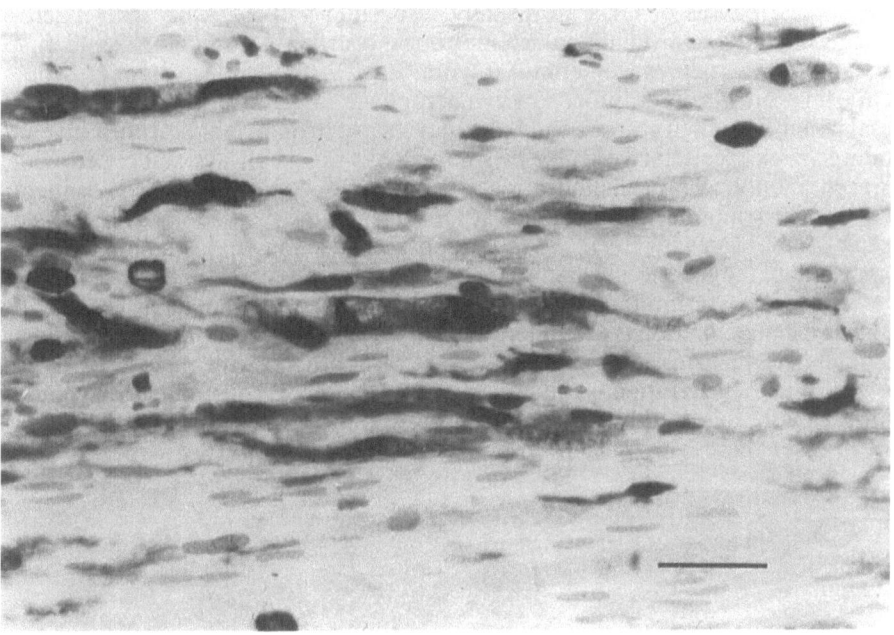

Fig. 4.10. Longitudinal section of a sural nerve biopsy from a patient with GBS showing numerous cells labelled with monoclonal antibody MAC 387 which is a marker for macrophages which have recently left the circulation. The marker is visualised with a second antibody tagged with peroxidase and developed with diaminobenzidine. Such macrophages are not seen in normal nerves and rarely seen in axonal neuropathy. Bar = 50 μm.

removed by macrophages which leave the Schwann cell basal lamina and can be seen migrating towards the endoneurial capillaries and perineurium. The Schwann cell proliferates and the processes envelop the axon, presumably stimulated by a signal from the denuded axon or one of the products of the breaking down myelin. At first the axon is merely wrapped by Schwann cell cytoplasm but within a few days the wraps condense to form new myelin lamellae. Not all the proliferating Schwann cells participate in the remyelination: some are relegated to a supernumerary position forming an extra wrap outside the axon basal lamina. These supernumerary Schwann cells probably disappear eventually but in CIDP recurrent episodes of demyelination may lead to formation of several wraps forming "onion bulbs". Post-mortem examinations on patients who died several years after having had GBS still showed sparse perivascular foci of mononuclear cells (Asbury et al. 1969). This might indicate long-lasting subclinical inflammation of the peripheral nervous system. In severe cases the axonal degeneration which develops in the acute stage causes permanent reduction of the myelinated nerve fibre density. There is corresponding muscle atrophy with fibre type grouping.

Summary

The earliest phases of GBS are usually associated pathologically with macrophage-associated demyelination in a multifocal distribution, sometimes affecting roots more than nerves, sometimes the opposite. With the electron microscope demyelination is seen to proceed via macrophage invasion of the Schwann cell basal lamina and then stripping and phagocytosis of myelin debris. Rarely myelin undergoes extracellular vesicular dissolution but this is always associated with adjacent mononuclear cell infiltration. In some autopsy cases there is prominent perivascular lymphocytic infiltration at all stages of the disease which has given rise to comparison with EAN and the proposal that cell-mediated immunity is an important pathogenetic mechanism. In other autopsy cases and most biopsy studies lymphocytic infiltration has been absent or sparse, although macrophage-mediated demyelination is prominent, which leads to the proposal that antibody-mediated mechanisms are important. In rare cases of GBS with explosively rapid onset axonal degeneration is the underlying pathological substrate. Demyelination is rapidly followed by Schwann cell proliferation and remyelination but in severe cases there is permanent loss of some nerve fibres.

References

Ammoumi AA, Pertschuk L, Daras M, Rosen AD (1980) Guillain–Barré syndrome. Results of direct immunofluorescent study. NY State J Med 80:1434–1435

Arstila AV, Reikinnen PJ, Rinne VK, Pelliniemi TT, Nev Alainen T (1971) Guillain–Barré syndrome. Neurochemical and ultrastructural study. Eur Neurol 5:257–269

Asbury AK, Arnason BG, Adams RD (1969) The inflammatory lesion in idiopathic polyneuritis. Its role in pathogenesis. Medicine 48:173–215

Asbury AK, Johnson PC (1978) Acute idiopathic polyneuritis and related disorders. Pathology of peripheral nerve. WB Saunders, Philadelphia, pp 120–135

Best PV (1985) Acute polyradiculoneuritis associated with demyelinated plaques in the central nervous system: a report of a case. Acta Neuropathol 67:230–234

Behar R, Penny R, Powell HC (1986) Guillain–Barré syndrome associated with Hashimoto's Hyroiditis. J Neurol 233:233–236

Brechenmacher C, Vital C, Laurentjoye L, Castaing Y (1981) Ultrastructural study of peripheral nerve in Guillain–Barré syndrome presence of mononuclear cells in axons. Acta Neuropathol 7 suppl:249–251

Brechenmacher C, Vital C, Deminiere C et al. (1987) Guillain–Barré syndrome: an ultrastructural study of peripheral nerve in 65 patients. Clin Neuropathol 6:19–24

Carpenter S (1972) An ultrastructural study of an acute fatal case of the Guillain–Barré syndrome. J Neurol Sci 15:125–140

Cornblath DR, Griffin DE, Chupp M, Griffin JW, McArthur JC (1987a) Mononuclear cell typing in inflammatory demyelinating polyneuropathy nerve biopsies. Neurology 37 suppl 1:253–254

Cornblath DR, McArthur JC, Kennedy PGE, Witte AS, Griffin JW (1987b) Inflammatory demyelinating peripheral neuropathies associated with human T-cell lymphotropic virus type III infection. Ann Neurol 21:32–40

Dehaene I, Martin JJ, Geens K, Cras P (1986) Guillain–Barré syndrome with ophthalmoplegia: clinicopathologic study of the central and peripheral nervous system, including the oculomotor nerves. Neurology 36:851–854

Feasby TE, Gilbert JJ, Brown WF et al. (1986) An acute axonal form of Guillain–Barré polyneuropathy. Brain 109:1115–1126

Finean JB, Woolf AL (1962) An electron microscope study of degenerative changes in human cutaneous nerve. J Neuropathol Exp Neurol 21:105–115

Hausmanowa-Petrusewicz I, Emeryk B, Rowinska-Marcinska K, Jedrzejowska H (1979) Nerve conduction in the Guillain–Barré–Strohl syndrome. J Neurol 220:169–184

Haymaker W, Kernohan JW (1949) The Landry–Guillain–Barré syndrome: a clinicopathologic report of fifty fatal cases and a critique of the literature. Medicine 28:59–141

Hays AP, Lee SSL, Latov N (1988) Immune reactive C3d on the surface of myelin sheaths in neuropathy. J Neuroimmunol 18:231–244

Honavar M, Tharakan JKJ, Hughes RAC, Leibowitz S, Winer JB (1990) A clinico-pathological study of Guillain–Barré syndrome. Brain (in press)

Julien J, Vital CL, Aupy G, Laugueny A, Darriet D, Brechenmacher C (1980) Guillain–Barré syndrome and Hodgkin's disease – ultrastructural study of a peripheral nerve. J Neurol Sci 45:23–27

Kanda T, Hayashi H, Tanabe H, Tsubaki T, Oda M (1989) A fulminant case of Guillain–Barré syndrome: topographic and fiber size related analysis of demyelinative changes. J Neurol Neurosurg Psychiatry 52:857–864

Koski CL, Sanders ME, Swoveland PT et al. (1987) Activation of terminal components of complement in patients with Guillain–Barré syndrome and other demyelinating neuropathies. J Clin Invest 80:1492–1497

Krücke W (1955) Die primär entzündliche Polyneuritis unbekannter Ursache. Lubasch O, Henke F, Rossle G (eds) Handbuch der speziellen pathologischen Anatomie und Histologie. Vol 13 Erkrankungen der peripheren Nerven. Springer, Berlin, pp 164–182

Luitjen JAFM, Faille-Kuyper EHB (1972) The occurrence of IgM and complement factors along myelin sheaths of peripheral nerves. An immunohistochemical study of the Guillain–Barré syndrome. J Neurol Sci 15:219–224

Mancardi GL, Cadoni A, Zicca A et al. (1988) HLA-DR Schwann cell reactivity in peripheral neuropathies of different origins. Neurology 38:848–852

Mei Liu H (1970) Ultrastructure of remyelination of peripheral nerves in Landry–Guillain–Barré syndrome. Acta Neuropathol 16:262–265

Miyakawa T, Murayama E, Sumiyoshi S et al. (1971) A biopsy case of Landry–Guillain–Barré syndrome. Acta Neuropathol 17:181–187

Nyland H, Matre R, Mork S (1981) Immunological characterisation of sural nerve biopsies from patients with Guillain–Barré syndrome. Ann Neurol 9 suppl:80–86

Phillips MS, Stewart S, Anderson JR (1984) Neuropathological findings in Miller–Fisher syndrome. J Neurol Neurosurg Psychiatry 47:492–495

Pollard JD, Baverstock J, McLeod JG (1987) Class II antigen expression and inflammatory cells in the Guillain–Barré syndrome. Ann Neurol 21:337–341

Prineas JW (1972) Acute idiopathic polyneuritis. An electron microscope study. Lab Invest

26:133–147

Prineas JW (1981) Pathology of the Guillain–Barré syndrome. Ann Neurol 9 suppl:6–19

Ramos-Alvarez M, Bessudo L, Sabin AB (1969) Paralytic syndromes associated with noninflammatory cytoplasmic or nuclear neuropathy. JAMA 207:1481–1492

Schober R, Itoyama Y, Sternberger NH et al. (1981) Immunocytochemical study of P_0 glycoprotein, P_1 and P_2 basic proteins, and myelin associated glycoprotein (MAG) in lesions of idiopathic polyneuritis. Neuropathol Appl Neurobiol 7:421–434

Tsukada N, Koh C-S, Inoue A, Yanagisawa N (1987) Demyelinating neuropathy associated with hepatitis B virus infection. Detection of immune complexes composed of hepatitis B virus surface antigen. J Neurol Sci 77:203–216

Van Zandycke M, Martin JJ, Gaer LV, Van den Heyning J (1982) Facial myokymia in the Guillain–Barré syndrome: a clinicopathologic study. Neurology 32:744–748

Winer JB, Hughes RAC, Osmond C (1988) A prospective study of acute idiopathic neuropathy. I. Clinical features and their prognostic value. J Neurol Neurosurg Psychiatry 51:605–612

Wisniewski H, Terry RD, Whitaker JN, Cook SD, Dowling PC (1969) The Landry–Guillain–Barré syndrome. A primary demyelinating disease. Arch Neurol 21:269–276

Chapter 5

Epidemiology

Introduction

The annual incidence of GBS is uniformly about one or two cases per 100 000 population (Table 5.1). Reports of large series of patients from the USA (Wiederholt et al. 1964; Pleasure et al. 1969), Scandinavia (Ravn 1967), Israel (Soffer et al. 1978), West Indies (Hanna 1979), Nigeria (Osuntokun and Agbebi 1973), Kenya (Bahemuka 1988), Singapore (Tong et al. 1979) and China (Zhao et al. 1981) indicate that the disease extends worldwide. GBS has been recognised for over a hundred years and data from Olmsted County in Minnesota indicate that the incidence has been stable during the past 40 years (Kennedy et al. 1978). Occasional reports of an unusual number of cases in relatively small communities, such as 19 cases in Lorimer County, Colarado in 1981–83 giving a crude annual incidence of 4.0 per 100 000 population per year (Kaplan et al. 1982), may represent chance occurrences. In every study the incidence of GBS in males has been slightly higher than that in females, by a factor of about 1.25 to 1 (Hurwitz et al. 1983). The slight male preponderance contrasts with the female preponderance in multiple sclerosis and in many autoimmune disorders such as Hashimoto's thyroiditis and myasthenia gravis. GBS was slightly more common in whites than blacks in one large American study (Hurwitz et al. 1983) but this difference might have been attributable to the method of ascertainment which depended on a system of sentinel neurologists to which whites might have had easier access. Most studies have shown that GBS is more common in the elderly and some have reported that children or young adults are more prone to develop GBS so that the age distribution of incidence is bimodal (Dowling et al. 1977; Schonberger et al. 1981). The largest body of data comes from 2575 cases detected by the national surveillance of GBS cases in the USA in the period from January 1978 to March 1981. The incidence distribution confirmed the increased risk in the elderly with an additional small increase in females aged between 15 and 30 (Fig. 5.1). The increased incidence in the elderly is compatible with the hypothesis that autoimmune disorders are more likely to occur in old age because of failure of normal suppressor mechanisms.

Swine Influenza Vaccine Epidemic

On 1 October 1976 the USA department of health began a massive campaign to immunise the American population against an influenza virus strain which was

Table 5.1. Incidence of Guillain–Barré syndrome

Reference	Year	Country	Region	No. of cases	Crude average annual incidence per 100 000
Kurland 1958	1945–1954	USA	Rochester, Minnesota	3	1
Gudmundson 1969	1954–1963	Iceland		13	0.7
Brewis et al. 1966	1955–1961	UK	Carlisle	3	0.6
Chen et al. 1968	1960–1966	USA	Guam	5	1.9
Lesser et al. 1973	1935–1968	USA	Olmsted County, Minnesota	29	1.6
Kennedy et al. 1978	1935–1976	USA	Olmsted County, Minnesota	40	1.7
Soffer et al. 1978	1969–1972	Israel		89	0.8
Hogg et al. 1979	1972–1976	USA	San Joaquim County, California	18	1.2
Larsen et al. 1985	1957–1982	Norway	Hordaland	109	1.19
Bremen and Hayner 1984	1 July 1976– 30 April 1977	USA	Michigan	79	1.87
Langmuir et al. 1984	1 Oct–19 Dec 1976	USA	Nationwide	260[a]	0.98
Haberman et al. 1982	1978	UK	North-West Thames Health Region	39	1.10
Hankey 1987	1980–1985	Australia	Western Australia	109	1.35

[a]Figure for associated population and for cases with "extensive" rather than "limited" paresis (see text).

expected to affect the nation that winter. In early December attention was drawn to the occurrence of GBS in some of the recipients of A/New Jersey/ 1976 or swine influenza vaccine. In mid-December, when 45 million immunisations had been given, further immunisations were stopped. A national surveillance programme was established and all cases of GBS encountered by sentinel neurologists throughout the USA were recorded from 1 October 1976 until 31 January 1977. The findings were the subject of preliminary reports (Schonberger et al. 1979; Langmuir 1979) but there was much initial confusion about what constituted a case. The most helpful summary of this episode is the evaluation of the epidemic by Langmuir et al. (1984). A panel of experts was commissioned by a court order to evaluate 1300 computerised summaries of cases of alleged GBS having their onset between 1 October 1976 and 31 January 1977. From the data available the panel identified 1098 cases who fulfilled the diagnostic and onset criteria and had a record of swine influenza vaccination. Of these cases 580 had "extensive" paresis of limbs and trunk or cranial nerve muscles whereas the remaining cases had "limited" paresis confined to the limbs or insufficient data for classification. A clear increase in incidence of "extensive" paresis, but not the other two categories, was demonstrated for at least six weeks and possibly eight weeks following immunisation (Langmuir et al. 1984). The increased incidence had a log-normal distribution which is characteristic of a point source epidemic (Fig. 5.2). The figure shows that the increase in risk rises sharply from the time of vaccination to reach a peak during the last half of the second week,

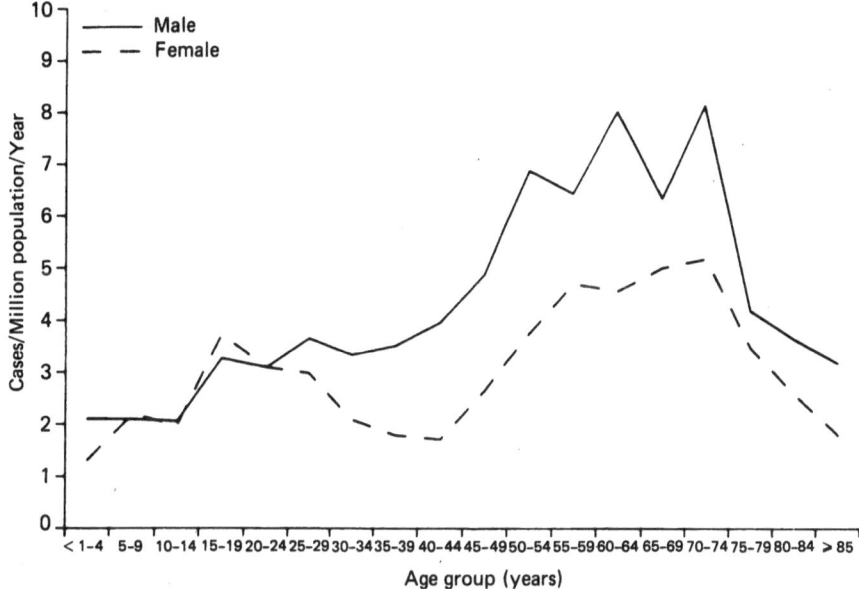

Fig. 5.1. Incidence of Guillain–Barré syndrome by age and sex in the United States from January to March 1981. Note that the absolute incidence is probably an underestimate but the relative sex and age distributions are the most reliable available. (From Kaplan et al. (1982), with permission.)

which is also the commonest time at which GBS is seen following infections. From these figures the relative risk of developing GBS during the six weeks following swine influenza vaccination was calculated as between four and eight times the background risk.

The explanation of this apparent epidemic of GBS following the 1976 swine influenza vaccination programme remains uncertain. It is possible that the increased incidence was more apparent than real because of ascertainment bias (Kurland et al. 1985). Kurland and Wiederholt (1986) have pointed out that there was no concomitant increase in reports of GBS among 1.7 million vaccinated members of the US Armed Forces. The cases included by Langmuir et al. (1984) had to be classified from a summary of limited information and were not the subject of formal follow-up and review as would be usual in a clinical paper. The possibility of successful litigation would have provided a much more powerful incentive for doctor and patient to diagnose and report GBS in vaccinated than unvaccinated persons. The reality of the increase in incidence shown in Fig. 5.2 awaits confirmation from a re-evaluation of the cases which was reported to be under way (Kurland and Wiederholt 1986). If the increased incidence was real it appears to have been unique for the "swine influenza" strain of vaccine. A few cases of GBS after other influenza vaccines have been reported but there is no evidence that these are more than a chance association (Winer et al. 1984; Winer et al. 1988c). Surveillance of the effects of other influenza vaccine campaigns, for instance in the USA in 1979–1980 and 1980–1981, has not detected any other epidemic (Kaplan et al. 1982). The cases of GBS included recipients of vaccine from different batches and different manufac-

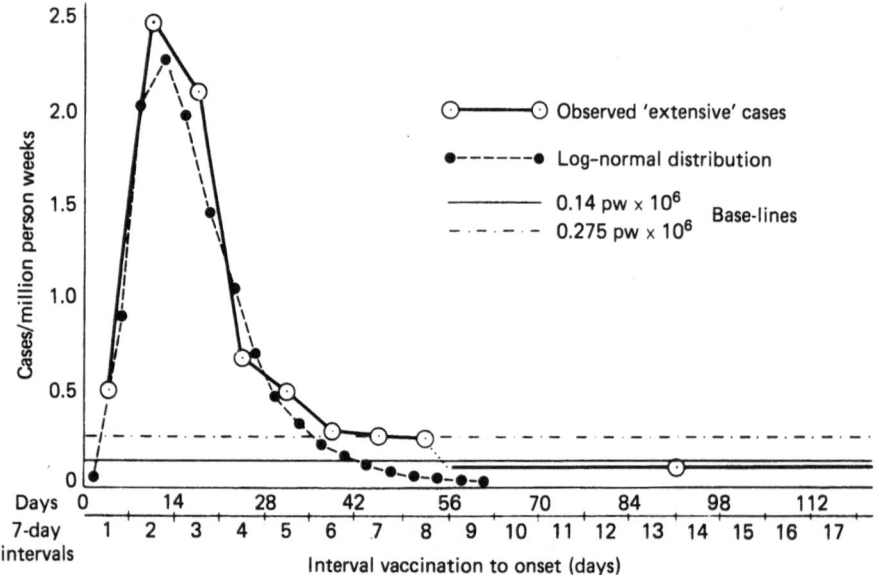

Fig. 5.2. Incidence of GBS with "extensive" paresis, i.e. affecting trunk or cranial nerves as well as limbs, following swine influenza vaccination in the United States between 21 October 1976 and 31 January 1977. The baseline estimate of 0.14 cases per million per week is derived from the unvaccinated cases in the USA in the same period and the baseline estimate of 0.275 from studies in Olmsted County, Minnesota and in Michigan. Some or even all of this apparent increase in incidence may be due to an artefact of reporting (see text). (From Langmuir et al. (1984), with permission.)

turers. Claims that the swine influenza vaccine might contain small amounts of chicken myelin P_2 basic protein (which causes EAN) have been refuted in particular by Brostoff and White (1982). A case-control study failed to detect either an increase in frequency of previous influenza vaccination or of allergy in the patients who developed GBS after swine influenza immunisation (Kaslow et al. 1987).

Attempts to reproduce disease by injecting experimental animals with influenza vaccine have not been successful. Ziegler et al. (1983) reported producing an EAN-like disease in rabbits with an emulsion of influenza vaccine, adjuvant and ganglioside. Such disease was predicted (Lennon 1986) and then shown to be due to the injection of ganglioside alone which produces some inflammation and demyelination in the peripheral nervous system of rabbits (Mizisin et al. 1987).

Leneman (1966) reported briefly that of 1100 cases of GBS in the literature 735 had an alleged precipitating event and 32 of these events were vaccinations. Winer et al. (1988b) undertook a case-control study of 100 patients and 100 age- and sex-matched hospital controls, who had been admitted to hospital for minor surgical conditions: six patients and five controls had been immunised during the previous 3 months; three patients and three controls had received tetanus toxoid. This case-control study and the published literature demonstrate that GBS only follows immunisation in a small percentage of cases and immunisation is rarely followed by GBS. The available studies do not rule out a modest increase in risk of GBS following immunisations since detection of a four- or fivefold increase in risk following a particular immunisation would require a national surveillance programme of the magnitude used to investigate the alleged increase in incidence following swine influenza immunisation in the USA.

Reports of GBS following typhoid vaccination (Miller and Stanton 1954; Cambier and Schott 1966) persuaded Arnason that the association was significant. In an early review there were three cases of GBS and nine of Landry's paralysis out of 50 cases with neurological complications following this vaccine (Miller and Stanton 1954). Meningitis and encephalitis were more common complications. I do not think we can decide whether this represents a real increase in risk following typhoid vaccination. Typhoid vaccination is certainly rarely followed by GBS and GBS is rarely preceded by typhoid vaccination.

Arnason (1984) doubted whether the rare reports in the literature of GBS following immunisation with poliomyelitis, measles, rubella, diphtheria, pertussis and tetanus toxoid exceed background incidence. I remain concerned that tetanus toxoid may be an occasional precipitant of GBS. Single cases have been reported following tetanus and diphtheria toxoid (Holliday and Bauer 1983) and tetanus toxoid alone (Newton and Janati 1987). Recurrent attacks of demyelinating neuropathy occurred following repeated immunisation with tetanus toxoid in one patient (Pollard and Selby 1978). The decision whether to avoid routine immunisation following an attack of GBS lacks adequate epidemiological data for scientific guidance. I think the concern is sufficient to omit routine tetanus toxoid booster injections following GBS but not to omit tetanus toxoid reimmunisation in the presence of a wound carrying a significant risk of inducing tetanus.

Possible increased evidence of GBS following oral poliomyelitis vaccination was noted recently in Finland. Ten patients were detected who had developed GBS within ten weeks after immunisation and the incidence was about three times the incidence expected in the population being monitored (Kinnunen et al. 1989; Uhari et al. 1989). Neurological complications used to be occasional complications of serum treatment, such as with antitetanus serum. Fifteen of 1100 cases of GBS from the literature had been given antitoxin shortly beforehand (Leneman 1966). The usual neurological complication of antiserum injection used to be brachial neuritis with an urticarial rash and fever. Out of 100 cases of neurological complications of serum treatment culled from the literature including five personal cases, only three were attributed to GBS but some others were labelled Landry's paralysis and might have represented the same pathological entity (Miller and Stanton 1954). Without autopsy or biopsy evidence it is impossible to exclude the possibility that these complications were due to vasculitis.

Post-Rabies Vaccine Neuritis

The neuroparalytic illnesses occasionally following rabies vaccination not uncommonly resemble GBS (Griffin 1988). The first vaccine introduced by Pasteur in 1875 consisted of dried formalised brain from a rabbit infected with rabies. This was later called Semple vaccine. Cases of encephalomyelitis which followed could be mimicked in animals by immunisation with rabbit brain alone, leading to subsequent research in EAE (Chapters 1 and 3). Appelbaum et al. (1953)

reported 44 cases of neuroparalytic disorders following Semple rabbit brain vaccine in New York State, an incidence of 1 per 2000 vaccinations. Four of these had GBS and some others had both CNS and PNS involvement. Hemachudha et al. (1987a) reported 32 cases of meningitis, encephalitis or myelitis in various combinations and four cases of GBS among recipients of Semple vaccine in Thailand, where the incidence of neuroparalytic complications was reported to be as high as 1 in 100 to 1 in 400. Since a tenth of the complications affected the PNS the risk of GBS following Semple rabies vaccine was in the order of 1 in 1000 to 1 in 4000.

In attempts to reduce the neuroparalytic complications of rabies vaccination a vaccine was produced from suckling mouse brain, which has a much lower myelin content than Semple vaccine. Unfortunately this vaccine is also followed by neuroparalytic disorders in which the proportion of GBS is higher. Toro et al. (1977) reported 21 neurological complications including 16 cases of GBS (six fatal). The overall incidence of neurological complications was 1 in 4615. Cabrera et al. (1987) describe five patients with GBS following suckling mouse brain vaccine in Peru: all had severe disease which was remarkable for prominent cranial nerve involvement, severe muscle wasting and a protracted course including severe ophthalmoplegia and wasting of the tongue. Detailed neurophysiological and histological studies have not been made. There is no theoretical reason why human diploid cell culture-derived rabies vaccine should precipitate GBS although cases of GBS following this vaccine have been reported (Boe and Nyland 1980; Knittel et al. 1989).

Recent immunological studies indicate that patients with neurological complications following Semple or suckling mouse brain vaccination have more antibody in their serum to myelin antigens than normal people or vaccinated persons without such complications. Patients with both CNS and PNS complications had increased reactivity of serum and CSF with crude CNS antigen, myelin basic protein, cerebroside or ganglioside detected by ELISA (Hemachudha et al. 1988; Hemachudha et al. 1987b). Antibody to myelin-associated glycoprotein could not be detected. Intrathecal synthesis of antibodies to MBP and cerebroside was detected in between a third to a half of patients in whom it was tested. Somewhat surprisingly increased antibody levels could not be detected against crude PNS antigens or P_2. Antibodies to a SV40 transformed human cell line were detected in the sera of patients with CNS and PNS complications following Semple vaccine but not in the sera of patients with GBS following suckling mouse brain or patients with sporadic GBS.

It does seem likely that the post-vaccine cases of GBS are caused by an EAN-like mechanism but it has to be admitted that the identity of the antigen remains elusive. The sera of patients with rabies vaccine-induced GBS do contain antibody to MBP but the distribution of MBP in both CNS and PNS makes it an unlikely target for a pathogenetically important autoimmune reaction in GBS.

Viral Infections as Antecedent Events

Some form of infection precedes nearly two-thirds of cases of GBS. In general neurological practice the precise infection is not clear from the history. Since the acute infection has usually subsided when the neuropathy begins, viral or bacterial cultures are usually negative and serological tests are rarely rewarding. Melnick and Flewett (1964) confirmed this association between GBS and an antecedent infection in a case-controlled study in Birmingham, UK. They compared 44 patients with GBS with 48 medical patients and 23 healthy hospital staff and found that 48% of the patients and only 18% of the controls gave a history of respiratory infection during the preceding month. In a retrospective review covering 1935–1976 from Olmsted County, Minnesota, Kennedy et al. (1978) found a history of infection in 28 of 37 patients within 4 weeks before the onset of GBS. This was more frequent than in age- and sex-matched control patients with Bell's palsy. This difference was largely due to the occurrence of gastrointestinal infections (12/37) in the GBS patients but not in those with Bell's palsy. In a prospective study conducted in south east England during 1983 and 1984, 99 patients and age- and sex-matched controls recruited from surgical wards were questioned about the occurrence of infections before the onset of GBS. The results were compared with the incidence in the corresponding period in the controls (Winer et al. 1988b). Respiratory infections had occurred in 38% of the patients and 12% of the controls, relative risk 4.1, and gastrointestinal infections in 17% of the patients and 3% of the controls, relative risk 7.5. Respiratory infections had usually started 1–2 weeks before the onset of the neuropathy and gastrointestinal infections usually only 1 week or less before (Fig. 5.3). These few case-control studies confirm the clinical impression that both respiratory and gastrointestinal infections precede GBS more commonly than would be expected by chance. The association with respiratory infection is more frequent but that with gastrointestinal infections emerges as more significant because these are less common in control populations.

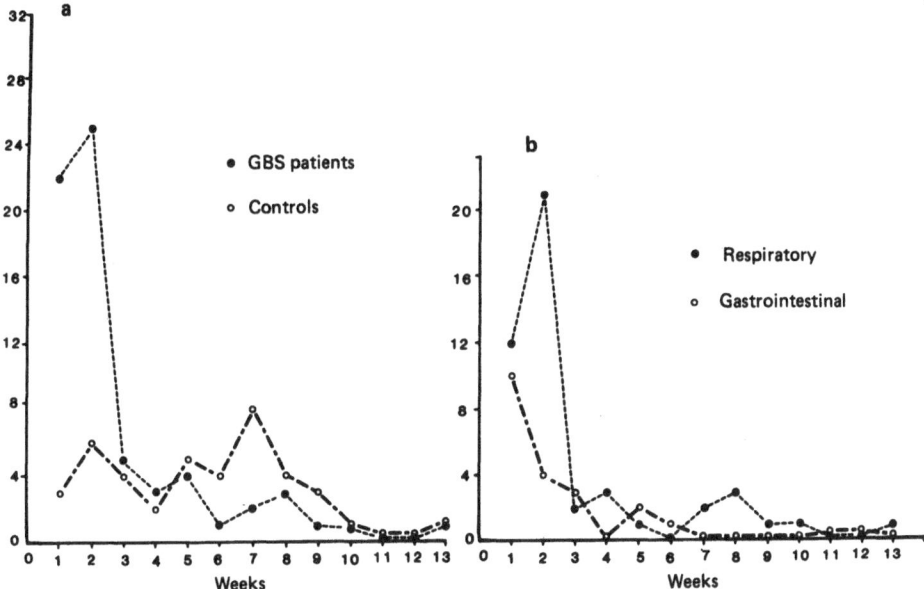

Fig. 5.3. **a** Incidence of infection at intervals before the onset of neuropathy in 99 patients with GBS and at the corresponding interval in 99 control subjects. **b** Incidence of respiratory or gastrointestinal infection in 99 patients with GBS at intervals before the onset of neuropathy. (From Winer et al. (1988b), with permission.)

In studies which have sought serological identification of the preceding infection the most commonly recognised has been cytomegalovirus (CMV). The association between CMV and GBS was first reported in 1967 (Klemola et al. 1967). In 1971 a collaborative study reported nine cases (Leonard and Tobin 1971). A German study reported evidence of CMV infection in 10% of 94 cases (Schmitz and Enders 1977). Dowling and Cook (1981) tested sera by immuno-fluorescence on infected fibroblasts for evidence of recent CMV infection and found positive results in 15% of sera from 220 patients with GBS seen in the New York region from 1967 to 1980. A study in south east England from 1983 to 1984 with a solid phase radioimmunoassay revealed serological evidence of CMV infection in 11 of 99 patients compared with only 1 of 99 age, sex-matched controls (Winer et al. 1988b). Other individual case reports are reviewed by Dowling and Cook (1981). Of the 33 patients with GBS associated with serological evidence of CMV infection reported by Dowling and Cook (1981) most had had a febrile illness usually affecting the upper respiratory tract but some had had no apparent antecedent illness. Most people develop antibodies to CMV at some time during their lives and the initial infection may be inapparent or may cause a glandular fever-like illness, especially in young women. In Dowling and Cook's series most of the sera with CMV IgM antibodies were from younger patients. This association between CMV and GBS might account for some or all of the "hump" in the age distribution of frequency of GBS observed in young women in the national survey of GBS in the USA from 1978 to 1981 (Fig. 5.1). Dowling and Cook (1981) commented that their GBS patients with CMV antibodies commonly had increased numbers of circulating lymphoblasts, elevated serum IgM, cold agglutinins and elevated liver enzymes in the plasma, all features of adult-onset CMV infection. CMV infection is particularly prone to occur in transplant subjects (Bale et al. 1980), following surgery and blood transfusion, and in acquired immunodeficiency syndrome (AIDS) (Bishopric et al. 1985), all situations in which GBS has been reported.

Although the association between CMV infection and GBS is established its explanation remains obscure. The possibility of molecular mimicry between viral protein and homologous sequences in myelin protein (Jahnke et al. 1985) seems unlikely because of the multiplicity of agents which may trigger GBS (Table 5.2). In CMV infection there is usually an increase in CD8+ lymphocytes and a reduction in CD4+ lymphocytes in the blood. A similar change in lymphocyte subsets in the blood has been observed in GBS following CMV infection although the changes were less marked than is usual with uncomplicated CMV infection (Winer et al. 1988a). It is possible that a perturbation of normal suppressor mechanisms permits the escape of autoimmune B and T cell responses directed against myelin antigens but this remains to be established.

Epstein–Barr virus (EBV), another herpes virus, has been reported to precede GBS on so many occasions that the association seems likely to be real. The first report was published in 1947 (Ricker et al. 1947) and numerous single case reports and small series have been published (Arnason 1984). Since most of the population become infected with EBV, IgG antibodies to the virus are ubiquitous and strict criteria for serological diagnosis of primary infection as opposed to reactivation must be adopted. False positive tests due to the presence of rheumatoid factor must also be eliminated by absorbing the serum with aggregated immunoglobulin before testing. Failure to adopt such criteria may explain why some early surveys suggested a very high incidence of recent EBV

Table 5.2. Antecedent events for which there is reason-
able evidence for a causative association with Guillain–
Barré syndrome

Vaccines
  Swine influenza (possible artefact of reporting)
  Suckling mouse brain rabies
  Semple rabbit brain rabies
  Vaccinia
Viruses
  Cytomegalovirus
  Epstein–Barr
  Varicella zoster
  Human immunodeficiency
Other agents
  Mycoplasma pneumoniae
  Campylobacter jejuni

---

virus infection in GBS. Dowling and Cook (1981) adopted reasonably strict criteria and found EBV specific IgM antibodies in the serum of 8 of 100 patients with GBS in New Jersey. Winer et al. (1988b) found only one probably and one definitely positive serum from 99 patients in south east England compared with two probably positive sera in 99 age- and sex-matched controls. The discrepancy might be accounted for by the age structure of the populations being studied, since the English patients were probably older than those in the New Jersey series. EBV infection may be complicated by a subclinical or even clinically evident encephalopathy which may cloud the diagnosis and prevent categorisation of an illness as GBS (Gautier-Smith 1965).

Varicella-zoster infection is another herpes infection which is probably a significant antecedent factor triggering GBS. The evidence for this is the gradual accumulation of case reports such that Sanders et al. (1987) were able to review 28 cases following varicella and 21 cases following zoster including personal cases and literature reports. Although probably significant the association is rare; only 10 of the 1100 cases of GBS reviewed by Leneman (1966) had had varicella-zoster beforehand and none of the 99 cases reported by Winer et al. (1988b). The incidence of GBS following varicella is of the order of 1 in 10000 or less (Arnason 1984).

Very few cases of proven active herpes simplex infection and GBS have been reported in the literature (Callaghan et al. 1974; Olivarius and Buhl 1975; Menonna et al. 1977; Morrison et al. 1979). In a serological study of 75 patients Dowling and Cook (1981) were only able to find one with herpes simplex-specific IgM antibodies. Although it is difficult to exclude such a ubiquitous virus from an aetiological role in GBS there is nothing in the literature to support such an idea.

It has recently become clear that both GBS and CIDP may occur in association with human immunodeficiency virus (HIV) infection. GBS usually occurs at (Piette et al. 1986) or soon after seroconversion and before the development of AIDS-related complex (ARC) or AIDS itself (Berger et al. 1987, five cases; Cornblath et al. 1987, three cases; Vendrell et al. 1987, one case). One patient with established AIDS has been reported who developed typical GBS with antibodies to myelin in the serum and subsequently improved (Mishra et al.

1985). Two patients with AIDS developed a progressive cauda equina syndrome and died of respiratory failure within a month. One had pneumocystis pneumonia. Both had marked CSF pleocytosis. One was shown at autopsy to have severe inflammatory cell infiltration and endoneurial vessel wall necrosis in the cauda equina. Endoneurial cells probably both Schwann cells and endothelial cells, contained cytomegalic inclusions and CMV infection was demonstrated with a biotinylated CMV DNA probe (Eidelberg et al. 1986). The histological picture was that of a vasculitis rather than a demyelinating process and nerve conduction studies were not reported. The clinical features of the series reported by Cornblath et al. (1987) were variable but consistent with the characteristic clinical pictures of CIDP (six patients) and GBS (three patients). The CIDP patients improved either spontaneously or following treatment with plasma exchange, prednisolone or ACTH. The GBS patients slowly improved following plasma exchange. Neurophysiological studies showed the expected combination of demyelination and then axon loss. The one distinguishing feature of these patients has been that the CSF usually contains an increased cell count (mean 23 cells/µl in the series of Cornblath et al. (1987)). One of the patients with GBS underwent nerve biopsy which showed internodal demyelination, extensive mononuclear cell infiltration and Wallerian degeneration. The endoneurial component of the infiltrate was shown with immunohistochemical staining techniques to include T lymphocytes. An unusual finding was prominent vacuolation of cells in the perineurium probably representing infiltrating macrophages. Viral nucleic acid could not be detected in any of the cells in the sural nerve biopsies of three CIDP or one GBS patient with a radiolabelled DNA probe capable of detecting HIV RNA in infected spleen cells (Cornblath et al. 1987). This makes it likely that inflammatory neuropathy associated with HIV infection is not directly related to the viral infection. The neuropathy is more likely to be caused in some way by the altered state of immunity in these patients who are also vulnerable to other putative autoimmune conditions including thrombocytopenia. The association between HIV and CIDP will be discussed further in Chapters 10 and 12.

Although GBS and CIDP usually occur at the time of seroconversion or before the development of other features of HIV infection 12 cases of multiple mononeuropathy or polyneuropathy have been reported in association with generalised lymphadenopathy, fever and night sweats (Lipkin et al. 1985). The CSF usually showed a pleocytosis and the IgG concentration was increased. EMG and biopsy studies showed axonal neuropathy and often demyelination as well. Three of six biopsies showed endoneurial and perivascular inflammation. Improvement occurred in some of the patients but others rapidly developed AIDS. In AIDS itself it is more common to encounter a progressive painful predominantly sensory neuropathy which is due to an axonopathy (Snider et al. 1983; Bailey et al. 1988).

Smallpox vaccination with live vaccinia virus used to be followed by GBS sufficiently frequently for Arnason (1984) to regard the association as established. Important reports were those of Lane et al. (1968) and Spillane and Wells (1964).

Several reports of GBS following measles have been published but it is clearly a rare complication, less common than encephalitis which is said to occur in about one per thousand cases (Miller et al. 1956; Lidin-Janson and Strannegard 1972). Only 11 of a retrospective series of 1100 (Leneman 1966) and none of the 100 prospectively studied cases of Winer et al. (1988a) had had measles.

Modern serological techniques now permit distinguishing the relationship between different forms of hepatitis and GBS. Some early reports might have been due to CMV or EBV infection both of which may cause hepatitis. Several reports now describe an association between GBS and serological evidence of acute hepatitis B (Detroyer 1975; Ng et al. 1975; Niermeijer and Gips 1975; Marti-Masso et al. 1979; Huet et al. 1980; Berger et al. 1981; Penner et al. 1982) or hepatitis A infection (Johnston et al. 1981; Dunk et al. 1982; Grover et al. 1986; Mares-Segura et al. 1986), and also non-A non-B hepatitis (Macleod 1987). These reports have included cases during the incubation, icteric and convalescent phases.

There have been at least 18 published reports of GBS occurring after mumps (Miller et al. 1956; Pollack et al. 1981). Similarly there are a few reports of rubella followed by GBS (Miller et al. 1956; Tomlinson 1975; Saeed and Lange 1978). It is difficult to know which of the reported associations has any aetiological significance because of the reporting bias. Many other virus infections have been implicated on the basis of one or two case reports. These include influenza A (Wells et al. 1959; Stevens et al. 1974; Arnason 1984), influenza B (Stevens et al. 1974; Arnason 1984), parainfluenza (Eisen and Humphreys 1974), ECHO virus types 6, 7 and 22 (Parker et al. 1960; Forbes et al. 1967; Urano et al. 1970; Eisen and Humphreys 1974), Coxsackie virus (Jackson 1961; Melnick and Flewett 1964; Eisen and Humphreys 1974; Usui et al. 1974), respiratory syncytial virus (Arrowsmith et al. 1985) and BK virus (Nordaa and Dillen 1977). Their identification depends on laboratory studies which are rarely performed.

Mycoplasma, Bacterial and Other Infections as Antecedent Events

There have been many reports of an association between Mycoplasma infection and GBS (Arnason 1984). In one series there was serological evidence of active Mycoplasma infection in five of 100 patients (Goldschmidt et al. 1980). This association is sufficiently frequent to deserve serious consideration as an aetiological link. Mycoplasma infection also induces antibodies to red cells, haemolytic anaemia, Stevens–Johnson syndrome and encephalitis.

There are reports of many bacterial infections preceding GBS and of these the most convincing concern *Campylobacter*. *Campylobacter jejuni* and *C. coli* are Gram-negative flagellated curved rods which are common causes of acute gastroenteritis, often with malaise, fever, headache and generalised aching. The incubation period is usually 3 days with a range of 1–7 days or longer and the severe symptoms usually subside spontaneously within 2–3 days. Recent infection can be identified serologically and during the 1980s this has permitted a plethora of reports of an association with GBS (Molnar et al. 1982; Rhodes and Tattersfield 1982; Constant et al. 1983; Pryor et al. 1984; Wroe and Blumhardt 1984; De Bont et al. 1986; Kohler and Goldblatt 1987; Speed et al. 1987; Sovilla et al. 1988; Ropper 1988). The significance of these reports is enhanced by two case-controlled studies. Of 56 (38%) patients with GBS admitted to an infectious diseases hospital in Victoria, Australia, 21, a self-selected sample, had evidence of anti-*Campylobacter* antibodies detected by ELISA and an immunoblot technique compared with none of a similar number of controls (Kaldor and Speed 1984; Speed et al. 1987). In a prospective study in England 14 of 99 (14%) patients with GBS had serological evidence of *Campylobacter* infection compared with two controls (Winer et al. 1988b). Only nine of the 14 had a history of gastroenteritis, so that the infection may go unnoticed and yet trigger GBS. There was a reduction of the blood CD8+ lymphocyte subset in patients with *Campylobacter* infection which was not evident in those who had had CMV infection (Winer et al. 1988c). A higher proportion of those infected with *Campylobacter* had a poor outcome than those not so affected (Ropper 1988; Winer et al. 1988c). The association between *Campylobacter* and GBS may account for part of the relationship between GBS and gastrointestinal infections noted in many series and the few case-control studies. However, it is probable that other bacterial intestinal infections can also trigger GBS: 19 cases were reported following an epidemic of gastroenteritis due to *Shigella boydii* in Jordan (Khoury 1978). The neuropathy began a median of seven days after the gastroenteritis. Several individual reports of GBS after typhoid have been published (Berger et al. 1986; Changdam and Waniganetti 1969; Donoso 1988; Samantray 1977). It is a rare complication since none of 959 Nigerian patients with typhoid developed typical GBS although three developed multiple mono-neuropathy and seven a mild sensory and motor neuropathy without albumino-cytological dissociation (Osuntokun and Agbebi 1973). Conversely even in countries where typhoid is common, GBS is rarely preceded by typhoid. For instance typhoid had occurred in only three of 302 cases of GBS in India (Samantray 1977). An interesting case of both botulism and demyelinating neuropathy has been reported following toxic enteritis due to *Clostridium botulinum* type F (Sonnabend et al. 1987).

The significance of isolated reports of GBS following some of the less common bacterial infections such as *Borrelia burgdorferi* (Bouma et al. 1989; Mancardi et al. 1989) paratyphoid, listeria, brucella, tularaemia (Mushinski et al. 1964; Syrjala et al. 1989) and chlamydia (Arnason 1984) is difficult to evaluate. If they were statistically valid associations they would emphasise the multiplicity of agents which may trigger GBS.

Other Possible Precipitants of GBS

The occurrence of GBS soon after a wide variety of surgical procedures has been reported (Arnason and Asbury 1968). Eight of 100 cases studied by Winer et al. (1988b) had had an operation during the previous 3 months: two also had serological evidence of infection, one with CMV and one with *Campylobacter*. Possible explanations of the relationship include surgical trauma to nerves releasing myelin antigens, precipitation of infection by surgery and accompanying blood transfusion, and interference with immunity by surgical stress possibly mediated by corticosteroid release. Some of these mechanisms might also apply to reported associations between GBS and head trauma (Duncan and Kennedy 1987) or myocardial infarction (McDonagh and Dawson 1987), although it seems more likely that these are coincidental. Fever treatment was once used for treating gonorrhoea and six of 1000 cases so treated developed GBS (Garvey et al. 1940). A single case of GBS has also been reported following hyperthermic treatment of cancer (Adam et al. 1987). Four cases of GBS following Hymenoptera stings were described in one report (Bachman et al. 1982) but this association has not been noted elsewhere (Winer et al. 1988b).

GBS has been reported in pregnancy or the puerperium in more than 30 cases (Ahlberg and Ahlmark 1978; McGrady 1987; Quinlan et al. 1988; Rodin et al. 1988). The neuropathic symptoms have started at all stages of pregnancy or in the puerperium. The disease has run its usual clinical course and the main problem is that weakness may hinder labour and increase the need for instrumental delivery. There is a possible increase in risk of premature birth but no instance of a baby being affected by neurological disorder at birth has been reported, even when delivery has occurred at the nadir of the mother's illness. This contrasts with the occurrence of neonatal myasthenia in the children of myasthenic mothers and argues against a humoral factor causing the disease. However, neonatal myelin may not express the appropriate epitopes for recognition by the antimyelin antibodies which have been described in GBS. It remains doubtful whether the occurrences of GBS in pregnancy or the puerperium exceed background levels. Of concern to the patient who has had GBS during pregnancy once is whether it is likely to recur with a further pregnancy. There is no clear answer to this question. On the one hand there are many instances in the literature and personal cases of uncomplicated pregnancies following GBS. On the other hand McCombe et al. (1987) reported that their patients with CIDP were more likely to experience worsening when pregnant rather than not.

Neuropathies following drug administration are usually acute axonal neuropathies involving dying back of the axons. Three reports of GBS following captopril have been published (Chakraborty and Ruddell 1987). Isolated case reports of GBS occurring during treatment with commonly used drugs have to be dismissed as probably coincidental. Four reports of GBS following streptokinase therapy have not been followed by the spate of reports which would be expected with the vastly increased use of streptokinase in myocardial infarction (Arrowsmith et al. 1985; Cicale 1987). An interesting series of cases of fever and myalgia followed by an acute polyradiculoneuropathy resembling GBS was encountered 6–17 days after starting treatment with zimeldine, a novel antidepressant (Fagius et al. 1985). Although pathological data were not obtained and the published neurophysiological data are limited the clinical descriptions fulfilled the standard criteria for GBS. This happened in the context of symptoms of generalised hypersensitivity which may have directly involved the nerves or indirectly triggered the process causing GBS.

Summary

GBS has a worldwide incidence of about 1–2 per 100000 population. It is slightly more common in males than females, in the elderly and also in young adult females. Immunisation with suckling mouse brain, rabies vaccine and infection with viruses (CMV, Epstein–Barr, varicella-zoster, HIV), Mycoplasma and bacteria (*Campylobacter*) are significant antecedent events, occurring one to two weeks before the onset of the neuropathy. A possible association with swine influenza vaccine may be an artefact of reporting. This wide range of precipitating agents suggests that they act by a non-specific triggering mechanism, possibly interfering with normal mechanisms for suppressing auto-immune responses.

References

Adam AM, Hughes RAC, Payan JA, McColl I (1987) Peripheral neuropathy and hyperthermia. Lancet ii:1270–1271

Ahlberg G, Ahlmark G (1978) The Landry–Guillain–Barré syndrome and pregnancy. Acta Obstet Gynecol Scand 57:377–380

Appelbaum E, Greenberg M, Nelson J (1953) Neurological complications following antirabies vaccination. JAMA 151:188–191

Arnason BGW (1984) Acute inflammatory demyelinating polyradiculoneuropathies. In: Dyck PJ, Thomas PK, Lambert EH, Bunge R (eds) Peripheral neuropathy. WB Saunders, Philadelphia, pp 2050–2100

Arnason BGW, Asbury AK (1968) Idiopathic polyneuritis after surgery. Arch Neurol 18:500–504

Arrowsmith JB, Milstein JB, Kuritsky JN, Murano G (1985) Streptokinase and the Guillain–Barré syndrome. Ann Intern Med 103:302

Bachman DS, Paulson GW, Mendell JR (1982) Acute inflammatory polyradiculoneuropathy following Hymenoptera stings. JAMA 247:1443–5

Bahemuka M (1988) GBS in Kenya: a clinical review of 54 patients. J Neurol 235:418–421

Bailey RO, Baltch AL, Venkatesh R, Singh JK, Bishop MB (1988) Sensory motor neuropathy associated with AIDS. Neurology 38:886–891

Bale JF, Rote NS, Bloomer LC, Bray PF (1980) Guillain–Barré like polyneuropathy after renal transplant: possible association with cytomegalovirus infection. Arch Neurol 37:784

Berger JR, Ayyar R, Sheremata WA (1981) Guillain–Barré syndrome complicating acute hepatitis B. A case with detailed electrophysiological and immunological studies. Arch Neurol 38:366–368

Berger JR, Ayyar DR, Kasovitz B (1986) Guillain–Barré syndrome complicating typhoid fever. Ann Neurol 20:649–650

Berger JR, Difini JA, Swerdloff MA, Ayyar DR (1987) HIV seropositivity in Guillain–Barré syndrome. Ann Neurol 22:393–394

Bishopric G. Bruner J, Butler J (1985) Guillain–Barré syndrome with cytomegalovirus infection of peripheral nerves. Arch Pathol Lab Med 109:1106–1108

Boe E, Nyland H (1980) Guillain–Barré syndrome after vaccination with human diploid cell rabies vaccine. Scand J Infect Dis 12:231–232

Bouma PAD, Carpay HA, Rijkema SGT (1989) Antibodies to *Borrelia burgdorferi* in Guillain–Barré syndrome. Lancet ii:739

Breman JG, Hayner NS (1984) Guillain–Barré syndrome and its relationship to swine influenza vaccination in Michigan, 1976–1977. Am J Epidemiol 119:880–889

Brewis M, Poskanzer DC, Rolland DC, Miller H (1966) Neurological disease in an English city. Acta Neurol Scand 42 suppl 24:1–89

Brostoff SW, White TM (1982) Absence of P_2 protein in swine flue vaccines. JAMA 247:495

Cabrera J, Griffin DE, Johnson RT (1987) Unusual features of the GBS after rabies vaccine prepared in suckling mouse brain. J Neurol Sci 81:239–246

Callaghan N, Flaherty T, McGarry J (1974) Case of GBS and herpes simplex virus. J Irish Med Assoc 67:541

Cambier J, Schott S (1966) Nosologie des polyradiculonévrites inflammatoires. Rev Neurol 115:811–842

Chakraborty TK, Ruddell WS (1987) Guillain–Barré neuropathy during treatment with captopril. Postgrad Med J 63:221–222

Changdam D, Waniganetti A (1969) Guillain–Barré syndrome associated with typhoid fever. Br Med J i:95–96

Chen KM, Brody JA, Kurland LT (1968) Pattern of neurologic diseases in Guam. Arch Neurol 573:577

Cicale MJ (1987) Guillain–Barré syndrome after streptokinase therapy. South Med J 80:1068

Constant OC, Bentley CC, Denman AM (1983) The Guillain–Barré syndrome following *Campylobacter enteritis* with recovery after plasmapheresis. J Infect 6:89–91

Cornblath DR, Griffin DE, Chupp M, Griffin JW, McArthur JC (1987) Mononuclear cell typing in inflammatory demyelinating polyneuropathy nerve biopsies. Neurology 37 suppl 1:253–254

De Bont B, Matthews N, Abbott K, Davidson GP (1986) Guillain–Barré syndrome associated with *Campylobacter enteritis* in a child. J Pediatr 109:660–662

Detroyer A (1975) Guillain–Barré syndrome in acute HBs antigen-positive hepatitis. Br Med J iv:732–734

Donoso R (1988) GBS following typhoid fever. Ann Neurol 23:627

Dowling PC, Cook SD (1981) Role of infection in Guillain–Barré syndrome: Laboratory confirmation of herpes viruses in 41 cases. Ann Neurol 9 suppl:44–45

Dowling PC, Menonna JP, Cook SD (1977) Guillain–Barré syndrome in Greater New York–New Jersey. JAMA 238:317–318

Duncan R, Kennedy PG (1987) Guillain–Barré syndrome following acute head trauma. Postgrad Med J 63:479–480

Dunk A, Jenkins WJ, Sherlock S (1982) Guillain–Barré syndrome associated with hepatitis A in a male homosexual. Br J Ven Dis 58:269–270

Eidelberg D, Sotrel A, Vogel H, Walker P, Keefield J, Crumpacker CS (1986) Progressive polyradiculopathy in AIDS. Neurology 36:912–916

Eisen A, Humphreys P (1974) The Guillain–Barré syndrome. A clinical and electrodiagnostic study of 25 cases. Arch Neurol 30:438–443

Fagius J, Osterman PO, Siden A, Wiholm B-E (1985) Guillain–Barré syndrome following zimeldine treatment. J Neurol Neurosurg Psychiatry 48:65–69

Forbes SJ, Brumlik J, Harding HB (1967) Acute ascending polyradiculoneuritis associated with Echo 9 virus. Dis Nerv Syst 28:537–540

Garvey PH, Jones N, Warren SL (1940) Polyradiculoneuritis (GBS) following the use of sulfanilamide and fever treatment. JAMA 115:1955–1959

Gautier-Smith PC (1965) Neurological complications of glandular fever (infectious mononucleosis). Brain 88:323–331

Goldschmidt B, Menonna J, Fortunato J, Dowling P, Cook S (1980) Mycoplasma antibody in Guillain–Barré syndrome and other neurological disorders. Ann Neurol 7:108–112

Griffin DE (1988) Post-infectious and post-vaccinal disorders of the central nervous system. Immunol Allergy Clin North Am 8(2):239–242

Grover P, Dalessandro L, Sanders JG, Walter MH, O'Meara TF, Redmond J (1986) Severe viral hepatitis A infection, Landry–Guillain–Barré syndrome, and hereditary elliptocytosis. South Med J 79:251–252

Gudmundson KR (1969) Prevalence and occurrence of some rare neurological diseases in Iceland. Acta Neurol Scand 45:114–118

Haberman S, Benjamin B, Capildeo R, Rose FC (1982) North West Thames registry of neurological disease. J R Soc Med 75:443–449

Hankey GJ (1987) Guillain–Barré syndrome in Western Australia, 1980–1985. Med J Austr 146:130–133

Hanna WJ (1979) Acute polyneuritis (Guillain–Barré syndrome) seen at the University Hospital of the West Indies, 1970–1975. West Indian Med J 28:164–171

Hemachudha T, Phanuphak P, Johnson RT, Griffin DE, Ratanvongsiri J, Siripramsomsup W (1987a) Neurologic complications of Semple-type rabies vaccine: clinical and immunologic studies. Neurology 37:550–556

Hemachudha T, Griffin DE, Giffels JJ, Johnson RT, Moser AB, Phanupak P (1987b) Myelin basic protein as an encephalitogen in encephalomyelitis and polyneuritis following rabies vaccination. N Engl J Med 316:369–374

Hemachudha T, Griffin DE, Chen WW, Johnson RT (1988) Immunologic studies of rabies vaccination-induced GBS. Neurology 38:375–378

Hogg JE, Kobrin DE, Schoenberg BS (1979) The Guillain–Barré syndrome, epidemiologic and clinical features. J Chron Dis 32:227–231

Holliday P, Bauer RB (1983) Polyradiculoneuritis secondary to immunisation with tetanus and diphtheria toxoids. Arch Neurol 40:56–57

Huet PM, Layrargues GP, Lebrun LH, Richer G (1980) Hepatitis B surface antigen in the cerebrospinal fluid in a case of Guillain–Barré syndrome. Can Med Assoc J 122:1158–1158

Hurwitz ES, Holman RC, Nelson DB, Schonberger LB (1983) National surveillance for Guillain–Barré syndrome: January 1978–March 1979. Neurology 33:150–157

Jackson AC (1961) A clinical study of the Landry–Guillain–Barré syndrome with reference to aetiology including the role of Coxsackie virus infection. S Afr J Lab Clin Med 7:121–137

Jahnke U, Fischer EH, Alvord EC (1985) Sequence homology between certain viral proteins and proteins related to encephalomyelitis and neuritis. Science 229:282–284

Johnston CL, Schwartz M, Wansbrough-Jones MH (1981) Acute inflammatory polyradiculoneuropathy following type A viral hepatitis. Postgrad Med J 57:647–648

Kaldor J, Speed BR (1984) GBS and *Campylobacter jejuni*: a serological study. Br Med J 288:1867–1870

Kaplan JE, Katona P, Hurwitz ES, Schonberger LB (1982) Guillain–Barré syndrome in the United States, 1979–1980 and 1980–1981. Lack of an association with influenza vaccination. JAMA 248:698–700

Kaslow RA, Sullivan-Bolyai JZ, Holman RC, Hafkin S, Dicker RC, Schonberger LB (1987) Risk factors for Guillain–Barré syndrome. Neurology 37:685–688

Kennedy RH, Danielson MA, Mulder DW, Kurland LT (1978) GBS: A 42-year epidemiologic and clinical study. Mayo Clin Proc 53:93–99

Khoury SA (1978) Guillain–Barré syndrome: epidemiology of an outbreak. Am J Epidemiol 197:433–438

Kinnunen E, Farkkila M, Hovi T, Juntenen J, Weckstrom P (1989) Incidence of Guillain–Barré syndrome during a nationwide oral poliovirus vaccine campaign. Neurology 39:1034–1036

Klemola E, Weckman N, Haltia K (1967) The Guillain–Barré syndrome associated with acquired cytomegalovirus infection. Acta Med Scand 181:603–607

Knittel T, Ramadori G, Mayet W-J, Lohr H, Meyer-zum-Buschenfelde (1989) Guillain–Barré syndrome and human diploid cell rabies vaccine. Lancet i: 1334–1335

Kohler PC, Goldblatt D (1987) Guillain–Barré syndrome following *Campylobacter jejuni* enteritis. Arch Neurol 44:1219

Kurland LT (1958) Descriptive epidemiology of selected neurologic and myopathic disorders with particular reference to a survey in Rochester, Minnesota. J Chron Dis 8:378–418

Kurland LT, Wiederholt WC (1986) Swine influenza vaccine and Guillain–Barré syndrome: lies, damn lies and . . . Arch Neurol 43:979–982

Kurland LT, Wiederholt WC, Kirkpatrick JW (1985) Swine influenza vaccine and GBS; epidemic or artefact. Arch Neurol 42:1089–1090

Lane JM, Ruben FL, Neff JM, Miller JD (1968) Complications of small-pox vaccinations. N Engl J Med 281:1201

Langmuir AD (1979) Guillain–Barré syndrome: the swine influenza virus vaccine incident in the United States of America, 1976–1977. Preliminary communication. J R Soc Med 72:660–669

Langmuir AD, Bregman DJ, Kurland LT, Nathanson N, Victor M (1984) An epidemiologic and clinical evaluation of GBS reported in association with administration of swine influenza vaccines. Am J Epidemiol 119:841–879

Larsen J, Kvale G, Nyland H (1985) Epidemiology of the Guillain–Barré syndrome in the county of Hordaland, Western Norway. Acta Neurol Scand 71:43–47

Leneman F (1966) The Guillain–Barré syndrome. Definition, etiology and review of 1000 cases. Arch Intern Med 118:139–144

Lennon VA (1986) Swine influenza vaccine and Guillain–Barré syndrome: lies damn lies and . . . Arch Neurol 43:981–982

Leonard M, Tobin JH (1971) Polyneuritis associated with cytomegalovirus infections – a report from various centres. Q J Med 40:435–442

Lesser RP, Hauser WA, Kurland LT, Mulder DW (1973) Epidemiologic features of Guillain–Barré syndrome. Neurology 23:1269–1272

Lidin-Janson G, Strannegard O (1972) Two cases of Guillain–Barré syndrome and encephalitis after measles. Br Med J ii:572–574

Lipkin WI, Parry G, Kiprov D, Abrams D (1985) Inflammatory neuropathy in homosexual men with lymphadenopathy. Neurology 35:1479–1483

Löffel NB, Rossi LN, Mumenthaler M (1977) The Landry–Guillain–Barré syndrome – complications, prognosis and natural history in 123 cases. J Neurol Sci 33:71–79

Macleod WN (1987) Sporadic non-A, non-B hepatitis and Epstein–Barr hepatitis associated with the Guillain–Barré syndrome. Arch Neurol 44:438–442

Mancardi GL, Del Sette M, Primavera A, Farinelli M, Fumarola D (1989) *Borrelia burgdorferi* infection and Guillain–Barré syndrome. Lancet ii:485–496

Mares-Segura R, Sola-Lamoglia R, Soler-Singla L, Pou-Serradell A (1986) Guillain–Barré syndrome associated with hepatitis A. Ann Neurol 19:100

Marti-Masso JF, Obeso JA, Cosme A et al. (1979) GBS associated with type B acute hepatitis. Med Clin (Barc) 73:447

Masucci EF, Kurtzke JF (1971) Diagnostic criteria for the Guillain–Barré syndrome. An analysis of 50 cases. J Neurol Sci 13:483–501

McCombe PA, McManis PG, Frith JA, Pollard JD, McLeod JG (1987) Chronic inflammatory demyelinating polyradiculoneuropathy associated with pregnancy. Ann Neurol 2:102–104

McDonagh AJ, Dawson J (1987) Guillain–Barré syndrome after myocardial infarction. Br Med J 294:613–614

McGrady EM (1987) Management of labour and delivery in a patient with Guillain–Barré syndrome. Anaesthesia 42:899

Melnick SC, Flewett TH (1964) Role of infection in Guillain–Barré syndrome. J Neurol Neurosurg Psychiatry 27:395–407

Menonna J, Goldschmidt B, Haidri N, Dowling P, Cook S (1977) Herpes simplex virus-IgM specific antibodies in Guillain–Barré syndrome and encephalitis. Act Neurol Scand 56:223–231

Miller HG, Stanton JB (1954) Neurological sequelae of prophylactic inoculation. Q J Med 23:1–27

Miller HG, Stanton JB, Gibbons JL (1956) Parainfections encephalomyelitis and related syndromes, a critical review of the neurological complications of certain specific fevers. Q J Med 25:427–453

Mishra BB, Sommers W, Koski CL, Greenstein JI (1985) Acute inflammatory demyelinating neuropathy in AIDS. Ann Neurol 18:131–132

Mizisin AP, Wiley CA, Hughes RAC, Powell HC (1987) Peripheral nerve demyelination in rabbits after inoculation with Freund's complete adjuvant alone or in combination with lipid hapten. J Neuroimmunol 16:381–395

Molnar GK, Mertgola J, Erkko M (1982) Guillain–Barré syndrome associated with *Campylobacter* infection. Br Med J 285:652

Morrison RE, Shatsky SA, Holmes GE, Top H, Martins AN (1979) Herpes simplex virus Type 1 from a patient with radiculoneuropathy. JAMA 241:393–394

Mushinski JF, Taniguchi RM, Stiefel JW (1964) GBS associated with tubero-glandular tularaemia. Neurology 14:877–879

Newton N, Janati A (1987) Guillain–Barré syndrome after vaccination with purified tetanus toxoid. South Med J 80:1053–1054

Ng PL, Powell LW, Campbell CB (1975) Guillain–Barré syndrome during the pre-icteric phase of acute Type 4 viral hepatitis. Aust NZ J Med 5:367–369

Niermeijer P, Gips CH (1975) Guillain–Barré syndrome in acute HBs antigen-positive hepatitis. Br Med J iv:732–733

Nordaa JVD, Dillen PW-V (1977) Rise in antibodies to human papova vi BK and clinical disease. Br Med J i:1471

Olivarius Bde F, Buhl M (1975) Herpes simplex virus and Guillain–Barré polyradiculitis. Br Med J i:192–193

Osuntokun BO, Agbebi K (1973) Prognosis of Guillain–Barré syndrome – the Nigerian experience. J Neurol Neurosurg Psychiatry 36:478–484

Parker W, Wilt JC, Dawson JW, Stackin W (1960) Landry–Guillain–Barré syndrome – isolation of an Echo Virus Type 6. Can Med Assoc J 82:814–815

Penner E, Maida E, Mamoli B, Gangl A (1982) Serum and cerebrospinal fluid immune complexes containing hepatitis B surface antigen in Guillain–Barré syndrome. Gastroenterology 82(3): 576–580

Piette AM, Tusseau F, Vignon D et al. (1986) Acute neuropathy coincident with sera conversion for anti LAV/HTLV III. Lancet i:852

Pleasure DE, Lovelace RE, Duvois RC (1969) The prognosis of acute polyradiculoneuritis. Neurology 18:1143–1148

Pollack S, Barr-On E, Enat R (1981) Guillain–Barré syndrome: association with mumps. NY State J Med 8:795–797

Pollard JD, Selby G (1978) Relapsing neuropathy due to tetanus toxoid. J Neurol Sci 37:113–125

Pryor W, Freiman JS, Gilles MA et al. (1984) GBS associated with *Campylobacter jejuni* infection. Aust NZ J Med 14: 687–688

Quinlan DJ, Moodley J, Lalloo BC, Nathoo UG (1988) Guillain–Barré syndrome in pregnancy. A case report. S Afr Med J 73:611–612

Ravn H (1967) The Guillain–Barré syndrome. A survey and clinical report of 123 cases. Acta Neurol Scand 43 suppl 30:1–164

Rhodes KM, Tattersfield AE (1982) Guillain–Barré syndrome associated with *Campylobacter* infection. Br Med J 285:173–174

Ricker W, Blumberg A, Peters CH, Widerman A (1947) The association of the Guillain–Barré syndrome with infectious mononucleosis with a report of two fatal cases. Blood 2:217–222

Rodin A, Ferner RE, Russell R (1988) Guillain–Barré syndrome in pregnancy and the puerperium.

J Obstet Gynecol 9:39–42

Ropper AH (1988) *Campylobacter* diarrhea and Guillain–Barré syndrome. Arch Neurol 45:655–666

Saeed AA, Lange LS (1978) Guillain–Barré syndrome after rubella. Postgrad Med J 54:333–334

Samantray SK (1977) Landry–Guillain–Barré–Strohl syndrome in typhoid fever. Aust NZ J Med 7:307–308

Sanders EACM, Peters ACB, Gratana JW, Hughes RAC (1987) GBS after varicella-zoster infection. Report of two cases. J Neurol 234:437–439

Schmitz H, Enders G (1977) Cytomegalovirus as a frequent cause of Guillain–Barré syndrome. J Med Virol 1:21–27

Schonberger LB, Bregman DJ, Sullivan-Bolynai JZ et al. (1979) Guillain–Barré syndrome following vaccination in the National Influenza Immunization program, United States 1976–1977. Am J Epidemiol 110:105–123

Schonberger LB, Hurwitz ES, Katona P, Holman RC, Bregman DJ (1981) Guillain–Barré syndrome: its epidemiology and associations with influenza vaccination. Ann Neurol 9 suppl:31–38

Snider WD, Simpson DM, Nielson G et al. (1983) Neurological complications of AIDS: analysis of 50 patients. Ann Neurol 14:403–418

Soffer D, Feldman S, Alter M (1978) Clinical features of the Guillain–Barré syndrome. J Neurol Sci 37:135–143

Sonnabend WF, Sonnabend OA, Grundler P, Ketz E (1987) Intestinal toxicoinfection by *Clostridium botulinum* type F in an adult. Case associated with Guillain–Barré syndrome. Lancet i:357–361

Sovilla JY, Regli F, Francioli PB (1988) Guillain–Barré syndrome following *Campylobacter jejuni* enteritis. Arch Intern Med 148:739–741

Speed BR, Kaldor J, Watson J et al. (1987) *Campylobacter jejuni/Campylobacter coli*-associated Guillain–Barré syndrome. Immunoblot confirmation of the serological response. Med J Aust 147:13–16

Spillane JD, Wells CEC (1964) The neurology of Jennerian vaccination. Brain 87:1–15

Stevens D, Burman D, Clarke SKR, Lamb RW, Harper ME, Sarafian AH (1974) Temporary paralysis in childhood after influenza B. Lancet ii:1254

Syrjala H, Koskela P, Kujala P, Myllyla V (1989) Guillain–Barré syndrome and tularemia pleuritis with high adenosine deaminase activity in pleural fluid. Infection 17:152–153

Tomlinson IW (1975) Rubella neuropathy. Postgrad Med J 51:30–32

Tong HI, Devathasar G, Wong PK (1979) The pattern of Guillain–Barré syndrome (acute polyradiculopathy) in Singapore – a critical analysis of 46 cases. Ann Acad Med Singapore 8:27–32

Toro G, Vergara I, Roman G (1977) Neuroparalytic accidents of antirabies vaccination with suckling mouse brain vaccine. Arch Neurol 34:694–700

Uhari M, Rantala H, Niemala M (1989) Cluster of childhood Guillain–Barré cases after an oral poliovaccine campaign. Lancet ii:440

Urano T, Kawase T, Kodaira K, Takeuchi Y, Kikuchi T, Kimura M (1970) GBS associated with ECHO virus type 7 infections. Pediatrics 45:294

Usui T, Hamada Y, Arita M (1974) A case of the Guillain–Barré syndrome associated with Coxsackie B5 virus infection. Tokushima J Exp Med 7:121–137

Vendrell J, Heredia C, Pujol M, Vidal J, Blesa R, Graus F (1987) Guillain–Barré syndrome associated with seroconversion for anti-HTLV-III. Neurology 37:544

Wells CEC, James WRL, Evans AD (1959) GBS and virus of influenza A (Asian strain): report of two fatal cases during the 1957 epidemic in Wales. Arch Neurol Psychiatry 81:699

Wiederholt HM, Mulder DW, Lambert EH (1964) The Landry–Guillain–Barré–Strohl syndrome of polyradiculoneuropathy – historical review report on 97 patients and present concepts. Mayo Clin Proc 49:427–451

Winer JB, Hughes RAC, Bradley GW, Scadding JW (1984) Guillain–Barré syndrome and influenza vaccine. Lancet i:1182

Winer JB, Gray IA, Gregson NA et al. (1988a) A prospective study of acute idiopathic neuropathy. III. Immunologic studies. J Neurol Neurosurg Psychiatry 51:619–625

Winer JB, Hughes RAC, Osmond C (1988b) A prospective study of acute idiopathic neuropathy. I. Clinical features and their prognostic value. J Neurol Neurosurg Psychiatry 51:605–612

Winer JB, Hughes RAC, Anderson MJ, Jones DM, Kangro H, Watkins RFP (1988c) A prospective study of acute idiopathic neuropathy. II. Antecedent events. J Neurol Neurosurg Psychiatry 51:613–618

Wroe SJ, Blumhardt LD (1984) Acute polyneuritis with cranial nerve involvement following

Campylobacter jejuni infection. J Neurol Neurosurg Psychiatry 48:593

Zhao B, Yinchang Y, Huifen H, Xiuqin L (1981) Acute polyradiculitis (Guillain–Barré syndrome): an epidemiological study of 156 cases observed in Beijing. Ann Neurol 9 suppl:146–148

Ziegler DW, Gardner JJ, Warfield DT, Walls HH (1983) Experimental allergic neuritis-like disease in rabbits after infection with influenza vaccines mixed with gangliosides and adjuvants. Infect Immun 42:824–830

Chapter 6

Clinical Features of Guillain–Barré Syndrome

Presentation

Guillain–Barré syndrome usually occurs after a preceding infection and begins with a combination of pain, sensory symptoms and weakness. The pain is usually a diffuse aching in the back, neck or limbs: it occurred in half of a series of 100 patients (Winer et al. 1988). It may be a prominent feature. Occasionally in children pain may cause meningism and rarely retrocollis. It is sometimes difficult to distinguish between the subsiding pain of the previous influenza like illness and the pain which accompanies the start of the neuropathic process. By definition weakness develops in all patients. It is usually first noted in the lower limbs as difficulty rising from a chair, climbing stairs, walking and standing. The upper limbs are often affected as well, usually less severely. Occasionally weakness is much more severe in or even confined to the lower limbs. Even more rarely weakness is more severe in the upper limbs. The weakness is usually proximal and distal and often more pronounced proximally than distally. It is usually approximately symmetrical but may be quite markedly asymmetrical especially at the onset. Loss of feeling and tingling in the extremities are characteristic early symptoms occurring in nearly 80% of patients (Winer and Hughes 1988). Sensory complaints are usually minor and little sensory deficit is discovered on conventional neurological examination although some patients have profound sensory loss and half lose joint position sense at the toes. The tendon reflexes are usually lost or at least diminished from an early stage and this is a required diagnostic criterion. Rare, usually mild, cases have an otherwise typical picture and neurophysiological evidence of demyelinating polyradiculoneuropathy but preserved tendon reflexes. If the reflexes are preserved the diagnosis should be regarded with suspicion. Cranial nerve involvement is common and usually accompanies limb involvement but rarely lags behind and may first become evident when the limb symptoms have reached a plateau. The facial nerves are the most commonly involved cranial nerves, being affected in nearly half of patients, whereas the bulbar nerves are compromised in a third and the ocular motor nerves in just over a tenth (Table 6.1). Visual impairment, deafness and anosmia do not occur but loss of taste is occasionally reported and may rarely be a prominent early feature. Respiratory muscle involvement occurs in about a quarter of patients and artificial ventilation is required in about one patient in seven. Although bladder involvement is not a presenting feature, bladder symptoms, usually difficulty with micturition sometimes leading to retention, occur in about a third of patients when the weakness is more advanced.

Clinical Course

The tempo of development of paralysis of events is quite variable. In some patients the weakness evolves rapidly to reach its worst in the course of a day whereas in others the symptoms evolve more gradually over 2 or 3 weeks. In a series of 100 patients who appeared at presentation to have acute neuropathy of the GBS type (Winer et al. 1988) 34% reached their nadir within 7 days, 70% within 14 days, and 84% within 21 days. Ninety-two per cent reached their nadir within one month, the artificial boundary which we have drawn to distinguish GBS from CIDP. Four patients continued to worsen until 44 days. Three pursued an even more chronic progressive course and one relapsed within the one year follow-up period.

The severity of the illness is similarly variable. A small proportion of the patients who are recognised as having GBS only have a trivial illness with numbness and weakness evolving over several days and never preventing walking or not even stopping them working. These mildly affected patients recover within a few weeks. Many such patients are never seen by a neurologist, probably never diagnosed and never reach the statistical database on which epidemiological studies depend. At the other end of the spectrum the disease represents one of the most alarming and life-threatening conditions with which modern medicine has to deal. GBS can completely paralyse a previously healthy person and require artificial ventilation in an intensive care unit within 24 hours. A recent survey reported the peak disability in 100 patients (Winer et al. 1988): 12% of patients remained able to walk unaided throughout their illness; 7% required support to walk; 47% became bed bound or chair bound; 33% required ventilation and 3% died without being ventilated.

The phase of worsening lasting up to 4 weeks is followed by a "plateau" phase

Table 6.1. Percentage occurrence of clinical features in 924 cases of Guillain–Barré syndrome

Clinical feature	%
Sensory presentation	46
Motor presentation	32
Motor and sensory presentation	21
Arms and legs involved	88
Legs alone	14
Arms alone	1
Cranial nerves involved total	52
Facial	40
Bulbar	30
Trigeminal	9
Ventilated	14
Sensory signs total	65
Proprioceptive loss	23
Sphincter disturbance	15

The percentages are derived from six large series (Marshall 1963; McFarland et al. 1966; Ravn 1967; Masucci and Kurtzke 1971; McLeod et al. 1976; Samantray et al. 1977). Reproduced from Hughes and Winer (1984) with permission.

of variable duration. Since it is difficult to define when worsening finishes and even more difficult to define when recovery begins precise measurements of the duration of the plateau are not available. At the extremes some patients begin to recover as soon as they have reached their worst whereas rare patients are left with devastating tetraplegia which does not begin to improve for several months. More typical plateau phases last 1–4 weeks. The use of plasma exchange significantly shortens the plateau and recovery phase (Chapter 8).

Once recovery begins improvement usually proceeds steadily. Most people remain bed bound one month after onset of the neuropathy and yet have begun to walk without aid between one and three months after onset (Fig. 6.1). The median time for return to work is between three and six months after onset. About half of patients have persistent symptoms after a year. These median figures disguise the facts that about 10% of patients die from GBS and 20% are left with some disability after a year. Although the persistent disability may be minor, at the end of a year three of the 100 patients studied by Winer et al. (1988) were bedbound, four required aid to walk and 14 were unable to work. The mortality in that series was 13%. A similar mortality has recently been reported in a multicentre study from France (Raphael et al. 1984). The conclusions of this prospective study support those which have been derived from previous large retrospective series (Table 6.2). Some large series report a much lower mortality, for instance 1.25% of 159 cases from the Massachusetts General Hospital (Ropper and Shahani 1984), possibly due to selection bias but more probably illustrating what can now be achieved with intensive care in a single centre of excellence. In that series residual deficit after one year was 23% (8% severe).

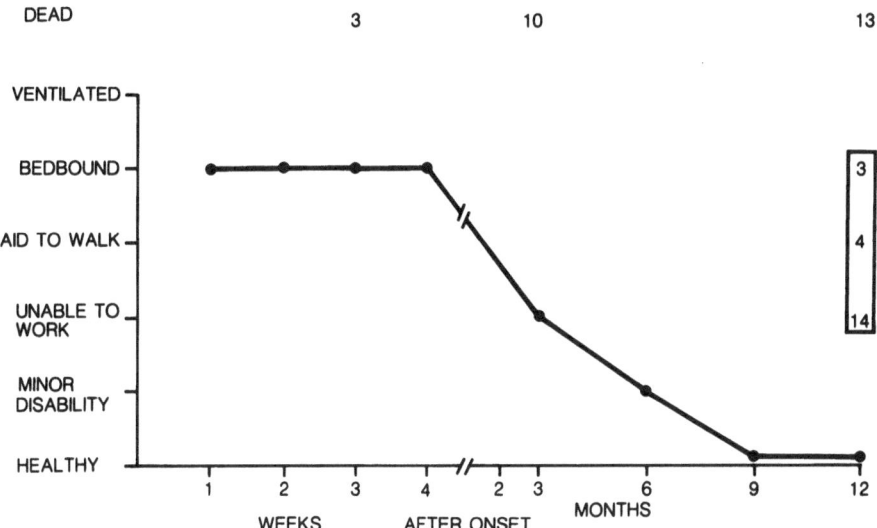

Fig. 6.1. Recovery from Guillain–Barré syndrome. The points on this diagram represent median disability grades of 100 patients. The figures on the top line represent the cumulative numbers of dead patients at those times. The figures in the box on the right represent the number of disabled survivors after one year. (Data from Winer et al. (1988), with permission.)

Table 6.2. Prognosis of Guillain–Barré syndrome: percentage recovery at intervals after onset

Interval from onset	Percentage recovery									Mean
	1[a]	2	3	4	5	6	7	8	9	
One month	19	–	–	–	–	–	–	–	–	19
Three months	27	–	–	–	–	–	–	20	–	24
Six months	–	38	23	–	–	–	96	57	–	54
One year	65	63	35	–	–	–	–	78	–	64
Two years	–	72	–	62	–	–	–	–	–	67
Total recovered[b]	65	77	65	62	72	94	96	78	78	77
Deaths	0	6	4	6	18	3	4	6	2	7

Data from Hughes and Winer (1984) with permission.

[a] 1, Eisen and Humphreys (1974) $n = 26$; 2, Wiederholt et al. (1964) $n = 97$; 3, Peterman et al. (1959) $n = 26$; 4, Pleasure et al. (1969) $n = 81$; 5, Ravn (1967) $n = 127$; 6, Moore and James (1981) $n = 33$; 7, Samantray (1977) $n = 302$; 8, Masucci and Kurtzke (1971) $n = 49$; 9, Löffel et al. (1977) $n = 123$.

[b] Figures are estimated on the basis of return to all normal activities and this does not exclude minor degrees of weakness and neurological signs.

Differential Diagnosis

The diagnosis of GBS is almost entirely dependent on the clinical history and examination. The cerebrospinal fluid characteristically shows an increased protein concentration and normal cell count after the first week of the illness but may remain normal (Chapter 9). However, the same albuminocytological dissociation may occur in other neurological conditions, including diphtheritic neuropathy, as pointed out by Guillain. Neurophysiological abnormalities may be difficult to find especially early in the disease even in quite severe cases, although slowed nerve conduction and conduction block are characteristic (Chapter 7). Reliable clinical immunological tests of autoimmunity to myelin have not yet been identified (Chapter 9). Fortunately the clinical picture is sufficiently characteristic that diagnosis is not usually difficult. However, the clinical picture is rather protean and although GBS is the commonest cause of acute neuromuscular paralysis in an otherwise healthy person, other possibilities must always be carefully considered and the diagnosis of GBS achieved by exclusion.

    The first step in the diagnosis is to establish that the paralysis is due to a peripheral neuropathy and many neurological conditions may sometimes mimic or be confused with GBS (Tables 6.3 and 6.4). It is not uncommon for the doctor who first sees an otherwise fit patient complaining of weakness to consider hysteria. I have seen several patients who had initially been turned away from accident and emergency departments only to return within 24 hours tetraplegic and in respiratory distress. It was noted in Chapter 1 that Landry's chief, Monsieur Gubler, initially thought that Landry's patient (who died) was hysterical. In some patients on the other hand a pretence of weakness may be so convincing as to require artificial ventilation (Hopkins and Clarke 1987). I have seen GBS causing very rapidly progressive tetraplegia with bulbar involvement misdiagnosed as a brainstem stroke. In Guillain–Barré syndrome the loss of re-

Table 6.3. Differential diagnosis of Guillain–Barré syndrome

Malingering
Hysteria

Locked-in syndrome
Brainstem infarction
Brainstem encephalomyelitis
Poliomyelitis
Spinal cord compression
Transverse myelitis
Acute necrotic myelopathy
Paralytic rabies
Cauda equina compression

Other causes of neuropathy (see Table 6.4)
Myasthenia gravis
Botulism
Tick bite paralysis
Pelagic paralysis

Acute myopathy
Hypokalaemia
Hyperkalaemia

Table 6.4. Other causes of acute neuropathy resembling Guillain–Barré syndrome

Toxins
  Alcohol
  Heavy metals: arsenic, lead, thallium, gold
  Organophosphate insecticides
  Hexacarbons: industrial solvents, glue sniffing
Drugs: nitrofurantoin, isoniazid, vincristine
Nutritional deficiency: thiamine
Vasculitis: polyarteritis nodosa and variants, systemic lupus erythematosus
Lymphomatous infiltration
Porphyria
Borreliosis (Lyme disease, Bannwarth's syndrome)
Diphtheria
Acute neuropathy of the critically ill
Acute sensory neuronopathy
Acute dysautonomia

flexes, preservation of bladder function at least in the early stages and normal alertness argue against an intrinsic brainstem lesion. However brainstem lesions do rarely cause depression of the tendon reflexes (Al-Din et al. 1982). This is further discussed in relation to a variant of GBS with ophthalmoplegia, ataxia and reflex loss called the Miller Fisher syndrome. In severe cases of GBS paralysis of all voluntary muscles including eye movements may occur producing a locked-in state which resembles that induced by brainstem infarction, GBS therefore enters the differential diagnosis of the locked-in state (Carroll and Mastaglia 1979).

The clinical picture of poliomyelitis is sufficiently different from polyneur-opathy that the distinction was made by Leyden over a century ago (Chapter 1). However, rare sporadic cases still cause confusion. Poliomyelitis is still common in some third world countries and occasional cases occur in supposedly vac-cinated communities and in non-immunised visitors to endemic regions. Charac-teristically a prodromal upper respiratory infection is followed by a meningitic illness with high fever: paralysis starts during the febrile period and is markedly asymmetrical. At this acute stage the CSF contains many cells including some polymorphonuclear leucocytes. Later the pleocytosis may disappear leaving only a raised protein as in GBS. Paralytic poliomyelitis is usually caused by poliomyelitis virus types 1 or 3 which can be cultured from the faeces. The diagnosis can be confirmed by demonstrating IgM antibodies and a rising titre of antibodies to the virus in paired sera. It can be difficult to make the diagnosis serologically because antibodies may be the result of previous immunisation and a rise in titre may be an anamnestic response. Other enteroviruses, especially Coxsackie, may produce a similar but usually milder disease.

In India and other developing countries where rabies is endemic the possi-bility of paralytic rabies has to be considered as a cause of ascending paralysis. A history of animal bite is not always forthcoming (Verma et al. 1985). In one series when the bite was remembered the mean interval from bite to onset of paralysis was about 50 days with a range from one week to three months (Chopra et al. 1980). The paralysis is usually preceded by tingling and pain at the site of the bite. Death occurs from respiratory failure or cardiac arrhythmia. The major pathological change is an encephalomyelitis with Negri inclusion bodies in neurons. There is also inflammation in the dorsal root ganglia and peripheral nerves which show a mixture of axonal degeneration and primary demyelination, the latter usually dominant (Chopra et al. 1980). The mental agitation and hydrophobia which are characteristic of the usual form of rabies may be absent.

Occasionally spinal cord or cauda equina compression will pass, usually fleet-ingly, through the differential diagnostic list of the worried clinician presented with a patient with back pain and rapidly progressive paraparesis. In cord com-pression the sharp sensory level, early bladder involvement, brisk reflexes and extensor plantar responses usually provide clear evidence of spinal cord involve-ment. Alternatively bladder involvement, saddle sensory loss, absent anal reflex and confinement of the deficit to the lower limbs draw attention to the cauda equina. Sometimes the diagnosis is much more clear in retrospect than in the heat of the moment and a radiculomyelogram or MRI scan may be necessary to exclude a compressive cause. In acute transverse myelitis the reflexes may be transiently depressed and in acute necrotic myelopathy they may be permanently lost. Occasional cases have clinical evidence of cord and peripheral nerve in-volvement (Brashear et al. 1985; Goode and Shearn 1982) and rarely the brain and optic nerves may also be involved. Until the pathogenesis of these dis-orders is understood they should be classified separately from GBS according to the distribution of the inflammatory demyelinating lesions, for instance encephalomyeloradiculoneuritis.

Theoretically any disorder of neuromuscular conduction might mimic GBS. In practice, the diagnosis of myasthenia gravis is usually evident from the history but myasthenia presenting with an acute myasthenia crisis can cause confusion with GBS. Botulism also presents a real problem in differential diagnosis and should always be considered in acute motor neuropathy. It is usually caused by the ingestion of home-canned food in which *Clostridium botulinum* has grown anaerobically and produced its toxin. Historically the common cause was eating poorly preserved sausage (hence the name from the Latin *botulus* = sausage). Soon after eating the toxin, abdominal pain and distension and nausea develop followed by constipation and ileus due to blockade of the myenteric plexus. The initial neurological symptoms are dry mouth and eyes. Other autonomic features seen later and in more severe cases are dilated pupils and postural hypotension. The cranial nerves are preferentially involved but the limb and respiratory muscles may also be affected. Sensory symptoms do not usually occur but vertigo and facial numbness have been reported in a well-documented case (Goode and Shearn 1982). The toxin acts presynaptically to block the release of acetylcholine in parasympathetic terminals, sympathetic ganglia and at the neuromuscular junction. Botulism may also occur following infection of wounds. Colonisation of the gut by *Clostridium* is rare because *Clostridium* is a fastidious strictly anaerobic organism: when it does occur it causes botulism, usually in neonates (Bradley et al. 1980). Confirmation of the diagnosis requires either culture of the organism from the suspect food, wound or faeces or detection of the toxin by injecting the patient's serum into mice. The disease is rare but early diagnosis is worthwhile because injection of anti-toxin may be life-saving. It should be suspected from the clinical features and neurophysiological evidence of pre-synaptic motor endplate transmission failure.

Tick bite paralysis is a rare form of ascending paralysis which has been reported from the USA and follows soon after being bitten by ticks. Removal of the ticks is usually followed by rapid recovery and the site of the neurophysiological lesion has not been determined. The favoured theory is that conduction is blocked close to but not at the neuromuscular junction (Donat and Donat 1981). The time course of the illness suggests that a toxin is the cause but none has been identified.

Having narrowed down the differential diagnosis to an acute peripheral neuropathy, a substantial list of diagnoses remains (Table 6.3). Most of the listed conditions can be eliminated by consideration of the probabilities based on the clinical history and examination. Careful enquiry about possible toxin exposure is mandatory since industrial, social and iatrogenic poisons can all induce an acute neuropathy. Examples include heavy metal poisoning with arsenic, lead or thallium or exposure to organophosphorus insecticides (Senanyave and Johnson 1982). Such exposures are usually evident from the history. Social poisons may not be volunteered. Glue sniffing may cause subacute neuropathy because of its content of the neurotoxic hexacarbon, n-hexane (LeQuesne 1984). The commonest social poison is alcohol which is usually combined with malnutrition, a combination which causes progressive sensory and motor neuropathy and also myopathy often in circumstances where the history is blurred. The list of drugs which may cause neuropathy is so long that it is essential to scrutinise any drug which a patient is taking for previous reports of an association with neuropathy (Argov and Mastaglia 1979). Common culprits are vincristine, isoniazid and nitrofurantoin (Toole and Parrish 1973). The neuropathy produced by drugs is usually axonal. Demyelinating neuropathy has been described with perhexiline (Lhermitte et al. 1976) and amiodarone (Martinez-Arizala et al. 1983) and a GBS-like syndrome occurred so frequently with an antidepressant, zimeldine, in Scandinavia that the drug was withdrawn (Fagius et al. 1985). There are two reports of a GBS-like syndrome following amitriptyline overdose (Leys et al.

1987). Any form of porphyria may present with an acute neuropathy and mimic GBS so closely that the urine should be screened for porphobilinogen in every case. In acute intermittent porphyria the illness is usually heralded by abdominal pain and confusion or psychotic disturbances. Autonomic involvement with prominent tachycardia is common. Bizarre patterns of sensory impairment in the trunk and proximal portions of the limbs are encountered. The neuropathy may evolve acutely or subacutely and be so severe as to require ventilation. Nerve conduction studies suggest an axonal neuropathy. Whereas some biopsy and autopsy studies suggest a dying back form of axonopathy, others also show some demyelination. Although acute attacks may be fatal some patients who have been nursed through the acute stage have recovered completely (Ridley 1984).

Vasculitis usually produces an acute or subacute multiple mononeuropathy but may present as a polyneuropathy (Chapter 13). Rare cases of leukaemia and lymphoma associated with neuropathy have been reported. Some are due to an association between the lymphoma and GBS, as with Hodgkin's disease, others are due to infiltration of the nervous system by lymphoma, usually a B cell lymphoma (Vital and Vital 1989).

Tick bite meningoradiculitis induced by the spirochaete *Borrelia burgdorferi* is worth further consideration in the differential diagnosis because it has been recognised relatively recently and may teach lessons relevant to the pathogenesis of GBS. Paralysis following tick bite was described in Europe by Garin and Bujadoux (1922). The onset is delayed and not like the rapid onset of tick bite toxin paralysis. In this case the tick bite is usually followed by a spreading erythematous skin lesion called erythema chronicum migrans. Bannwarth (1941) described associated chronic lymphocytic meningitis and polyneuritis with arthralgia. A similar constellation of symptoms was described in the town of Lyme in Connecticut, USA and shown to be caused by a spirochaete which was injected into the victims by the bite of a tick *Ixodes burgdorferi* (Benach et al. 1983). The European disease was soon shown to be caused by a similar spirochaete (Pfister et al. 1984; Weber 1984). The occurrence of clinical disease in patients at risk may be quite low: of 40 forestry workers bitten by ticks in the New Forest in southern England only 10 had anti-*Borrelia* antibodies; of these only two had erythema chronicum migrans and none had neurological symptoms (Guy et al. 1989).

Vallat et al. (1987) give a clear description of the clinical picture and nerve biopsy findings in 10 cases, all of which occurred in the summer or autumn. Nine of the ten patients recalled a tick bite followed, within one to 15 days, by erythema chronicum migrans. After an average of three weeks (range 2–8 weeks) pain close to the site of the bite heralded severe radicular and back pain and then proximal leg weakness, radicular sensory loss in the distribution of the pain and sometimes cranial, especially facial, nerve palsies. The CSF cell count was always raised and the protein concentration usually. The diagnosis was confirmed by finding IgG and IgM antibodies to Lyme spirochaete antigens in the blood. Neurophysiological studies in this series demonstrated denervation and only minor slowing of nerve conduction more suggestive of axonal degeneration than demyelination. However, in some cases neurophysiological studies have suggested demyelination (Sterman et al. 1982). Nerve biopsies in the series of Vallat et al. (1987) showed widespread perivascular infiltration of lymphocytes and plasma cells round epineurial, perineurial and endoneurial vessels without vessel wall necrosis. In no case was active macrophage-mediated demyelination discovered, whereas that is the usual finding in GBS. The authors were unable to identify spirochaetes in the nerves but suggest that the inflammatory changes might be due to an immunological reaction directed against the spirochaete. The difference from GBS is striking and reminiscent of the inflammatory changes without demyelination induced by a tuberculin reaction in the nerve of a sensitised animal (Chapter 2). Parenteral penicillin appears of benefit in about half the patients treated (Steere et al. 1985).

Immunisation programmes have caused the almost complete disappearance of diphtheria from Europe and North America. However, the level of immunity has dropped in some communities and cases are being reported again (Rappuoli et al. 1988). In developing countries it remains a problem in differential diagnosis. Swift and Rivener (1987) provide a helpful review. The incidence of neuropathy is greater in children who have had more severe faucial diphtheria. In diphtheria palatal palsy develops during the first week; three to five weeks later loss of ocular accommodation and eye muscle, especially lateral rectus, palsies develop. More generalised weakness develops after that and may mimic GBS. Nerve conduction studies demonstrate demyelination. The CSF usually shows a pleocytosis as well as increased protein concentration. The neuropathy is due to demyelination caused by the toxin secreted by *Corynebacterium diphtheriae*. The toxin has two peptides, B which attaches to cell surface receptors and A which inhibits cell protein synthesis. The toxin is probably excluded from the brain by the blood–nerve barrier and has a predilection for sites, such as nerve roots, where the blood–nerve barrier is least but the explanation for its toxicity for myelin rather than other membranes such as axolemma is not clear. Early treatment with antitoxin will prevent progression of the disease but by the time neuropathy has set in such treatment may be too late. Nevertheless, if the patient is supported through the critical illness recovery will ensue (Christie 1987). The diagnosis should have been obvious before the polyneuropathy developed. If the diagnosis was not made earlier the tonsillar or nasopharyngeal exudate will usually have disappeared and if treatment has been given throat swabs are unlikely to grow *Corynebacterium*. The Schick test in which toxin is injected into the skin gives a positive red reaction in non-immune subjects. It should become negative in immunised people and would be expected to have become negative by the stage of diphtheritic neuropathy so it will not be a great help in diagnosis.

The title "critical illness polyneuropathy" has recently been applied to patients who emerge from a period of respiratory support following a variety of procedures complicated by sepsis (Bolton et al. 1984; Zochodne et al. 1987). About one month after the initial intubation these patients are found to be unable to wean from the ventilator and have a more or less severe tetraparesis with areflexia. Neurophysiological studies show evidence of denervation in the muscles and relatively well-preserved conduction velocities suggesting an axonal neuropathy. The CSF is relatively normal. Biopsy and autopsy studies have confirmed widespread axonal degeneration. The underlying cause is unclear but the potential number of toxic factors which might be harmful is legion. Factors complicating assessment may be toxic encephalopathy and catabolic myopathy in these very sick patients.

In some patients with GBS sensory symptoms are unusually prominent but cases of acute purely sensory demyelinating neuropathy resembling GBS either do not occur or are extremely rare. They have to be distinguished from a rare syndrome of acute sensory neuronopathy. In the initial paper describing this syndrome three adults developed numbness, painful paraesthesiae and sensory ataxia acutely a few days after an infection treated with antibiotics. Sensory nerve action potentials were absent or delayed and the CSF protein was elevated. The sensory deficit, especially of large fibre modalities, did not recover so that fall out of sensory neurons and not demyelination was probably the primary pathology (Sterman et al. 1986). This syndrome occurs with Sjogren's syndrome and inflammatory changes have been identified in dorsal root ganglia (Kennett and Harding 1986; Laloux et al. 1988; Griffin et al. 1988).

Occasionally symptoms suggesting acute sensory neuronopathy are associated with acute autonomic dysfunction. At autopsy one such case had very marked loss of dorsal root ganglion cells and corresponding loss of myelinated nerve fibres from dorsal roots, dorsal columns and peripheral nerves. The ventral roots were only mildly affected. There were small clusters of inflammatory cells in the dorsal roots only (Fagius et al. 1983). Although the authors drew a comparison between their patient's disorder and GBS, such cases were closer to the syndrome of acute sensory neuronopathy.

Patients with GBS may have prominent autonomic involvement but rare patients have a syndrome of acute pure pandysautonomia which may recover completely. Although several cases of transient dysautonomia have been reported, only one well-documented example of acute pure dysautonomia with recovery is available (Young et al. 1975). The deficit evolved over a few weeks and recovery took a year. The symptoms included postural hypotension, blurred vision, dry eyes and mouth, loss of sweating, urinary incontinence and impotence. Clinical and pharmacological tests showed severe dysautonomia. The site of the lesion can only be surmised since no pathological information was available apart from a sural nerve biopsy showing a marginal reduction of small calibre unmyelinated nerve fibres. However, the authors discuss the possibility that the transient pandysautonomia was due to a form of GBS confined to the autonomic nerves.

Miller Fisher Syndrome

Three patients with acute external ophthalmoplegia, sluggish pupil reflexes, ataxia and absent tendon reflexes were described by Miller Fisher (1956). Two patients had no weakness and in the third there was a unilateral facial palsy and questionable limb weakness. All three patients recovered spontaneously. In one patient the CSF protein became markedly raised, in one it remained normal and in one it was not examined. Because of the similar ophthalmoplegia in some cases of GBS, the elevated CSF protein in one case and the absence of mental impairment and long tract signs Miller Fisher proposed that his three cases had a disorder akin to GBS affecting the peripheral parts of the ocular motor nerves. He commented on the resemblance of the ataxia to that seen in cerebellar disorder and could not account for this deficit by position sense loss since this was not found in any of his cases. Ropper and Shahani (1984) described a similar case with ataxia consisting of irregular large amplitude side to side movements of the upper limbs which were not worsened by eye closure. The lower limbs were very slightly tremulous and the gait was slightly ataxic. Although their patient did have some loss of position sense they thought that the characteristics of the tremor were cerebellar and not like the clumsiness usually associated with a sensory neuropathy. They proposed that the ataxia was due to a mismatch of information from muscle spindles and other proprioceptors to the cerebellum. This proposal has received some experimental support. In a small series of patients ataxia was prominent when there was a mismatch between position sense impairment estimated clinically and loss of the late component of the silent period of the electromyogram following stimulation of a motor nerve during voluntary contraction of the muscle which it supplied (Jamal and Donaghy 1989). The late component of the silent period is considered to be dependent on afferent input to the cerebellum. In a neurophysiological study of two cases, sensory nerve action potentials were small or absent in the early stages despite relatively normal motor nerve conduction, consistent with a predominantly sensory axonal neuropathy or neuronopathy in the Miller Fisher syndrome (Guiloff 1977). This conclusion has been supported by subsequent electrophysiological studies including a series of ten cases and literature review of nine others (Fross and Daube 1987). The pure Miller Fisher syndrome lacks any weakness or cutaneous sensory impairment and is rather uncommon. Overlap syndromes with weakness and some sensory improvement resemble the usual case of GBS. Rare recurrent cases of Miller Fisher syndrome have been described: one patient had a first attack with weakness and a second attack without weakness emphasising the similarity between GBS and Miller Fisher syndrome (Schapira and Thomas 1986).

In most cases of pure Miller Fisher syndrome or GBS with ophthalmoplegia and ataxia subjected to MRI scans or post-mortem examinations no CNS lesion has been demonstrated (Ropper 1983; Phillips et al. 1984; Landau et al. 1987). It has sometimes been proposed that Miller Fisher syndrome may be caused or at least mimicked by brainstem encephalitis. The cases which have been used to support this proposal have mostly shown features which were conspicuously absent in Miller Fisher's cases, such as drowsiness, brisk reflexes, CSF pleocytosis or extensor plantar responses (Shuaib and Becker 1987). On the other hand mesencephalic and upper pontine reticular formation lesions can depress tendon reflexes. Two cases of brainstem encephalitis with radiologically or autopsy proven lesions had tendon areflexia as well as drowsiness and extensor plantar responses (Al-Din et al. 1982). Two cases somewhat resembling Miller Fisher syndrome with a low attenuation CT scan lesion in the mid-brain or medulla have been reported (Derakshan et al. 1979; Al-Din et al. 1982).

Although some of the physical signs of the Miller Fisher syndrome can be mimicked by a brainstem lesion, the typical case has never been documented radiologically or pathologically as having a CNS lesion. Certain physical signs occur in Miller Fisher syndrome which are normally associated with CNS lesions but can all in fact be produced by peripheral lesions. Preservation of Bell's phenomenon may be encountered in Miller Fisher syndrome but is also seen occasionally in other clearly peripheral disorders, such as myasthenia or botulism, reflecting the much stronger stimulus to upgaze by forced opening of the closed eye. Similarly gaze palsy and ataxic nystagmus though usually disorders of supra or intranuclear pathways may also occur with peripheral disorders. Following these arguments and the proposal that the tremor and ataxia are due to a mismatch between ordinary position sense and peripheral cerebellar sensory input it becomes possible to explain all the clinical features of Miller Fisher syndrome on the basis of a peripheral neuropathy (Ropper 1983).

Autonomic Dysfunction

Involvement of the autonomic nervous system is common and an important cause of complications and death. Recognition of autonomic dysfunction dates back to Osler (1982) who described "paralysis of the heart" occurring in what he called "acute febrile polyneuritis". Attention was drawn to the frequency and severity of dysautonomia in a study of 28 consecutive patients admitted to Long Island Jewish Medical Center from 1965 to 1969 (Lichtenfeld 1971). Four of the six deaths in that series were attributed to autonomic causes, probably cardiac arrhythmias. The frequency of autonomic involvement has been confirmed in subsequent studies (Tuck and McLeod 1981; de Jager et al. 1985; Winer and Hughes 1988). Most patients have a sinus tachycardia, even in the absence of an infective or circulatory cause. While anxiety may be a contributory factor, sinus tachycardia is a feature of the denervated heart, occurring for instance in the neuropathy associated with porphyria. It would be expected to result from relative overactivity of the sympathetic compared with vagal nerve supply to the heart. Loss of the normal sinus arrhythmia, present in about half of patients with GBS might be due to vagal involvement affecting the afferent or efferent limbs of the reflex (Persson and Solders 1983; de Jager et al. 1985). Sustained or paroxysmal hypertension occurs in about 70% of patients. Abrupt fluctuations of blood pressure are not uncommon and may be the harbingers of an arrhythmia and death (Lichtenfeld 1971; Winer and Hughes 1988). Sometimes the sources of hypertension are sufficient to cause hypertensive encephalopathy (McQuillan and Bullock 1988). Plasma catecholamine and urine vanilylmandelic acid excretion in patients with hypertension have sometimes been raised (Mitchell and Meilman 1967; Lichtenfeld 1971) whereas plasma renin levels have been increased in two patients (Stapleton et al. 1978; Laufer et al. 1981). The causes of hypertension in GBS remain unclear and may well vary from patient to patient. In addition to catecholamine or renin angiotensin-mediated mechanisms the loss of sensory input from the glossopharyngeal fibres derived from the baroreceptor in the carotid sinus has been proposed as a mechanism which would release blood pressure from control. Plasma atrial natiuretic factor, which increases in acute hypertension and hypervolaemia, has been shown to rise and fall in response to fluctuations in blood pressure in a single patient with GBS (Saxenhofer et al. 1988). Postural hypotension occurs in about 20% of patients: it is sometimes a prominent early symptom but more commonly becomes a nuisance during the phase when the patient is beginning to be mobilised into sitting and standing after a prolonged period supine in an intensive care unit.

The most worrying autonomic disturbance is the occurrence of cardiac arrhythmias which may be tachycardia, bradycardia or asystole. The most common problem is sinus bradycardia or asystole often occurring during or just after tracheal suction. Seven of 100 patients in a recent series had such complications and in two this was documented after tracheal suction (Winer and Hughes 1988). This may be due to hypersensitivity of the vagal reflex or absence of the normal sympathetic activity which keeps that reflex in check. Atropine may control such episodes but endocardial pacing may be necessary. It must also be recalled that prolonged tracheal suction without prior hyperoxygenation may cause hypoxia which may itself cause bradycardia (Pace 1976). Management of autonomic complications is discussed in Chapter 8. Tachyarrhythmias of many sorts have been described, including atrial fibrillation (Stewart 1973; Dalos et al. 1988; Winer and Hughes 1988), atrial tachycardia (Lichtenfeld 1971), ventricular tachycardia and ventricular fibrillation (Winer and Hughes 1988). Requirement for ventilation, presence of hypertension and loss of sinus arrhythmia are factors which predict an increased likelihood of cardiac arrhythmia (Table 6.5). Unfortunately although a requirement for ventilation is the strongest predictive factor, patients who do not require ventilation are not free from risk since asystole may occur before or after respiratory failure occurs or after it has recovered (Fig. 6.2).

Various forms of paroxysmal autonomic disturbances occur in severely affected patients. Sometimes these suggest sympathetic overactivity, such as attacks of tachycardia, pupil dilatation and agitation, or episodes of sweating and peripheral vasoconstriction. Lichtenfeld (1971) described two patients who died during such episodes. Lability of pulse and blood pressure preceded dangerous cardiac arrhythmias in 6 of 11 patients in the series of Winer and Hughes (1988). Cardiac arrhythmias may be associated with marked electrocardiographic changes including T wave inversion and in one case S-T segment elevation (Fig. 6.3) (Palferman et al. 1982). These changes are usually transient and the T wave inversion may be reversed temporarily by atropine (Lichtenfeld 1971).

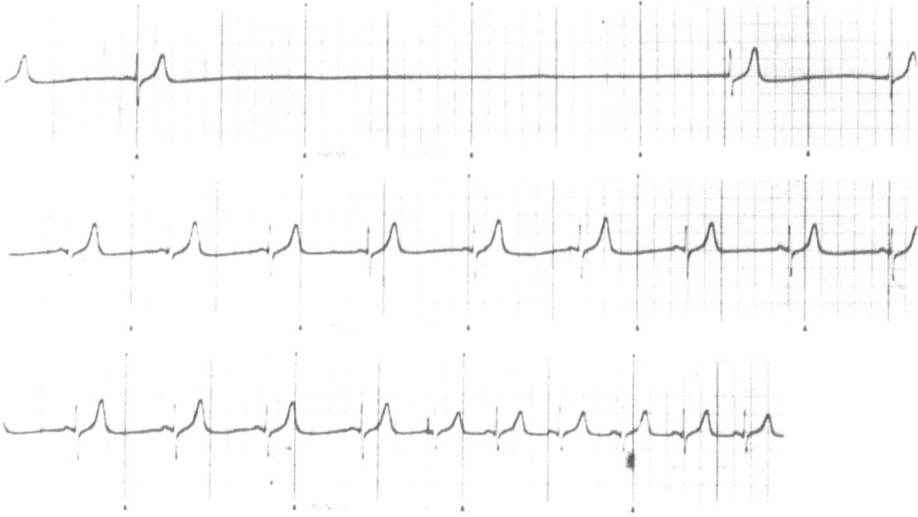

Fig. 6.2. Electrocardiogram recorded from a man with GBS who developed this episode of asystole before developing respiratory failure. Following this record a demand driven endocardial pacemaker was inserted and several more episodes of asystole occurred justifying its insertion. Twelve days later the pacemaker was removed, no further episodes occurred and the patient recovered.

Table 6.5. Frequency of autonomic dysfunction in Guillain–Barré syndrome patients with and without subsequent arrhythmias

	With (%)	Without (%)	Odds ratio (95% CI)
Sinus tachycardia	5/11 (45)	15/87 (17)	4.0 (14.8–1.1)
Systolic hypertension	8/11 (73)	20/87 (23)	8.9 (36.8–2.2)**
Reduced R-R variation	9/11 (82)	36/82 (44)	5.8 (28.2–1.2)*
Postural hypotension	1/4 (25)	0/56 (0)	
Excessive sweating	3/11 (27)	26/89 (29)	0.9 (3.7–0.2)

From Winer and Hughes (1988), with permission.
$*P<0.05$. $**P<0.005$.
Postural hypotension was not assessed in severely affected patients who could not stand or be tilted.

Excessive sweating was noted in 29% of 100 patients in one series (Winer and Hughes 1988) and profuse sweating may form part of the dangerous paroxysmal autonomic disturbances described above. However, extensive anhidrosis on the limbs was found in all of seven patients studied in detail by Tuck and McLeod (1981). Most of the autonomic disturbances in GBS could be explained by demyelinating lesions blocking conduction in the parasympathetic and sympathetic nerve trunks where they contain myelin. Autopsy studies have shown demyelination lesions in the glossopharyngeal and vagus nerves and sympathetic chains and white rami (Roseman and Aring 1941; Matsuyama and Haymaker 1967; Asbury et al. 1969). The paroxysmal disturbances are more difficult to understand. Three explanations seem possible. First, there might be denervation hypersensitivity of receptors in blood vessel walls and the heart to circulating catecholamines (Krone et al. 1983). Second, there might be paroxysmal increases in firing rate of sympathetic neurons, akin to myokymia in other nerves, due to inflammatory lesions. Experimental evidence supports a third and more likely explanation that the paroxysmal disturbances are due to loss of damping of autonomic reflexes by damage to myelinated fibres in the afferent limb of the arc. Microelectrode recordings show that sympathetic discharges were increased in muscle branches of the peroneal nerves of three GBS patients with sympathetic overactivity (Fagius and Wallin 1983). Similar increased sympathetic discharges occur in the nerves of muscles of patients to whom a negative pressure is applied to the abdomen. More direct proof was obtained by experiments in which the authors bravely anaesthetised their own vagus and glossopharyngeal nerves with local anaesthetic and induced tachycardia and severe hypertension: during this period sympathetic nerve activity was increased in the motor branches of the peroneal nerve at rest and sympathetic reflex responses to trivial stimuli were enhanced (Fagius et al. 1985). This suggests that reduction of baroreceptor afferent activity, as would occur with inflammatory lesions in these nerves, causes increased sympathetic activity.

Although sphincter involvement is conspicuous by its absence during the early stages of GBS and the bladder may remain unaffected in many cases, there are some patients in whom the bladder does become involved at the nadir of the disease. A catheter is so often inserted as part of the intensive care of the severely ill patient that it is impossible to give an accurate figure for the frequency of bladder involvement. During the acute stage of the disease it is not uncommon that patients have difficulty with absorbing nasogastric feeds which may be

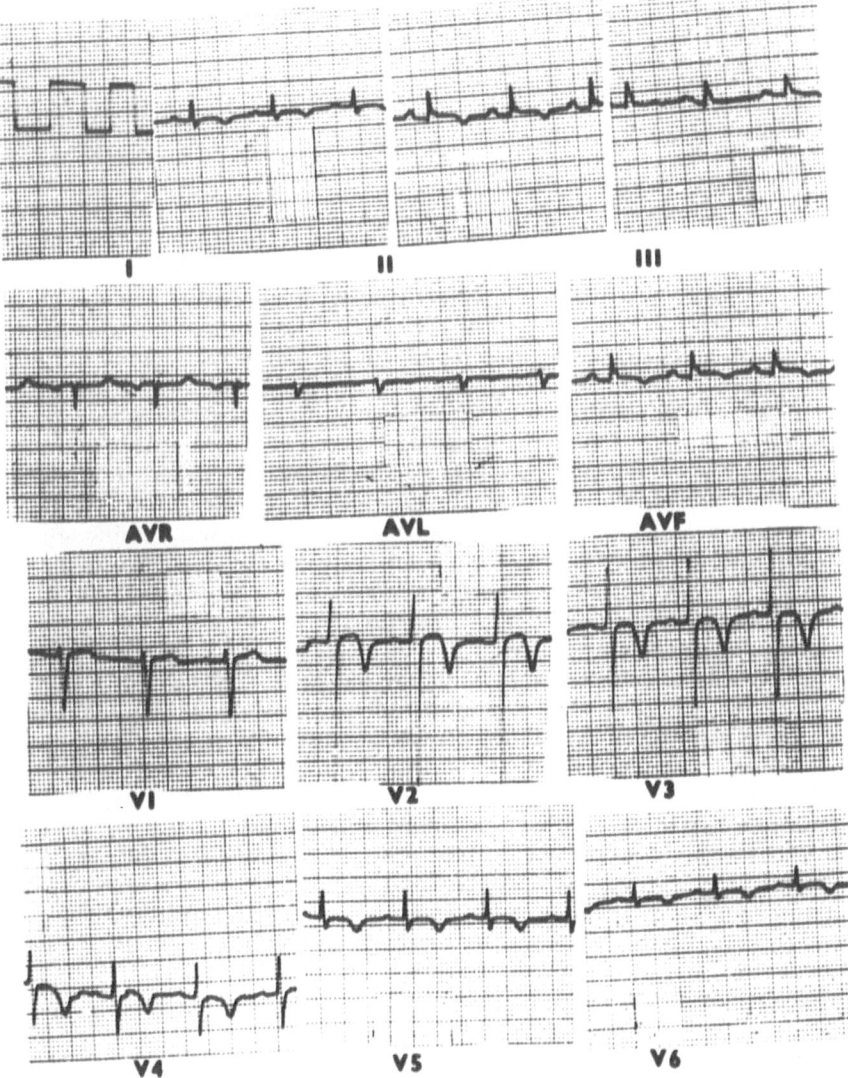

Fig. 6.3. This ECG was recorded from a woman with acute, severe GBS in the absence of chest pain or alteration of cardiac enzymes. The abnormalities disappeared spontaneously after two days. (From Palferman et al. (1982), with permission.)

due to reduced gastric mobility. Similarly constipation is a common problem both from weakness of the abdominal wall preventing voluntary straining and from reduction in gut mobility.

The occurrence of hyponatraemia is a well-recognised though uncommon complication of GBS (Posner et al. 1967). Its cause has rarely been thoroughly investigated. In one patient inappropriate antidiuretic hormone (ADH) secretion occurred at a lower plasma osmolality than usual. The authors suggested that an apparent resetting of the osmotic pressure receptor had occurred as a result of damage to afferent pathways by the inflammation in the PNS (Penney et al. 1979).

Papilloedema

Papilloedema is a rare complication of GBS, first described by Gilpin et al. (1936). A series of 31 cases, comprising all those from the early literature and four personal cases, was comprehensively reviewed by Morley and Reynolds (1966). They found that there was little difference between the clinical features of patients with GBS who developed papilloedema and those who did not, except that there was a tendency for patients with papilloedema to have a more prolonged progressive phase or a recurrent course. In most patients with papilloedema, but not all, the CSF protein has been increased, often markedly. More constant was the relationship between papilloedema and raised intracranial pressure which has usually been raised when it has been measured and might have been raised in some of the remainder had it been measured at the appropriate time. There has been no relationship between the presence of papilloedema and the severity or course of the neuropathy: in fact in many cases the papilloedema has been noted after the neuropathic symptoms have reached their worst or are beginning to improve.

A priori swollen optic discs might be due to papillitis, systemic hypertension, cerebral oedema or hydrocephalus. Since the visual acuity remains normal, visual fields show only enlargement of the blind spot and optic atrophy does not occur, with rare exceptions (Perniola and Torelli 1968), papillitis cannot be the explanation. Rare cases of optic neuritis associated with GBS require separate consideration (Behan et al. 1981). Systemic hypertension has been reported in association with GBS and papilloedema on several occasions (Davidson and Jellinek 1977) but is a very inconstant association. Controversy still exists concerning the two remaining hypotheses, cerebral oedema and hydrocephalus. In favour of cerebral oedema are observations of normal ventricle size on pneumo-encephalograms in five cases reviewed by Morley and Reynolds (1966) and normal absorption of radioactively labelled albumin, injected into the CSF space (Davidson and Jellinek 1977; Behan et al. 1981). In addition Joynt (1958) claimed that cerebral cortical tissue removed from his case showed intracellular oedema. It is not immediately obvious how GBS would cause cerebral oedema and the alternative proposal of communicating hydrocephalus is more attractive. Denny Brown (1952) described a patient who developed raised intracranial pressure and papilloedema: a ventriculogram was normal and the papilloedema subsided with repeated lumbar punctures. The CSF obtained clotted in the test tube and the proposal that similar clotting might obstruct the arachnoid villi and interfere with CSF absorption was said to be confirmed at autopsy. This hypothesis has been confirmed with modern radiological techniques. A girl developing papilloedema six weeks after the onset of GBS had a high CSF pressure, increased CSF protein concentration, enlarged ventricles on CT scan and impaired absorption of technetium pertechnetate from the CSF: following improvement six months later the ventricle size on the CT scan had returned to normal (Farrell et al. 1981). A similar case was reported by Reid and Draper

(1980). It seems likely that communicating hydrocephalus is the most common cause of papilloedema in GBS.

Other Complications

Glomerulonephritis has occasionally been reported in association with GBS and this has sometimes caused nephrotic syndrome. In one series renal biopsies showed evidence of subclinical glomerulonephritis in a high proportion of cases. In most instances in which glomerulonephritis has been symptomatic the underlying neuropathy has had the time course of progressive CIDP (Chapter 9).

Venous thrombosis and pulmonary embolism are important in GBS and are considered in Chapter 7.

Summary

Guillain–Barré syndrome presents with a combination of numbness and tingling, weakness and pain which reach their nadir within about 2 weeks and, by definition, in not more than 4 weeks. There is a large differential diagnosis including psychological and CNS disorders, myasthenia and specific causes of acute neuropathy. The Miller Fisher variant of GBS consists in its pure form of ophthalmoplegia, areflexia and ataxia which are due to demyelinating neuropathy affecting both nerves to the external ocular muscles and sensory afferents serving the monosynaptic tendon reflex and cerebellar input. The autonomic nervous system is commonly affected in GBS: autonomic complications include sinus tachycardia, dangerous arrhythmias, both fast and slow, rapid fluctuations of blood pressure, and abnormalities of sweating. The CNS is not affected in GBS but papilloedema is a rare complication and is usually due to impaired absorption of CSF probably related to the increased CSF protein concentration.

References

Al-Din AN, Anderson M, Bickerstaff ER, Harvey I (1982) Brainstem encephalitis and the syndrome of Miller Fisher: a clinical study. Brain 105:481

Argov Z, Mastaglia F (1979) Drug-induced peripheral neuropathies. Br Med J i:663–664

Asbury AK, Arnason BG, Adams RD (1969) The inflammatory lesion in idiopathic polyneuritis. Its role in pathogenesis. Medicine 48:173–215

Bannwarth A (1941) Chronische lymphocytäre Meningitis, entzündliche Polyneuritis und "Rheumatismus". Arch Psychiatr Nervenkr 113:284–376

Behan PO, Harrington H, Sekoni G (1981) Papilloedema in the Landry–Guillain–Barré syndrome. Eur Neurol 20:62–63

Benach JL, Bosler EM, Hanrahan JP et al. (1983) Spirochetes isolated from the blood of two patients with Lyme disease. N Engl J Med 308:740–742

Bolton CH, Gilbert JJ, Hahn AF, Sibbold WJ (1984) Polyneuropathy in critically ill patients. J

Neurol Neurosurg Psychiatry 47:1223–1231

Bradley WG, Shahani BT, Hyslop NE (1980) Rapidly progressive neurologic disorder following gastrointestinal symptoms. N Engl J Med 303:1347–1355

Brashear HR, Bonnin JM, Login IS (1985) Encephalomyeloneuritis simulating Guillain–Barré syndrome. Neurology 35:1146–1151

Carroll WM, Mastaglia FL (1979) Locked-in coma in postinfective polyneuropathy. Arch Neurol 36:46–47

Chopra JS, Benerjee AK, Murthy MJK, Pal SR (1980) Paralytic rabies: a clinicopathological study. Brain 103:789–802

Christie AB (1987) Diphtheria. In: Weatherall DJ, Ledingham JGG, Warrell DA (eds) Oxford textbook of medicine. Oxford University Press, Oxford, pp 5.164–5.165

Dalos NP, Borel C, Hanley DF (1988) Cardiovascular autonomic dysfunction in Guillain–Barré syndrome. Therapeutic implications of Swan-Ganz monitoring. Arch Neurol 45:115–117

Davidson DLW, Jellinek EH (1977) Hypertension and papilloedema in the Guillain–Barré syndrome. J Neurol Neurosurg Psychiatry 40:144–148

de Jager AE, Op de Coul AA, Lambregts PC (1985) Cardiovascular dysfunction in Guillain–Barré syndrome. Neurology 35:1805

Denny Brown DE (1952) The changing pattern of neurological medicine. N Engl J Med 246:839–846

Derakshan I, Lotfi J, Kaufman B (1979) Ophthalmoplegia, ataxia and hyporeflexia (Fisher's syndrome) with a midbrain lesion demonstrated by CT scanning. Eur Neurol 18:361–366

Donat JR, Donat JF (1981) Tick paralysis with persistent weakness and electromyographic abnormalities. Arch Neurol 38:59–62

Eisen A, Humphreys P (1974) The Guillain–Barré syndrome. A clinical and electrodiagnostic study of 25 cases. Arch Neurol 30:438–443

Fagius J, Wallin BG (1983) Microneurographic evidence of excessive sympathetic outflow in the Guillain–Barré syndrome. Brain 106:589–600

Fagius J, Westerberg C-E, Olsson Y (1983) Acute pandysautonomia and severe sensory deficit with poor recovery. A clinical neurophysiological and pathological case study. J Neurol Neurosurg Psychiatry 46:723–725

Fagius J, Osterman PO, Siden A, Wiholm B-E (1985) Guillain–Barré syndrome following zimeldine treatment. J Neurol Neurosurg Psychiatry 48:65–69

Farrell K, Hill A, Chauang S (1981) Papilloedema in Guillain–Barré syndrome – a case report. Arch Neurol 38:55–58

Fisher M (1956) Syndrome of ophthalmoplegia, ataxia and areflexia. N Engl J Med 255:57–65

Fross RD, Daube J (1987) Neuropathy in the Miller Fisher syndrome: clinical and electrophysiologic findings. Neurology 37:1493–1498

Garin C, Bujadoux C (1922) Paralysie par les Tignes. J Med Lyon 3:765–767

Gilpin SF, Moersch FP, Kernohan JW (1936) Polyneuritis. A clinical and pathological study of a special group of cases frequently referred to as instances of neuronitis. Arch Neurol Psychiatry 35:937–953

Goode GB, Shearn DL (1982) Botulism. A case with associated sensory abnormalities. Arch Neurol 34:55

Griffin JW, Cornblath DR, Alexander E (1988) Sensory ganglionitis associated with connective tissue disorders. Neurology 38:243

Guiloff RJ (1977) Peripheral nerve conduction in Miller Fisher syndrome. J Neurol Neurosurg Psychiatry 40:801–807

Guy EC, Martyn CN, Bateman DE, Heckels JE, Lawton NF (1989) Lyme disease: prevalence and clinical importance of Borrelia burgdorferi specific IgG in forestry workers. Lancet i:484–485

Hopkins A, Clarke C (1987) Pretended paralysis requiring artificial ventilation. Br Med J 294:961–962

Hughes RAC, Winer JB (1984) Guillain–Barré syndrome. In: Matthews WB, Glaser GH (eds) Recent advances in clinical neurology, 4th edn. Churchill Livingstone, Edinburgh, pp 19–49

Jamal J, Donaghy M (1989) A peripheral mechanism for the ataxia associated with the Miller Fisher syndrome of acute ophthalmoplegia, ataxia and areflexia. J Neurol Neurosurg Psychiatry 52:1210

Joynt RJ (1958) Mechanism of production of papilloedema in the Guillain–Barré syndrome. Neurology 8:8–12

Kennett RP, Harding AE (1986) Peripheral neuropathy associated with the sicca syndrome. J Neurol Neurosurg Psychiatry 49:90–93

Krone A, Reuther P, Fuhrmeister U (1983) Autonomic dysfunction in polyneuropathies: a report on 106 cases. J Neurol 230:111–121

Laloux P, Brucher JM, Guerit JM, Sindic CJM, Laterre EC (1988) Subacute sensory neuropathy associated with Sjögren's sicca syndrome. J Neurol 235:352–354

Landau WM, Glenn C, Dust G (1987) MRI in Miller Fisher variant of Guillain–Barré syndrome. Neurology 37:1431

Laufer J, Passwell J, Keren G, Brandt N, Cohen DE (1981) Raised plasma renin activity in the hypertension of the Guillain–Barré syndrome. Br Med J 282:1272–1273

LeQuesne PM (1984) Toxic neuropathies. In: Asbury AK, Gilliatt RW (eds) Peripheral nerve disorders. Butterworths, London, pp 184–204

Leys D, Pasquier F, Lamblin MD, Dubois F, Petit H (1987) Acute polyradiculoneuropathy after amitriptyline overdose. Br Med J 294:608

Lhermitte F, Fardeau M, Chedru F, Mallecourt J (1976) Polyneuropathy after perhexiline maleate therapy. Br Med J 1:1256–1257

Lichtenfeld P (1971) Autonomic dysfunction in the Guillain–Barré syndrome. Am J Med 50: 772–780

Löffel NB, Rossie LN, Mumenthaler M (1977) The Landry–Guillain–Barré syndrome – complications, prognosis and natural history in 123 cases. J Neurol Sci 33:71–79

Marshall J (1963) The Landry–Guillain–Barré Syndrome. Brain 86:56–66

Martinez-Arizala A, Sobol SM, McCarty GF, Nichols BR, Rarita L (1983) Amiodarone neuropathy. Neurology 33:643–665

Masucci EF, Kurtzke JF (1971) Diagnostic criteria for the Guillain–Barré syndrome. An analysis of 50 cases. J Neurol Sci 13:483–501

Matsuyama H, Haymaker W (1967) Distinction of lesions in the Landry–Guillain–Barré syndrome with emphasis on involvement of the sympathetic system. Acta Neuropathol 8:230–241

McFarland HR, Heller GL, Arbor A (1966) Guillain–Barré disease complex. Arch Neurol 14: 197–201

McLeod JG, Walsh JC, Prineas JW, Pollard JD (1976) Acute idiopathic polyneuritis – a clinical and electrophysiological follow-up study. J Neurol Sci 27:145–162

McQuillan JJ, Bullock RE (1988) Extreme labile blood pressure in Guillain–Barré syndrome. Lancet ii:172–173

Mitchell PL, Meilman E (1967) The mechanism of hypertension in the Guillain–Barré syndrome. Am J Med 42:986–995

Moore P, James O (1981) Guillain–Barré syndrome. Incidence, management and outcome of major complications. Crit Care Med 7:549–555

Morley JB, Reynolds EH (1966) Papilloedema and the Guillain–Barré syndrome. Brain 89:205–222

Osler W (1982) Acute ascending (Landry's) paralysis. In: Young J (ed) Principles and practice of medicine. Pentland, Edinburgh, London

Pace NL (1976) Cardiac monitoring and demand pacemaker in Guillain–Barré syndrome. Arch Neurol 33:374

Palferman TG, Wright I, Doyle DV, Amiel S (1982) Electrocardiographic abnormalities and autonomic dysfunction in GBS. Br Med J 284:1231–1232

Penney MD, Murphy D, Walters G (1979) Resetting of osmoreceptor response as cause of hyponatraemia in acute idiopathic polyneuritis. Br Med J 2:1474–1476

Perniola T, Torelli D (1968) Papilloedema in Guillain–Barré syndrome. Lancet ii:919

Persson A, Solders G (1983) R-R variations in Guillain–Barré syndrome – a test of autonomic dysfunction. Acta Neurol Scand 67:294–300

Peterman AF, Daly DD, Dion FR, Keith HM (1959) Infectious neuronitis (Guillain–Barré syndrome) in children. Neurology 9:533–539

Pfister H, Walter E, Karl P, Mursic V, Wilske B, Schierz G (1984) The spirochaetal aetiology of lymphocytic meningoradiculitis of Bannwarth. J Neurol 231:141–144

Phillips MS, Stewart S, Anderson JR (1984) Neuropathological findings in Miller Fisher syndrome. J Neurol Neurosurg Psychiatry 47:492–495

Pleasure DE, Lovelace RE, Duvois RC (1969) The prognosis of acute polyradiculoneuritis. Neurology 18:1143–1148

Posner JB, Ertel NH, Kossman RJ, Scheinberg LC (1967) Hyponatraemia in acute polyneuropathy. Arch Neurol 17:530–541

Raphael JC, Masson C, Morice V, Bronel D, Goulon M (1984) Le syndrome de Guillain–Barré: étude retrospective de 233 observations. Sem Hôp Paris 60:2543–2546

Rappuoli R, Perugini M, Falsen E (1988) Molecular epidemiology of the 1984–1986 outbreak of diphtheria in Sweden. N Engl J Med 318:12–14

Ravn H (1967) The Guillain–Barré syndrome. A survey and clinical report of 123 cases. Acta Neurol Scand 43 suppl 30:1–164

Reid AC, Draper IT (1980) Pathogenesis of papilloedema and raised intracranial pressure in GBS.

Br Med J 281:1393–1394

Ridley A (1984) Porphyric neuropathy. In: Dyck PJ, Thomas PK, Lambert EH, Bunge R (eds) Peripheral neuropathy. WB Saunders, Philadelphia, pp 1704–1716

Ropper AH (1983) The CNS in Guillain–Barré syndrome. Arch Neurol 40:397–398

Ropper AH, Shahani BT (1984) Diagnosis and management of acute areflexic paralysis with emphasis on Guillain–Barré syndrome. In: Asbury AK, Gilliatt RW (eds) Peripheral nerve disorders. A practical approach. Butterworths, London

Roseman E, Aring CD (1941) Infectious polyneuritis. Medicine 20:463–494

Samantray SK (1977) Landry–Guillain–Barré–Strohl syndrome in typhoid fever. Aust NZ J Med 7:307–308

Samantray SK, Johnson SC, Mathao KU, Pulimood BM (1977) Landry–Guillain–Barré syndrome. A study of 302 cases. Med J Aust 2:84–91

Saxenhofer H, Weidman P, Shaw S, Sulzer M, Siegrist P, Staubli M (1988) Atrial natriuretic factor in the Landry–Guillain–Barré syndrome. N Engl J Med 319:448

Schapira AHV, Thomas PK (1986) A case of recurrent idiopathic ophthalmoplegic neuropathy (Miller Fisher syndrome). J Neurol Neurosurg Psychiatry 49:463–464

Senanayave N, Johnson MK (1982) Acute polyneuropathy after poisoning by a new organophosphate insecticide. N Engl J Med 306:155–157

Shuaib A, Becker WJ (1987) Variants of Guillain–Barré syndrome: Miller Fisher syndrome facial diplegia and multiple cranial nerve palsies. Can J Neurol Sci 14:611–616

Stapleton FB, Skoglund RR, Daggett RB (1978) Hypertension associated with the Guillain–Barré syndrome. Pediatrics 62:588–590

Steere AC, Green J, Schoen RT et al. (1985) Successful parenteral penicillin therapy of established Lyme arthritis. N Engl J Med 312:869–874

Sterman AB, Nelson S, Barclay P (1982) Demyelinating neuropathy accompanying Lyme disease. Neurology 32:1302–1305

Sterman AB, Schaumburg HH, Asbury AK (1986) The acute sensory neuronopathy syndrome: a distinct clinical entity. Ann Neurol 7:354–358

Stewart IM (1973) Arrhythmias in the Guillain–Barré syndrome. Br Med J 2:665–667

Swift TR, Rivener MH (1987) Infectious diseases of nerve. In: Matthews WB (ed) Neuropathic handbook of clinical neurology, Vol 7. Elsevier, Amsterdam, pp 179–184

Toole JF, Parrish ML (1973) Nitrofurantoin polyneuropathy. Neurology 23:554–559

Tuck RR, McLeod JG (1981) Autonomic dysfunction in Guillain–Barré syndrome. J Neurol Neurosurg Psychiatry 44:983–990

Vallat JM, Hugon J, Lubeau M, Leboutet MJ, Dumas M, Desproges-Gutteron R (1987) Tick bite meningoradiculitis: clinical electrophysiologic and histologic findings in 10 cases. Neurology 37:749–753

Verma AK, Masheshwari MC, Chawdhary C, Tickoo S (1985) Acute ascending motor paralysis due to rabies: a clinicopathological report. Eur Neurol 24:160–162

Vital C, Vital A (1989) T cell leukemic lymphomatosis. Neurology 39:1272

Weber K (1984) Lymphocytic meningoradiculitis of Bannwarth and erythema migrans disease. J Neurol 231:281–282

Wiederholt HM, Mulder DW, Lambert EH (1964) The Landry–Guillain–Barré–Strohl syndrome of polyradiculoneuropathy – historical review report on 97 patients and present concepts. Mayo Clin Proc 49:427–451

Winer JB, Hughes RAC (1988) Identification of patients at risk of arrhythmia in the Guillain–Barré syndrome. Q J Med 68:735–739

Winer JB, Hughes RAC, Osmond C (1988) A prospective study of acute idiopathic neuropathy. I. Clinical features and their prognostic value. J Neurol Neurosurg Psychiatry 51:605–612

Young RR, Asbury AK, Corbett JL, Adams RD (1975) Pure pandysautonomia with recovery – description and discussion of diagnostic criteria. Brain 98:613–636

Zochodne DW, Bolton CF, Wells AG et al. (1987) Critical illness polyneuropathy. A complication of sepsis and multiple organ failure. Brain 110:819–842

Chapter 7

Clinical Neurophysiology

Introduction

Neurophysiological studies usually show some abnormality of nerve conduction in GBS and may eventually show evidence of denervation in severe cases. They are an important adjunct in the diagnosis but even extensive studies may be normal at first in occasional clinically typical cases. Neurophysiological techniques have been particularly valuable in probing the underlying pathophysiology of GBS. Biopsies give information about only a tiny sample of a particular cutaneous sensory nerve which is usually relatively unaffected, while autopsy studies have their own limitations. Albers (1987) has provided an excellent review to which must be added interesting recent studies concerning the pattern of conduction failure (Van der Meche et al. 1988) and the relationship between small distally evoked action potentials and poor prognosis (Cornblath et al. 1988; Winer et al. 1988).

Frequency of Abnormalities of Motor Nerve Conduction

Several substantial series of patients with GBS have been studied neurophysiologically (Bannister and Sears 1962; Lambert and Mulder 1964; McLeod et al. 1976; Hausmanowa-Petrusewicz et al. 1979; McLeod 1981; Albers et al. 1985; Albers 1987). In one series of 49 patients studied in the first three weeks of the illness maximum motor nerve conduction velocity was slowed to less than 70% of the normal mean in 61%: prolonged distal motor latencies were the only abnormality in 25% and 14% of examinations were normal (Lambert and Mulder 1964). In another study of 50 patients at rather variable times in the illness severe slowing of nerve conduction was found in only 50% of patients and normal results were obtained in 14% (McLeod et al. 1976). In this study the criteria for demyelination were median or ulnar nerve motor conduction velocity <40 m/s in the forearm segment, a distal motor latency >7.0 ms or a common peroneal motor conduction velocity <30 m/s. These criteria had been shown to be characteristic of segmental demyelination in a study correlating neurophysiological results with biopsy findings (McLeod et al. 1973). The impairment of conduction varied from nerve to nerve in an individual patient implying a multifocal rather than diffuse demyelinating process. No relationship was identified between motor conduction velocity and severity of illness at the time of the neurophysiological study. This is in keeping with the finding that patients with demy-elinating forms of hereditary motor and sensory neuropathy may have normal strength despite very slow nerve conduction slow velocities.

Another study of 20 patients with GBS during the acute stage (Hausmanowa-Petrusewicz et al. 1979) led to similar conclusions and added two further obser-vations. First, longer nerves were affected earlier and more severely than shorter nerves, a difference which is easily explained by their greater chance of being af-fected by a multifocal demyelinating process. Second, nerve conduction velocities may become paradoxically slower during the early stages of clinical recovery. This might be explained if the eloquent lesions producing the neurological defi-cit are proximally situated and conduction block at those sites is improving while new lesions are developing in the forearm and leg segments in which con-duction velocity is measured. Incidentally histological evidence of mononuclear cell infiltration was reported in all of six biopsies in this study, an unusually high incidence (Chapter 4).

In GBS the compound muscle action potential amplitude (CMAP) evoked by a proximal stimulus is often significantly smaller than that evoked by a distal stimulus. A difference in which the proximally evoked CMAP is less than 70% of the distally evoked amplitude is usually considered significant (Albers 1987). This decrement may be caused by block of conduction in some of the axons between sites of stimulation (Fig. 7.1) or by dispersion of the volley. The poten-tial recorded from each axon is triphasic and with dispersion the negative com-ponents of one potential may cancel the positive components of another. Con-duction block detected electrophysiologically has been shown to correlate with histological evidence of demyelination in teased fibres (Feasby et al. 1985). As

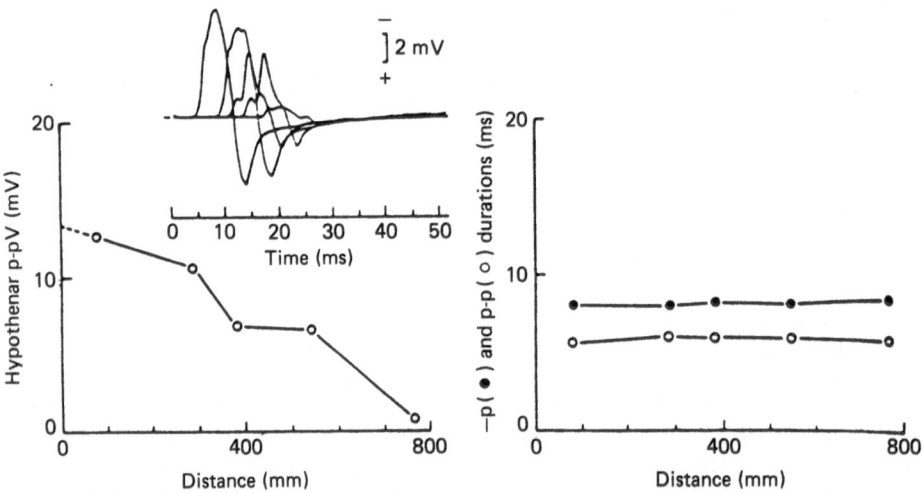

Fig. 7.1. Patient with GBS examined seven days after the onset. The changes in the maximum hypothenar M potentials elicited by stimulation at progressively more proximal sites, namely the wrist, just distal to the cubital tunnel, just proximal to the elbow, the proximal upper arm and Erb's point are shown. The recorded potentials are shown in the inset and the plots illustrate the changes in the M potential amplitude in mV and duration in ms. There was more than 90% decline in am-plitude accompanied by less than 15% increase in duration between the wrist and Erb's point. (From Brown and Feasby (1984), with permission.)

in other types of demyelinating neuropathy conduction is preferentially delayed at sites of compression in GBS. In a series of 25 patients motor conduction was most commonly slowed in the elbow segment of the ulnar nerve, the distal portion of the median nerve and across the fibular head in the peroneal nerve (Brown and Feasby 1984). This study sought conduction block in the common peroneal nerve between the popliteal fossa and ankle, in the median nerve between the proximal arm and the wrist and in the ulnar nerve between Erb's point and the wrist. Conduction block was identified in at least one nerve in 13 of 19 patients tested within the first 14 days. Of the six other patients, four showed progressive increases in the duration of the CMAPs as the stimulus was moved proximally preventing assessment of conduction block. In one patient only two axons could be excited in one ulnar nerve at the wrist and none in the median and peroneal nerves. In one patient there was neither temporal dispersion nor conduction block.

Time Course of Nerve Conduction Abnormalities

A comprehensive neurophysiological study of GBS confirmed these observations and contained sufficient observations at different intervals after onset to give a much clearer temporal profile of the incidence of abnormalities of motor and sensory conduction (Albers et al. 1985). According to strict criteria (Table 7.1) only half the patients had evidence of demyelination in the first two weeks of the illness, 85% in the third week and about two-thirds thereafter. Abnormalities of motor nerve function were already present in nearly 90% of patients in the first two weeks whereas sensory nerve conduction remained within normal limits in most patients at this early stage (Fig. 7.2). It is not surprising that the more thorough studies identify more abnormalities. Only one from this series of 70 patients had no abnormality during the first five weeks of the illness. The greatest abnormalities occurred between three and four weeks after onset, which is likely to have been just after the time at which clinical disability was at its worst. Abnormalities of motor or sensory nerve conduction

Table 7.1. Criteria suggestive of demyelination in the electrodiagnostic evaluation of acute inflammatory demyelinating polyradiculoneuropathy

Demonstrate at least three of the following in motor nerves (exceptions noted below):
1. Conduction velocity less than 90% of lower limit of normal if amplitude exceeds 50% of lower limit of normal: less than 80% if amplitude less than 50% of lower limit of normal (two or more nerves).[a]
2. Distal latency exceeding 115% of upper limit of normal if amplitude normal: exceeding 125% of upper limit of normal if amplitude less than lower limit of normal (two or more nerves).[b]
3. Evidence of unequivocal temporal dispersion or a proximal to distal amplitude ratio less than 0.7.[b,c]
4. F-response latency exceeding 125% of upper limit of normal (one or more nerves).[a,b]

From Albers and Kelly (1989), with permission.
[a] Excluding isolated ulnar or peroneal nerve abnormalities at the elbow or knee, respectively.
[b] Excluding isolated median nerve abnormality at the wrist.
[c] Excluding the presence of anomalous innervation (e.g. median to ulnar nerve crossover).

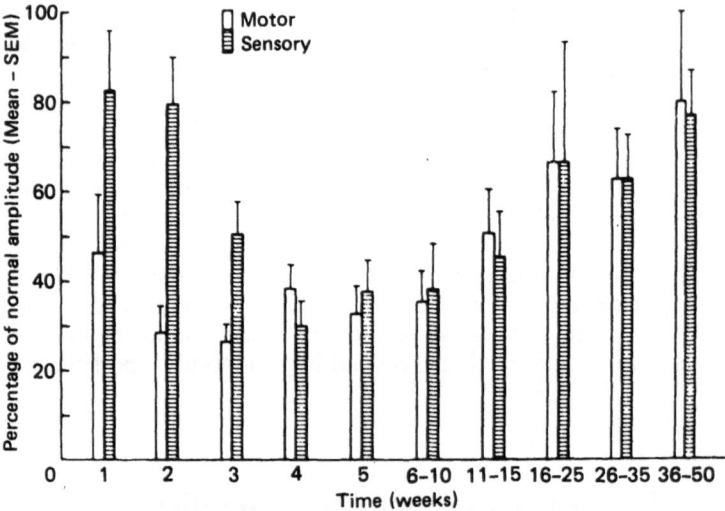

Fig. 7.2. Motor-evoked (*white bar*) and sensory-evoked (*striped bar*) response amplitudes expressed as a percentage of the normal mean as a function of time after disease onset in patients with acute inflammatory demyelinating polyradiculoneuropathy. The responses are significantly ($P<0.05$) different for weeks 1, 2 and 3. No significant differences exist thereafter. (From Albers et al. (1985), with permission.)

were still present in most of 12 patients studied 6–9 months after onset and in three of five studied a year after onset.

Direct measurement of proximal motor nerve conduction velocity was undertaken in two patients with GBS by Brown and Feasby (1984) and shown to be abnormal in both (Fig. 7.3). These measurements involved stimulating the lumbar roots and sciatic nerve at the sciatic notch with needle electrodes, techniques which are unlikely to gain acceptance in a routine laboratory. Mills and Murray (1985) used an electrical stimulation technique applied to the neck to excite cervical roots. The CMAP elicited by this neck stimulation was abnormal in 15 of 21 patients with GBS. In five of the 15 patients with abnormal results neck stimulation resulted in an abnormally small CMAP whereas the potential elicited from more distal sites remained normal. This suggested a proximal site of conduction block. However, the result could also be explained by a random multifocal process blocking conduction throughout the length of the nerve which would cause the compound muscle action potential to decline as an exponential function of the length of the nerve (Van der Meche et al. 1988).

In clinical practice the most frequently used measure of proximal conduction is the F wave latency. Stimulation of, for instance, the median nerve at the wrist normally elicits an orthodromic M wave in abductor pollicis brevis and, about 25 ms later, a smaller inconstant F wave of variable latency which has been shown to represent an antidromic impulse conducted proximally and causing the anterior horn cell to backfire and stimulate the muscle again. Only a small proportion of motor neurons backfire and the pool that backfires varies from stimulus to stimulus and may not always represent the fastest conducting fibres. Allowance has to be made for the length of the nerve (usually in proportion to the height of the patient) and the refractory period of the proximal axon (about 1 ms). The technique provides a useful, non-invasive method of measuring

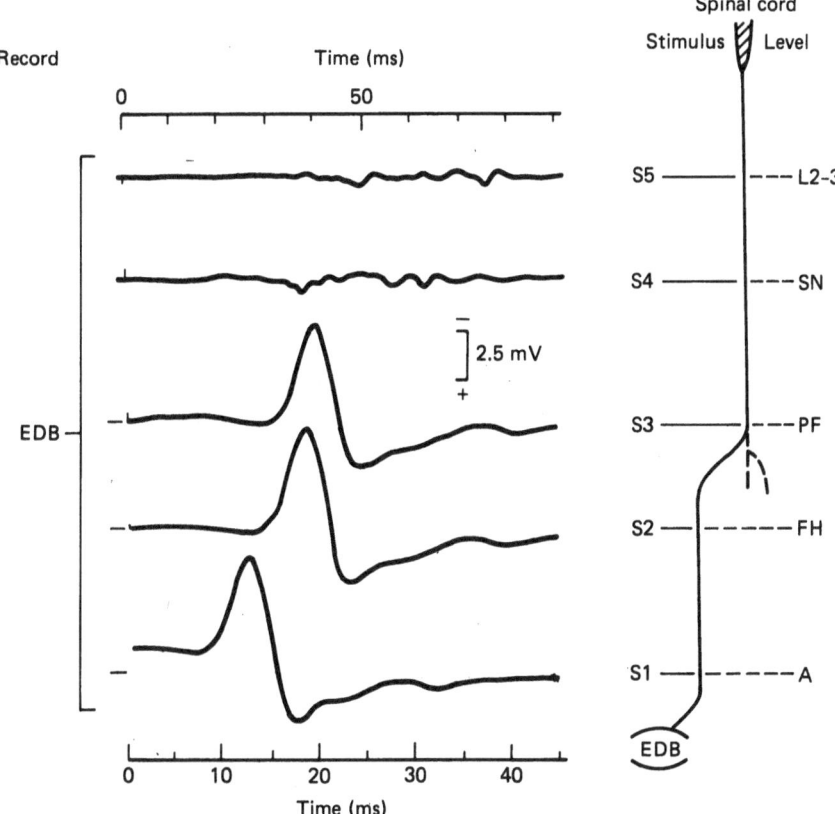

Fig. 7.3. Changes in the maximum M response in EDB on stimulation at the ankle (A), fibular head (FH), popliteal fossa (PF), sciatic notch (SN) and L2–3 roots. Even though there was more than 90% reduction in the EDB M potential between the sciatic notch and popliteal fossa, this was accompanied by a doubling of the total duration of the M potential (note change in time scales). The latter indicated a substantial degree of temporal dispersion, especially between the popliteal fossa and sciatic notch stimulus sites. Maximum motor conduction velocity between L2–3 and the sciatic notch, the sciatic notch and fibular head, and the fibular head and ankle were 37, 46 and 42 m/s respectively. The intermediate and proximal conduction velocities were much less than two standard deviations below the mean of control subjects. (From Brown and Feasby (1984), with permission.)

proximal motor conduction velocity. In four of nine patients (Kimura and Butzer 1975) and two of 11 patients (King and Ashby 1976) proximal motor conduction velocity was shown to be slowed by this technique in cases where forearm conduction velocity remained normal or relatively normal. It is possible to compare proximal with distal nerve conduction by stimulating the median or ulnar nerve at the elbow or the tibial nerve at the knee and expressing the proximal conduction latency (with a 1.0 ms allowance for the refractory period) as a ratio of the distal conduction latency. This so-called F ratio remained normal in half of 45 patients with GBS, was increased in a quarter and decreased in another quarter (Kimura 1978). In other words half the patients had similar abnormalities in proximal and distal nerve segments indicating a diffuse process, a quarter had proximally predominant and a quarter distally predominant slowing of conduction. F waves were often absent early in the development of

GBS. When F waves could be recorded they reached their longest latency 4–5 weeks after disease onset (Albers et al. 1985).

Sensory Nerve Conduction

Although positive and negative sensory symptoms are common at or soon after the onset of GBS, weakness is the dominant clinical sign. In the early stages sensory nerve action potential (SNAP) amplitudes are usually better preserved in both amplitude and latency than CMAPs. Furthermore, median SNAPs are more commonly abnormal (72%) than sural SNAPs (23%) (Albers et al. 1985). This accords with the frequency with which sural nerve biopsies are normal (Chapter 4). In generalised polyneuropathies sural SNAPs are more commonly abnormal than median SNAPs. This differing result in GBS re-emphasises the fact that the underlying pathological process is multifocal.

There is also evidence for proximal (nerve root) delay of sensory conduction in GBS. The latency of the potentials evoked from the brachial plexus and cervical cord by distal stimulation of the median nerve were abnormal slightly more frequently than F wave latencies in a study of 17 patients (Walsh et al. 1984). Conduction delay between Erb's point and the cervical cord was demonstrated in 10 of 11 patients within 2 weeks of onset of GBS when distal sensory nerve conduction velocity remained normal. These delayed latencies are probably attributable to slowed conduction in sensory nerve fibres in the cervical roots or brachial plexus. The results do firmly suggest that sensory conduction is impaired proximally more than distally, consistent with pathological involvement in or near the dorsal root ganglion, a site where the blood–nerve barrier is relatively impaired.

The results of a single case studied in detail conflict with this conclusion (Wexler 1983). A young woman with typical features of GBS developing over 28 days was studied on seven occasions from 17 to 95 days after onset. Median SNAPs (digit–wrist) and tibial (ankle–knee) and "sciatic" (knee-T12-L1 epidural interspace) nerve action potentials were recorded. At the first examination 17 days after onset the proximal (conus medullaris) action potential remained normal while the median and tibial nerve action potentials were dispersed and irregular. Over the next 11 days as the patient's clinical state worsened all the potentials became more dispersed including the conus potential. During recovery the conus potential reappeared first and eventually became normal whereas the median sensory and tibial nerve potentials remained delayed. This pattern indicated a centripetal pattern of involvement beginning peripherally and spreading proximally with recovery in the reverse order. As the author acknowledges, this single case may not be representative of GBS patients as a group and the conclusion is at variance with that derived from the sensory evoked potential studies.

A recent thought-provoking study distinguished two groups of GBS patients (Van der Meche et al. 1988). One group developed progressive conduction block of motor nerve fibres while SNAPs were usually preserved. This corresponds to the characteristic pattern of GBS recognised by other authors. In this group sensory impairment was variable and tendon reflexes were preserved until quite severe (MRC grade 3) weakness had supervened. The second group exhibited early uniform diminution of both distal and proximal CMAPs and SNAPs. The neurophysiological observations were more likely explained by distal conduction block than axonal degeneration since early recovery occurred. Clinically this group of patients had sensory as well as motor deficit and the tendon reflexes were lost at an early stage. It will be important to discover whether these patterns turn out to be homogeneous in different nerves within the same patient and whether similar patterns can be discovered in larger series.

Axonal Degeneration in GBS

Electromyographic studies have shown that fibrillation potentials and positive sharp waves develop in many patients with severe GBS between two and four weeks after disease onset. Such spontaneous activity occurs in proximal and distal muscles at about the same time, consistent with multifocal axonal damage. By contrast denervation changes in porphyric symmetrical polyneuropathy progress from proximal to distal, consistent with an origin of the axonal damage in the anterior roots (Albers 1987). The occurrence of axonal degeneration in patients with severe GBS who develop pronounced wasting and persistent weakness is evident from the clinical examination alone. A computerised electromyographic study of 17 patients with GBS demonstrated persistent reduction of motor unit numbers in the extensor digitorum brevis muscle several months after the disease had occurred. The remaining motor units had larger than normal amplitudes due to regenerating collateral axonal sprouts reinnervating the muscle fibres which had lost their attachment to the degenerated axons (Martinez-Figueroa et al. 1977). Rare patients with rapidly progressive tetraparesis and areflexia show the electrophysiological characteristics of an acute axonal neuropathy with inexcitable peripheral nerves and later profuse fibrillation potentials and no evidence of demyelination. Such a patient was recently described in which nerve biopsy confirmed the presence of axonal degeneration without primary demyelination or inflammatory cell infiltration (Feasby et al. 1986). A similar personal case was a child who had two episodes of acute neuropathy: in the first neurophysiological studies showed slowing of motor nerve conduction consistent with demyelination; following the second the peripheral nerves became inexcitable and biopsy and subsequent autopsy showed uniform axonal degeneration without significant lymphocytic infiltration in the peripheral nervous system. Such neurophysiological studies support evidence that the pathology of GBS is heterogeneous, usually demyelinating but rarely axonal. In the axonal form of the disorder it is particularly important to exclude toxic causes and porphyria.

Myokymia in GBS

Clinical myokymia, a spontaneous quivering of the facial muscles is occasionally seen in the acute stage of GBS (Daube et al. 1979). Subclinical electrophysiological myokymic discharges also have been recorded from limb muscles (Albers 1987). The myokymic discharges consist of spontaneous semirhythmic bursts of normal-appearing muscle action potentials. They occur in bursts of about 3–10 potentials at 30–60 Hz separated by intervals of 0.5–3 s. Although it has been proposed that the bursts might be caused by ephaptic transmission at a focus of demyelination it is difficult to see how this would lead to bursts of activity in an individual axon. The alternative proposal that partial axonal damage sometimes produces an unstable oscillating potential generator seems more likely.

Evoked Potentials

Mention has already been made of the delay in conduction of the sensory evoked potentials (SEPs) between Erb's point and the cervical cord commonly demonstrated in GBS. Studies of conduction time in central sensory conduction pathways are sometimes impossible because of proximal conduction block. Central conduction times of median or lower limb SEPs were usually normal, when they could be measured, in a study of 21 patients with acute GBS (Ropper and Chiappa 1986).

Brainstem and visual evoked potentials have been used as a non-invasive means of investigating possible CNS involvement in GBS (Ropper and Chiappa 1986). Of 21 patients 18 had normal brainstem auditory evoked potentials: two patients had increased I–III interwave latencies, one unilateral, the other bilateral, suggesting impaired conduction between the proximal portion of the eighth nerve and lower pons; one patient had a marginal unilateral delay of the III–V interwave latency. Studies of visual evoked responses were normal in nine of ten patients but showed bilateral delay in one patient with intracranial hypertension. In general such studies support the conclusion that the CNS is not directly affected in GBS.

Electrophysiological Evidence and Prediction of Prognosis

It is only to be expected that patients with denervation potentials indicating axonal degeneration will be more likely to be severely affected and take longer to recover. This expectation was fulfilled in the first thorough attempt to correlate neurophysiological evidence and prognosis. Among 50 patients studied in Vellore, India, 19 had profuse fibrillation detected in the first four weeks of the illness and only six improved whereas the 25 of the 31 without fibrillation recovered rapidly and well (Raman and Taori 1976). Other authors have confirmed the adverse prognostic implications of fibrillation (Eisen and Humphreys 1974; McLeod 1981). Since fibrillation may not appear until between two and three weeks after an axonal insult this sign is of limited use in assessing prognosis at the very early stage when treatment has to be decided. Recent papers have agreed that the amplitude of distally evoked muscle action potentials is a helpful prognostic indicator. In a prospective study of 100 patients with GBS 94 were assessed electrophysiologically. An absent or small (<1 mV) CMAP elicited by stimulating the median nerve at the wrist was associated with a delayed or incomplete recovery, or both (Table 7.2) (Winer et al. 1988). Analysis of neurophysiological results obtained during a large North American collaborative study of plasma exchange in GBS strongly supported a similar conclusion. The amplitudes of the distally evoked muscle action potentials were expressed as a percentage of the lower limit of normal of the measurement in each laboratory and all the values for an individual patient were averaged to provide a mean for that person. The distribution of the means for 210 patients for whom data were available was bimodal. One group had a mean distally evoked muscle action potential less than 20% of the lower limit of normal and the other group had a modal value about 60%. Only a fifth of the patients had a distally evoked CMAP amplitude of less than 20% of the lower limit of the normal value and these patients had a worse prognosis than those with larger potentials. Furthermore, this measure was a more powerful predictor of prognosis than the other variables examined, although proximally evoked CMAP amplitude and motor conduction velocity were also related to prognosis (Cornblath et al. 1988; McKhann et al. 1988).

There are conflicting reports of the relationship between conduction velocity at the acute stage of GBS and ultimate prognosis. Some of the earlier reports found an approximate correlation between slowing of motor conduction and eventual outcome (Eisen and Humphreys 1974; Takeuchi et al. 1984; Hausmanowa-Petrusewicz et al. 1979) but others did not (Raman and Taori 1976; McLeod et al. 1976). Slowing of phrenic nerve conduction was associated with a greater likelihood of death or poor recovery in a series of 18 patients (Gourie-Devi and Ganapathy 1985). In the recent larger studies severe slowing

Table 7.2. Electrophysiological measurements in the acute stage of GBS compared with outcome about 3 and 12 months after onset

Outcome	3 months		12 months	
	Poor[a]	Good	Poor	Good
APB MAP				
<1mV	10/18**	8/53**	11/24*	7/45*
absent	6/18***	1/53***	6/24*	1/45*
Median CV max				
<48 m/s	7/10	31/58	12/18	26/48
<40 m/s	6/10	19/58	11/18*	14/48*
L limb CV max				
<41 m/s	7/9	31/48	10/14	28/43
<30 m/s	1/9	10/48	2/14	9/43
Denervation	4/15	10/40	5/20	8/35

After Winer et al. (1988), with permission.
$*P<0.05$; $**P<0.01$; $***P<0.001$
[a] Poor prognosis after three months was defined as being bedbound or worse, good prognosis as ability to walk with assistance or better. Poor prognosis after 12 months was defined as inability to undertake manual work, good prognosis as either complete recovery or persistence of minor symptoms or signs.

of motor nerve conduction was associated with a greater likelihood of poor recovery (Table 7.2) (Winer et al. 1988; Cornblath et al. 1988).

Mechanisms

In GBS, conduction block is usually the main cause of weakness and is readily reversible. Slowing of nerve conduction is due to demyelination. In severe cases axonal degeneration occurs and is associated with delayed and often incomplete recovery. Our understanding of the precise neurophysiological mechanisms remains incomplete. For instance, why does the damage have a predilection for motor rather than sensory fibres? This predilection may extend to individual nerves. Thus progressive reduction of the proximally/distally evoked CMAP ratio of the median nerve of a patient during the acute stage of GBS was recorded while the corresponding ratio of the sensory nerve action potential remained normal (Fig. 7.4) (Van der Meche et al. 1988). This observation neatly indicated that motor nerve fibres were being blocked in the same segment of nerve in which sensory nerve fibres continued to conduct normally. A possible neurophysiological explanation is that the sensory nerve fibres largely responsible for the SNAP measured have a larger diameter than the motor fibres. The immune or inflammatory response causing the impairment of conduction may affect smaller nerve fibres preferentially, perhaps because fibres with thicker myelin sheaths have a greater safety factor for saltatory conduction from one internode to the next. Experimental support for this possibility has been provided. Anti-galactocerebroside serum applied to spinal root blocks conduction in smaller diameter axons before that in larger (Sumner 1981). Although there is little evidence of anti-galactocerebroside antibodies in the serum of patients with GBS it is possible that antibodies to unrecognised myelin or Schwann cell antigens might exert a similar effect. The hypothesis that the motor fibres are the primary target of the attack and the sensory fibres are affected as bystanders in untenable because macrophage-mediated demyelination is frequently observed in biopsies of the purely sensory sural nerve. A more plausible explanation for the discrepancy between motor and sensory nerve fibre conduction is untenable because macrophage-mediated demyelination is frequently observed sensory nerve myelin. Although slight differences in the myelin composition between dorsal root and ventral root myelin including a lower content of P_2 protein in the dorsal root (Kadlubowski et al. 1984) have already been described, none seem sufficient to explain the predominance of motor deficit in GBS. The fact that EAN can be induced with both dorsal and ventral root myelin is strong evidence against this argument.

Summary

Neurophysiological evidence of demyelination, consisting of marked slowing of motor nerve conduction, prolonged latencies of distally evoked CMAPs, or delayed F wave latencies appear in most patients with GBS during the first 3 weeks of the illness. In the third week 85% of patients show at least one of these

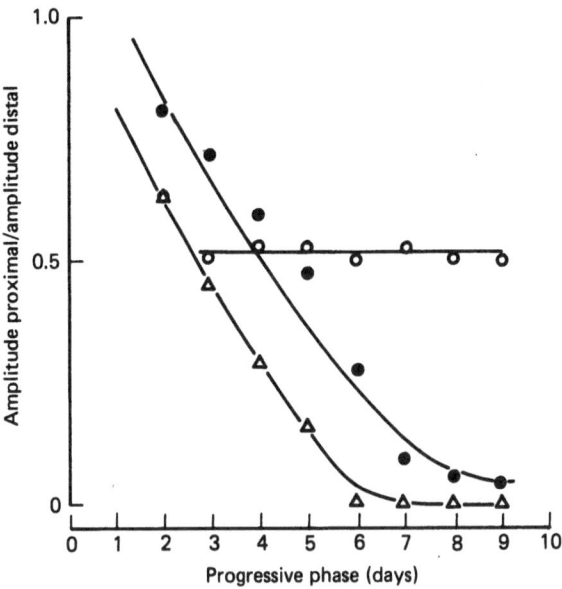

Fig. 7.4. Ratio of amplitudes with elbow and wrist stimulation and knee and ankle stimulation obtained during clinical deterioration in GBS. A length-dependent amplitude reduction is characterised by a decrease in the ratio as seen in the motor fibres of the median and peroneal nerves. No length-dependent reduction can be observed in the sensory fibres of the median nerve. For sensory nerves a ratio of 0.5 is within normal limits, the normal ratio for motor nerves is above 0.9. CMAP from abductor pollicis brevis (*filled circles*), CMAP from extensor digitorum brevis (*open triangles*), SNAP – second digit (*open circles*). (From Van der Meche et al. (1988), with permission.)

abnormalities. In the early stages weakness is largely due to conduction block. Upper limb SNAPs become reduced in amplitude slightly later and lower limb SNAPs are less commonly affected. Sensory nerve fibres may remain spared in the same nerve segment in which motor conduction is progressively blocked. The pattern of neurophysiological involvement is multifocal. Severely affected patients develop widespread fibrillation about three weeks after the onset and the fibrillation affects proximal and distal muscles simultaneously. If the amplitude of the distally evoked CMAP is small (less than 20% of the lower limit of normal) there is a significantly greater chance of a slow and incomplete recovery. The explanation of the predominantly motor involvement in GBS remains unclear but might be related to the size distribution of motor fibres or unidentified immunological differences between myelin of motor and sensory nerve fibres. Evoked potential studies of central sensory, auditory and visual pathways usually give normal results.

References

Albers JW (1987) Inflammatory demyelinating polyradiculoneuropathy. In: Brown WF, Bolton CF (eds) Clinical electromyography. Butterworths, Boston, pp 211–244
Albers JW, Kelly JJ (1989) Acquired inflammatory demyelinating polyneuropathies: clinical and

electrodiagnostic features. Muscle Nerve 12:435–451

Albers JW, Donofrio PD, McGonagle TK (1985) Sequential electrodiagnostic abnormalities in acute inflammatory demyelinating polyradiculoneuropathy. Muscle Nerve 6:504–509

Bannister RG, Sears TA (1962) The changes in nerve conduction in acute idiopathic polyneuritis. J Neurol Neurosurg Psychiatry 25:321–328

Brown WF, Feasby TE (1984) Conduction block and denervation in Guillain–Barré polyneuropathy. Brain 107:219–239

Cornblath DR, Mellits ED, Griffin JW et al. (1988) Motor conduction studies in Guillain–Barré syndrome: description and prognostic value. Ann Neurol 23:354–359

Daube JR, Kelly JJ, Martin RA (1979) Facial myokymia with polyradiculoneuropathy. Neurology 29:662–669

Eisen A, Humphreys P (1974) The Guillain–Barré syndrome. A clinical and electrodiagnostic study of 25 cases. Arch Neurol 30:438–443

Feasby TE, Brown WF, Gilbert JJ, Hahn AF (1985) The pathological basis of conduction block in human neuropathies. J Neurol Neurosurg Psychiatry 48:239–244

Feasby TE, Gilbert JJ, Brown WF et al. (1986) An acute axonal form of Guillain–Barré polyneuropathy. Brain 109:1115–1126

Gourie-Devi M, Ganapathy GR (1985) Phrenic nerve conduction time in Guillain–Barré syndrome. J Neurol Neurosurg Psychiatry 48:245–249

Hausmanowa-Petrusewicz I, Emeryk B, Rowinska-Marcinska K, Jedrzejowska H (1979) Nerve conduction in the Guillain–Barré–Strohl syndrome. J Neurol 220:169–184

Kadlubowski M, Hughes RAC, Gregson NA (1984) Spontaneous and experimental neuritis and the distribution of the myelin protein P_2 in the nervous system. J Neurochem 42:123–129

Kimura J (1978) Proximal versus distal slowing of motor nerve conduction velocity in the Guillain–Barré syndrome. Ann Neurol 3:344–350

Kimura J, Butzer JF (1975) F wave conduction velocity in Guillain–Barré syndrome. Arch Neurol 32:524–529

King D, Ashby DW (1976) Conduction velocity in the proximal segments of a motor nerve in the Guillain–Barré syndrome. J Neurol Neurosurg Psychiatry 39:538–544

Lambert EH, Mulder DW (1964) Nerve conduction in the Guillain–Barré syndrome. Electroencephalogr Clin Neurophysiol 17:86–93

Martinez-Figueroa A, Hansen S, Ballantyne JP (1977) A quantitative electrophysiological study of acute idiopathic polyneuritis. J Neurol Neurosurg Psychiatry 40:156–161

McKhann GM, Griffin JW, Cornblath DR et al. (1988) Plasmapheresis and Guillain–Barré syndrome: analysis of prognostic factors and the effect of plasmapheresis. Ann Neurol 23:347–353

McLeod JG (1981) Electrophysiological studies in the Guillain–Barré syndrome. Ann Neurol 9:20–27

McLeod JG, Prineas JW, Walsh JC (1973) The relationship of conduction velocity of pathology in peripheral nerves. In: Desmedt JE (ed) New developments in electromyography and clinical neurophysiology, Vol 2. Karger, Basel, pp 238–258

McLeod JG, Walsh JC, Prineas JW, Pollard JD (1976) Acute idiopathic polyneuritis – a clinical and electrophysiological follow-up study. J Neurol Sci 27:145–162

Mills KR, Murray NMF (1985) Proximal conduction block in early Guillain–Barré syndrome. Lancet ii:659

Raman PT, Taori GM (1976) Prognostic significance of electrodiagnostic studies in the Guillain–Barré syndrome. J Neurol Neurosurg Psychiatry 39:163–170

Ropper AH, Chiappa KH (1986) Evoked potentials in Guillain–Barré syndrome. Neurology 36:587–590

Sumner J (1981) The physiological basis for symptoms in Guillain–Barré syndrome. Ann Neurol 9 Suppl:28–30

Takeuchi H, Takahishi M, Kang J, Keno S, Yamada A, Miki H (1984) The Guillain–Barré syndrome; clinical and electroneuromyographic studies. J Neurol 231:6–10

Van der Meche FG, Meulstee J, Vermeulen M, Kievit A (1988) Patterns of conduction failure in the Guillain–Barré syndrome. Brain 111:405–416

Walsh JC, Yiannikas C, McLeod JG (1984) Abnormalities of proximal conduction in acute idiopathic polyneuritis: comparison of short latency evoked potentials and F waves. J Neurol Neurosurg Psychiatry 47:197–200

Wexler I (1983) Sequence of demyelination–remyelination in Guillain–Barré disease. J Neurol Neurosurg Psychiatry 46:168–174

Winer JB, Hughes RAC, Osmond C (1988) A prospective study of acute idiopathic neuropathy. I Clinical features and their prognostic value. J Neurol Neurosurg Psychiatry 51:605–612

Chapter 8

Treatment of Guillain–Barré Syndrome

Prognosis

Treatment required for GBS ranges from the full panoply of modern intensive care in severely paralysed ventilated patients to gentle reassurance in an out-patient setting in patients with minor sensory symptoms and signs. It would be desirable to predict at an early stage which patients were likely to become severely affected and take a long time to recover. Treatment could then be concentrated on this poor prognosis group, leaving patients who were destined to make a reasonably rapid and complete recovery untreated. Unfortunately assessing prognosis at an early stage is difficult. Attempts have been made to define a pure form of GBS with a uniformly benign prognosis by insisting on absence of sensory deficit and sphincter disturbance (Osler and Sidell 1960). In other studies the presence of sensory deficit and sphincter disturbance bore no relation to outcome (Marshall 1963). In general and not surprisingly the more severe the motor deficit at the nadir of the disease the greater has been the risk of residual disability (Peterman et al. 1959; Pleasure et al. 1969; Löffel et al. 1977). Patients whose weakness is so severe that they require ventilation have a distinctly worse prognosis than those who do not. The outcome of 71 carefully documented patients who had taken part in therapeutic trials was studied: 20 died or were left with persistent deficit after 12 months and of these 16 (80%) had required ventilation. On the other hand 51 recovered to be able to work again and only 13 (26%) had required ventilation ($P<0.001$) (Winer et al. 1985). Similarly in a prospective study 59% of 32 patients who required ventilation and only 22% of 64 who did not were left with persistent deficit ($P<0.001$) (Winer et al. 1988) (Table 8.1).

Muscle wasting has been noted more often in patients who were left disabled (Löffel et al. 1977; McLeod 1981). Wasting usually indicates the occurrence of axonal degeneration so that it is not surprising that electrophysiological signs of denervation in the acute phase should be association with a poor outcome (McLeod 1981 and Chapter 7). In studies in which electromyography has been undertaken widespread profuse fibrillation and positive sharp waves have been found to predict a poor outcome (Albers et al. 1985). In other studies a small muscle action potential amplitude evoked by distal stimulation of a nerve has also correlated with poor outcome. In one study (Winer et al. 1988) 71% of 24 patients who had abductor pollicis brevis muscle action potential amplitudes less than 1 mV had a poor outcome compared with only 18% of 45 who had larger muscle action potential amplitudes ($P<0.0001$). Analysis of a North American trial of plasma exchange in 245 patients with GBS showed that a small distally

Table 8.1. Analysis of relationship between clinical features at initial assessment and outcome after 12 months

Clinical feature	Percentage of poor outcome patients who had the feature in the acute stage ($n = 32$)	Percentage of good outcome patients who had the feature in the acute stage ($n = 64$)
Ventilation	59***	22***
Age >40 years	88**	53**
Bedbound	100**	73**
Bedbound within 4 days	75*	46*
Muscle wasting <28 days	52**	25**
Some tendon reflexes retained	0	25**
Time to improvement >21 days	59	45

After Winer et al. (1988), with permission.
* $P<0.05$; ** $P<0.01$; *** $P<0.001$
Poor outcome was defined as inability to do manual work including housework, shopping or gardening after one year. Good outcome was any state better than that. The poor outcome group included 13 who died.

evoked muscle action potential was the most predictive clinical or electrophysiological variable (Cornblath et al. 1988). Markedly slowed motor nerve conduction velocity also predicted poor outcome but less powerfully in both studies. Although measurement of the distally evoked muscle action potential has been found useful as a prognostic factor, the physiological significance of a small or absent distally evoked muscle action potential is ambiguous. It could be due to distal conduction block or to axonal degeneration. In the studies in which this relationship has been found between a small distally evoked muscle action potential and prognosis the studies had been performed early in the course of the disease (within two weeks in 80% of the cases of Cornblath et al. (1988)). The distally evoked muscle action potential becomes reduced in more and more patients later in the course of GBS (Albers et al. 1985) and it is not known whether reduction or loss of the distally evoked muscle action potential at a later stage has the same adverse effect on prognosis.

Just as GBS becomes more common with increasing age, so the prognosis becomes worse: patients over 40 fared significantly worse than those under 40 in the study by Winer et al. (1988) and older patients did worse in the North American plasma exchange trial (McKhann et al. 1988). Although children have fared well in some series with little residual disability and low mortality (Rossi et al. 1976; Cole and Matthew 1987), a recent comparison found that the outcome of 18 children was similar to that of 50 adults seen at a single Dutch secondary referral centre. The duration of ventilation and hospitalisation was the same and two children died (Kleyweg et al. 1989). Male or female sex does not influence prognosis (Winer et al. 1988; McKhann et al. 1988).

My clinical impression that some patients with an explosive onset of their disease are more likely to develop axonal degeneration, recover slowly and be left with disability has been difficult to substantiate, although this impression is shared by others (Ropper 1986). In a retrospective analysis of 20 good and 51

Table 8.2. Relative risks of poor outcome after 12 months attributable to the four most discriminating features

	Relative risk	95% confidence intervals
Age > 40 years	4.4	1.2–16.4
Disability grade 4 in <4 days	3.1	1.0–9.6
Requiring ventilation	2.6	0.9–7.6
APB MAP[a] (mV)	3.8	1.0–14.0

From Winer et al. (1988), with permission.
[a] APB MAP, abductor pollicis brevis muscle action potential evoked by stimulating the median nerve at the wrist at the initial assessment.

poor outcome patients 35% of the poor outcome patients had reached their worst state in less than 4 days compared with 14% of the good outcome group, but these proportions are not significantly different (Winer et al. 1985). In a larger prospective study (Winer et al. 1988) a rapid onset, defined as becoming bedbound within four days, was associated with poor outcome (Table 8.1).

Many other features have been examined for a possible relationship with outcome and none has been found. The catalogue includes CSF cell count, CSF protein concentration, previous infection, immunisation or operation, sensory deficit, facial weakness or bladder involvement (Ravn 1967; McKhann et al. 1988; Winer et al. 1988). Mildly affected patients in whom some tendon reflexes were retained had a better prognosis in one study (Table 8.1).

To summarise, old age, rapid onset, severe disease requiring ventilation, and small distally evoked muscle action potentials are all adverse prognostic factors (Table 8.2). The relative risks of a poor outcome have been deduced from these four features and can be used to estimate the prognosis of untreated patients (Table 8.3). This predictive model has not yet been validated on a second population of patients and can only be used as a general guide. A similar predictive model has been constructed from the North American Guillain-Barré steroid trial data (McKhann et al. 1988).

Table 8.3. Predicted probabilities of poor outcome at 12 months in percentages using the four most discriminating variables

	Bedbound 0–4 days		Bedbound >4 days	
	Age <40 yr	Age >40 yr	Age <40 yr	Age >40 yr
APB MAP[a] <1 mV				
Not ventilated	36	71	15	44
Ventilated	59	86	31	67
APB MAP >1 mV				
Not ventilated	12	39	4	17
Ventilated	27	62	11	34

From Winer et al. (1988), with permission.
[a] APB MAP, abductor pollicis brevis muscle action potential evoked by stimulating the median nerve at the wrist at the initial assessment.

General Care

Good general medical and nursing care are a *sine qua non* without which plasma exchange or other forms of intervention will not be successful. From the time of presentation while the diagnostic process is being worked through it is essential to consider the factors which might cause permanent or fatal complications. The wide range of mortality quoted in different series, from 2%–3% in North America to 13% in the UK and France (Chapter 6), may partly reflect the standard or availability of intensive care. The causes of death in 13 patients from a series of 100 in a UK prospective study are listed in Table 8.4. There will always be a residue of fatal complications occurring in the context of a serious paralysing illness which cannot be avoided but careful attention must be paid to recognising respiratory failure, avoiding and treating atelectasis and chest infections, monitoring and dealing with cardiac arrhythmias and preventing pulmonary embolism (Ferner and Hughes 1987).

Early respiratory failure due to neuromuscular paralysis is deceptive but its prompt recognition is essential (Ropper and Shahani 1984). Doctors and nurses are used to seeing patients who are cyanosed, wheezing and obviously distressed by breathlessness from airway obstruction and lung disease. Patients with weakness of the neuromuscular bellows action of the chest wall may rapidly decompensate and have a respiratory arrest. Therefore, deteriorating breathing capacity must be monitored and artificial ventilation instituted prophylactically. As the respiratory muscles weaken the tidal volume and vital capacity fall and the patient becomes incapable of the sighing, intermittent deep breaths, which normally keep the peripheral alveoli expanded. Coughing becomes feeble. If a bulbar palsy is present, as it often is in GBS, the ability to protect the airway from inhaled secretions is compromised and food and drink make the patient splutter ineffectually. Atelectasis of peripheral alveoli occurs causing arterio-

Table 8.4. Causes of death within 12 months of onset of Guillain–Barré syndrome; prospective study of 100 cases in the UK in 1983–4

Age (years)	Major cause of death	Interval from onset (weeks)
59	Cardiac arrest	52
73	Cardiac arrest	4
62	Chest infection	7
74	Pulmonary embolus	1
78	Chest infection	5
72	Renal failure	2
78	Respiratory failure	4
76	Cardiac arrest	5
58	Cardiac arrest	3
66	Chest infection	8
67	Aortic aneurysm	2
63	Suicide	40
50	Pancreatic carcinoma	49

From Winer et al. (1988), with permission.

venous shunting and hypoxia. The respiratory rate increases to compensate for the hypoxia. The chest radiograph remains normal. The respiratory muscles have to work harder to compensate for the hypoxia, and if they are affected by the neuropathic process they may become fatigued. By this stage the patient may be feeling breathless and perspiring. The respiratory rate is increased, the accessory respiratory muscles including the muscles of the external nares and the sternocleidomastoid come into play. The abdominal muscles are sucked paradoxically inwards with each inspiration because of diaphragmatic weakness. Sentences are interrupted to take breaths and coughing sounds feeble and becomes ineffective.

A patient demonstrating all these symptoms and signs urgently needs intubation and respiratory support. The most useful laboratory aid to the assessment of respiratory failure is measurement of the vital capacity. Waiting for the blood gases to deteriorate before initiating ventilatory support is a recipe for disaster since hypoxia may develop precipitously and cause a respiratory arrest because of progressive atelectasis, sputum plugging or inhalation of fluid. Measurements of peak expiratory flow rate are often incorrectly substituted because of the ready availability of peak expiratory flow meters: it is not slowed airflow but inadequate amounts of air breathed which are a risk to life in GBS. The vital capacity can be measured with any available apparatus: a vitalograph or Wright's respirometer are both adequate but the author uses a small portable respirometer with a digital display. Technique of vital capacity measurement is important. The vital capacity manoeuvre has to be carefully explained to the patient. Measurements are prone to inter-observer error. If the patient has facial weakness attempts to get an adequate lip seal are difficult but measurements can be made by attaching the respirometer to a facemask. By the time the vital capacity has fallen from the normal of about 90 ml/kg to 15 ml/kg (i.e. 1000 ml in the average adult) there is a critical danger of respiratory muscle fatigue and respiratory arrest so that the patient should be intubated and ventilated. The decision to ventilate has to be taken on the strength of all the clinical findings and the vital capacity readings have to be interpreted in the context of other signs of respiratory failure and the speed of deterioration.

Patients are usually more comfortable intubated with a nasotracheal rather than an orotracheal tube. Suxamethonium should not be used as the paralysing agent during intubation because of reports of ventricular tachycardia caused by a rise in plasma potassium when it has been used in patients with denervated muscles due to chronic or recurrent polyneuropathy (Fergusson et al. 1981). It is usually appropriate to pass a nasogastric tube at the same time as the intubation: this should have a large enough bore to permit aspiration of the stomach contents since gastric stasis and ileus are common in the early stages.

Ventilation should be performed with a volume cycled machine set to deliver large tidal volumes which will help to prevent atelectasis. The ventilator should be set to support the patient's own breathing by using the synchronised intermittent mandatory ventilation (SIMV) mode. Volumes of 90 ml/kg delivered at six breaths of 15 ml/kg per minute have been recommended (Ropper and Shahani 1984). The settings must be adjusted to maintain normal blood gases: hyperventilation would cause respiratory alkalosis and exacerbate anxiety. The use of SIMV mode facilitates weaning from the ventilator when the ventilator rate may be gradually reduced. The vital capacity should be monitored daily while the patient is on the ventilator as an index of progress. There is no difficulty about taking the patient off the ventilator briefly, attaching a vital capacity meter to the endotracheal tube, and asking the patient to perform the vital capacity manoeuvre. Most patients can cooperate with this procedure and many modern ventilators have an appropriate meter in their circuit. By the time the vital capacity has risen again to 15 ml/kg it should be possible to start weaning the patient off the ventilator.

The median time of ventilation was 24 days even in the patients treated with plasma exchange in the North American trial (Guillain–Barré Syndrome Study Group 1985). Since a nasotracheal or orotracheal tube will only be tolerated for a maximum of about 12 days because of necrosis of the nasal mucosa and oedema of the vocal cords, it is usually appropriate to arrange a tracheostomy at an early stage. Patients are usually much less distressed after a tracheostomy has been substituted for the oro- or nasotracheal tube.

Meticulous ventilator care and chest physiotherapy are essential to avoid atelectasis and sputum plugging. Every hour the patient should be taken off the ventilator and the chest fully inflated with several large breaths with a hand-held bag using 100% oxygen. The trachea should then be gently sucked clear of sputum with a soft blunt tipped catheter. Such suction, though necessary to prevent sputum plugging, carries the risk of stimulating the vagus and causing asystole. This risk is reduced by the hyperoxygenation. Regular turning also helps to reduce the risk of segmental lung collapse since the weight of the viscera will tend to cause atelectasis of the dependent segments in any particular position. This can lead to the "elephant lung" syndrome in which first one and then another lobe collapses depending on the patient's position. The term arises from the grave problems which arise when anaesthetising an elephant, a species which is particularly prone to lung collapse during anaesthesia for precisely the same reason! Antibiotics should be used as appropriate, guided by both clinical signs of chest infection and purulent sputum containing organisms.

Because of the risk of cardiac arrhythmias, especially but not only in ventilated patients, all patients should have continuous electrocardiographic monitoring from admission until the nadir of deficit has been reached or well after the patient has been weaned off the ventilator. The most dangerous of the cardiac arrhythmias are the episodes of bradycardia and even sinus arrest which most commonly occur during the tracheal toilet. These can be prevented to some extent by previous hyperoxygenation and gentle suction but in some patients they persist and are the cause of occasional deaths in the acute stage (Winer and Hughes 1988). Tracheal suction provides a massive vagal stimulus which is normally checked by sympathetic activity brought into play by information transmitted via myelinated afferent nerve fibres from the baroreceptors. In GBS this afferent limb of the reflex is often damaged and the vagal stimulus is unchecked. Atropine is the obvious antidote. However, the dose of atropine to block cardiac vagal activity is so large that it may cause gastroparesis, ileus, drying of respiratory secretions and confusion. Accordingly if episodes of profound bradycardia or sinus arrest do occur during tracheal suction it is the author's practice to insert a temporary endocardial pacemaker. This can be left in situ for about 12 days by which time the episodes of bradycardia associated with tracheal suction have usually stopped.

Tachycardias are usually best monitored closely and left untreated unless the circulation becomes compromised. Sinus tachycardia is extremely common and rarely harmful on its own. The use of betablockers to control tachycardia and hypertension is discouraged because of their tendency to aggravate sinus brady-cardia. Episodes of paroxysmal atrial fibrillation or other supraventricular tachy-cardias need to be treated appropriately. Hypertension can be controlled by calcium channel antagonists. Swings of hypertension and hypotension can occur abruptly and be difficult to manage: they may herald cardiac arrest (Winer and Hughes 1988). Such swings may be due to inadequate sedation or occur in response to conventional doses of sedatives to which GBS patients are some-times supersensitive. In seriously ill ventilated patients intra-arterial pressure and central venous pressure are always monitored continuously. Since right atrial pressures do not always reflect left atrial pressure, additional monitoring of pulmonary capillary wedge pressure has been recommended (Dalos et al. 1988).

Another occasional cause of death is venous thrombosis leading to pulmonary embolism (Bredin 1976; Leese 1976). To avoid this subcutaneous heparin 5000 units twice daily is used routinely in addition to regular passive movements of the lower limbs and anti-embolism stockings. Also, if possible, the femoral route is not used for plasma exchange because of the risk that the femoral vein puncture will lead to thrombosis. An anticoagulant regime is not necessary in children in whom venous thrombosis and pulmonary embolism are rare.

Although guarding against the fatal respiratory, cardiac and thrombotic com-plications is essential, every aspect of intensive care is important in helping patients with severe GBS through the acute stage. Perhaps the most important is reassurance: the disease needs to be explained clearly so that the patient under-stands the probable deterioration, eventual improvement and hoped for, but not invariable, complete recovery. This reassurance needs to be repeated constantly and consistently and advice about the probable duration and extent of improve-ment has to remain guarded in patients in whom a poor outcome is relatively unlikely (Table 8.2). Over-optimistic estimates of time to wean off the ventilator or recover independent function lead to loss of confidence and morale. Visits from someone who has recovered from GBS are often appreciated and are best supervised by a member of the team looking after the patient. Sedation and analgesia in ventilated patients can be achieved with intravenous infusions of papavaretum and additional bolus injections of midazolam. Patients with GBS may show exaggerated responses to barbiturates, opiates and other sedatives so that doses must be titrated carefully.

Preserving two-way communication can be difficult in an intubated patient with a facial and bulbar palsy. However, most patients are left with some ability to nod and communication is often best maintained with a simple signal for *yes* and *no* and then a game of "twenty questions". It is essential for doctors, nurses and relatives to use the same code or else confusion will reign and frustration will intensify. More elaborate phrase boards with common wants such as "please may I have a drink", "please can I empty my bladder", "I have a pain" or letter boards are available. More sophisticated messages can be communicated by pointing with the eyes at letters on an alphabet board. Such messages require perseverance and intelligence on the part of patient and attendant: one or other (usually the patient!) usually has these admirable qualities and both usually resort to the "twenty questions" guessing game. A nasotracheal tube leaves the lips free to mouth words more easily than with an orotracheal tube. A trac-heostomy is more comfortable than either. Several different models of cuffed tracheotomy permitting phonation have been introduced (Editorial 1987). They have an extra channel opening above the cuff to which an external humidified air supply is connected. These devices might help GBS patients who have preserved bulbar or vocal cord function.

In patients with bulbar palsy and impairment of swallowing, a nasogastric tube has to be passed. It is best to start with a conventional nasogastric tube which will permit aspiration of gastric contents because gastroparesis and ileus are quite common in severely affected patients who have been heavily sedated during their first period on the ventilator. During this initial period parenteral feeding via a subclavian line may be necessary and can be achieved with specially prepared bags which contain appropriate calories, electrolytes, and vitamins (Table 8.5). Fortunately nasogastric feeds such as Clinifeed will usually be tolerated quickly. Feeds are best started at half strength in small volumes of 50 ml every 2 hours and then gradually increased to about 100 ml every hour with the aim of giving 1600 kcal/day for basal metabolism or more for a hyper-metabolic patient with infection. Extra iron to compensate for venepunctures and potassium to compensate for losses associated with increased aldosterone secretion in response to stress need to be given. Plasma exchange may cause transient hypocalcaemia requiring extra calcium. Hypophosphataemia may be induced by administration of large amounts of intravenous glucose and need correction. Fortunately, kidney and liver function usually remain normal and the metabolic problems associated with GBS are not great. An exception is the occasional patient with profound and persistent hyponatraemia which is prob-ably due to inappropriate secretion of ADH. This can usually be managed by restricting water intake and giving saline. Rare cases may need demeclocycline.

Correct positioning, turning, an appropriate bed and physiotherapy are important to minimise discomfort. Severely paralysed patients cannot make the constant changes in position which we all make subconsciously to keep ourselves comfortable. To overcome this and prevent the development of pressure sores, repositioning from side to back to side at hourly intervals is necessary. Com-pletely paralysed patients prefer large cell air beds (such as Mediscus) or special resin filled beds (such as Clinitron) which reduce the need for turning. As strength returns an ordinary bed becomes more suitable again so that the patient can have a firm base on which to move around. A programme of passive physio-

Table 8.5. Total parenteral nutrition

Constituents of a 2.5 litre bag infused over 24 hours via a central line		
Volume	2.5 litres	Additives:
Nitrogen	14.0 g	Folic acid 15 mg 1 × week
Non-nitrogen energy	2200 kcals	Water-soluble vitamins
Glucose	300 g	B complex, C (5 × week)
Lipid	100 g	Fat-soluble vitamins
Sodium	73 mmol	A, D, K (2 × week)
Potassium	60 mmol	
Phosphate	30 mmol	
Calcium	7.5 mmol	
Magnesium	14 mmol	
Zinc	90 µmol	
Manganese	40 µmol	
Chloride	103 mmol	
Acetate	100 mmol	

therapy must be organised to put all joints through a full range of movement several times each day to prevent contractures. Careful positioning of the limbs by day and night must be designed to prevent shortening of muscles, especially gastrocnemius, and to avoid pressure on nerves especially the ulnar nerve at the elbow and the common-peroneal at the fibular head. As recovery begins assisted active movements can be started and the patient gradually helped towards sitting, standing and then walking with support under the supervision of a physiotherapist.

Pain can be a problem throughout the course of GBS. Painful paraesthesiae and pain from muscles which are becoming paralysed are quite common during the acute stage. Discomfort from not being able to move and pain from joints which are wrenched accidentally during turning manoeuvres is common at the height of paralysis. Painful paraesthesiae replace anaesthesia during recovery and provide the final insult. Careful reassurance that pain is a usual part of the disease is an essential basis for its treatment but is certainly not enough. Many of the points made about positioning, beds and physiotherapy are directed towards the relief of pain. During the acute stage in the intensive care unit paparavetum infusions are appropriate and usually sufficient. Conventional analgesics, such as paracetamol or codeine, or non-steroidal anti-inflammatory agents, such as ibuprofen, are all worth trying. Quinine has been recommended (Nixon 1978) but I have never found it helpful. Severe pain has been reported to be dramatically relieved by a single intramuscular dose of methylprednisolone (Ropper and Shahani 1984).

In severely affected patients urine retention may occur at the nadir of the disease and even in those without bladder symptoms it may be appropriate to insert an indwelling catheter for convenience of nursing. In the male this is less trouble and less uncomfortable than a penile sheath drainage device. In the female it is the only practical proposition. Obviously an indwelling catheter carries the risk of infection but with meticulous care this can be reduced. Urine specimens should be cultured at intervals during and after catheterisation and infection treated appropriately if the urine contains pus cells and bacteria.

Steroids

Although steroids are the most effective anti-inflammatory and immunosuppressive agents available their efficacy in GBS has not been established. The first report of the use of cortisone was discouraging in that the authors reported an apparent lack of response in three cases (Shy and McEachern 1951). Shortly afterwards the first of several reports appeared claiming a beneficial effect of ACTH and cortisone (Stillman and Ganong 1952). The first three years experience was summarised by the statement that the response to steroids was occasional, unpredictable and possibly only coincidental (Plum 1953). A later review of the early literature found that 50 of 68 patients treated with steroids had "responded" and only two had died, which seemed a better record than the mortality ranging between 9%–42% in previous series (Jackson et al. 1957). The pendulum of opinion about the benefit of steroids continued to swing backwards and forwards as more series were published. Eighteen patients treated with ACTH or cortisone were considered to have recovered more quickly than 17 not so treated but treatment was not randomised and details of the untreated group were not given (Heller and de Jong 1963). Thirty-seven patients treated with 30–50 mg prednisolone daily began to improve more quickly than 29 non-steroid treated patients but this benefit was offset by a longer hospital stay in the steroid-treated group and again treatment was not randomly allocated (Frick and Angstwurm 1968). Another study reached a similar conclusion: the plateau phase was shortened in 22 steroid-treated patients compared with 68 non-steroid treated patients but the overall disease duration was not altered and one steroid patient had osteoporosis and another gastrointestinal haemorrhage (Wiederholt et al. 1964). Twenty-three patients treated with steroids had a longer duration of disease than 13 not so treated, and a subgroup with mild-to-moderate disease had significantly longer disease (Goodall et al. 1974). Other series reported no difference between patients treated with cortisone (Löffel et al. 1977) or prednisolone (Samantray et al. 1977) and controls. The early reports of steroid usage have been reviewed by Hughes et al. (1981). All commentators agreed that a controlled trial was essential to determine whether steroids are helpful in GBS.

Unfortunately the available controlled trials of steroids have been too small to identify a modest beneficial effect. In the first, eight patients with mild or moderate GBS were randomised to receive ACTH 100 units intramuscularly daily for 10 days and eight patients to receive placebo (Swick and McQuillen 1976). Although there was no obvious immediate improvement the duration of disease in the ACTH-treated patients was 4.4 ± 2.5 months compared with 9.0 ± 5.2 months in the non-treated patients. This difference was considered significant by the authors but the analysis was not conducted on an intention to treat basis and the significance would have been lost if a patient in the ACTH group who left hospital against advice and died had been retained in the analysis.

The largest available randomised controlled trial tested the effect of oral prednisolone 60 mg daily for one week, 40 mg daily for four days, 30 mg daily for three days and then prednisolone continued or reduced at discretion against placebo (Hughes et al. 1978). The trial was conducted double-blind and used a six-point disability scale which has formed the basis for several subsequent trials (Table 8.6). At randomisation the 21 eligible prednisolone-treated patients did not differ significantly from the 19 control patients in sex ratio, age, duration of disease (median 7.5 and 8 days in the two groups), disability grade, CSF cell count or CSF protein concentration. The only outcome comparison which favoured the prednisolone-treated group was that improvement began slightly earlier (Table 8.7). There was even a suggestion that those patients who were

Table 8.6.   Scale of disability grades used in randomised trial of prednisolone

---

0 = Healthy
1 = Minor signs or symptoms of neuropathy but capable of manual work
2 = Able to walk without support of a stick but incapable of manual work
3 = Able to walk with a stick, appliance or support
4 = Confined to bed or chairbound
5 = Requiring assisted ventilation
6 = Dead

From Hughes et al. (1978).

Table 8.7. Comparisons of groups in the randomised trial of prednisolone

Findings	Prednisolone ($n = 21$)	Control ($n = 19$)
Median time to onset of improvement (days)	8	10
Number of patients with residual minor signs	17	13[a]
Number of patients with residual inability to do manual work	7	2[a]
Number of patients who relapsed within 1 year	3	0
Number of deaths	1	1[a]

From Hughes et al. (1978), with permission.
[a] Excluding one suicide during convalescence.

Table 8.8. Improvement in disability grade 3 months after entry in the randomised trial of prednisolone

Delay (days) from onset to entry	n	Prednisolone mean improvement	n	Control mean improvement
<7	10	0.9*	6	2.5*
8–14	11	1.0	14	2.0
>15	10	0.9	5	1.2

From Hughes et al. (1978).
* $P < 0.05$.

randomised to receive prednisolone early, within the first week after the onset of neuropathic symptoms, fared worse than the placebo-treated patients. The average disability grade improvement was actually less in the prednisolone than the placebo-treated patients (Table 8.8). Another source of concern was that there were three relapses among the prednisolone-treated patients and none among the controls. In a series of 256 prednisolone-treated patients six had relapses compared with none of 46 non-steroid-treated patients (Samantray et al. 1977). These reports of relapses in steroid-treated patients are a source for continued concern but the numbers are too small to be significant. In a more recent even smaller trial plasma exchange and prednisolone 100 mg daily for 10 days was given to 13 patients and 12 patients received neither treatment but there was no difference in outcome (Mendell et al. 1985).

Anecdotal reports have appeared of dramatically successful treatment with higher doses of steroids. One patient with rapidly progressive GBS was given methylprednisolone intravenously daily and improved unusually rapidly and similar but less dramatic improvement occurred in 2 of 4 other patients (Dowling et al. 1980). Possible benefit from high dose steroids was also reported in another small series (Haass et al. 1988). In one case intravenous methylprednisolone appeared effective when oral prednisolone had not (Brumback 1980). An international multi-centre trial of intravenous methylprednisolone 500 mg daily for five days versus placebo is as present being conducted. It is planned to recruit

240 patients randomised between the two groups and to measure the outcome on a modified version of the disability grade scale (Table 8.9) and on an arm disability grade scale (Table 8.10). In the meantime the conclusion has to be that the small trials available make it unlikely that conventional doses of oral steroids have a major beneficial effect but the power of the trials has not been sufficient to identify a modest effect. Theoretical arguments suggest that large doses given early might damp down the inflammatory response and reduce the amount of demyelination and axonal degeneration. Since steroids have serious side-effects they should not be used in the acute stage of acute GBS except in a clinical trial. The situation is different in chronic idiopathic demyelinating polyradiculoneuropathy where steroids do have a place. There may be a place for a single injection of steroids to relieve pain during recovery from GBS. The patients who continue to progress for more than the four weeks which is arbitrarily allowed for GBS present a special problem. As progression moves into its fifth or sixth week I find myself increasingly tempted to use steroids on the ground that the course is beginning to resemble CIDP rather than GBS. The situation is far from satisfactory and clear guidelines cannot be given.

Plasma Exchange

Even 40 years after the introduction of steroids the definitive controlled trial is still awaited. By contrast within 6 years of the first report of possible benefit from plasma exchange in GBS a controlled trial had established its value. Brettle et al. (1978) reported a single case in which plasma exchange appeared beneficial. This single report was followed by many other individual cases of which most re-

Table 8.9. Disability grade scale being used in multicentre Guillain–Barré syndrome steroid trial

0 = Healthy: no signs or symptoms due to GBS

1 = Minor symptoms or signs and capable of running

2 = Able to walk 5 metres across an open space without assistant, walking frame or stick, but unable to run

3 = Able to walk 5 metres across an open space with the help of one person and waist level walking frame, stick or sticks

4 = Chair bound/bed bound: unable to walk as in 3

5 = Requiring assisted ventilation (for at least part of day or night)

Table 8.10. Arm disability grade scale being used in multicentre Guillain–Barré syndrome steroid trial

0 = Normal

1 = Minor symptoms or signs but able to put hand on top of head when sitting with head upright and able to oppose thumb to each fingertip

2 = Either able to put hand on top of head when sitting with head upright or able to oppose thumb to each fingertip, but not both

3 = Some movement but unable to perform either of the tasks in 2

4 = No movement

Table 8.11. Early reports of plasma exchange in GBS

Reference	No. treated	No. improved
Brettle et al. (1978); Gross et al. (1982)	4	2
Ropper et al. (1980)	4	3
Mark et al. (1980)	1	0
Dureux et al. (1980); Schooneman et al. (1981); Gerard (1981)	10	10
de Jager et al. (1981)	5	3
Rumpl et al. (1981)	8	8
Durward et al. (1981)	5	5
Kennard et al. (1982)	12	4
Osterman et al. (1982)	9	4
Vedeler et al. (1982)	3	2
Tindall (1982); Cook et al. (1980)	7	3
Valbonesi et al. (1980); Zerbi et al. (1981)	4	4
Fiorini et al. (1982)	1	1
Total	73	49

ported benefit (Rumpl et al. 1981; Schooneman et al. 1981) but some did not
(Kennard et al. 1982) (Table 8.11).

A small British controlled trial showed no significant benefit to the group
which received plasma exchange (Greenwood et al. 1984). Fourteen patients
were randomised to receive plasma exchange on average 13.5 days after onset of
neuropathy and 15 patients were randomised not to receive it on average 11.9
days after onset. The changes in disability grade (Table 8.4) favoured the plasma
exchange group, especially soon after the exchange (Table 8.12) but the eventual outcome was similar in both groups (Table 8.13). The sample sizes in the

Table 8.12. Change in disability grade in treated and control patients in the British randomised
trial of plasma exchange

	n	Entry grade	Change in grade			
			2 weeks	4 weeks	12 weeks	52 weeks
All patients						
Exchange	14	4.43±0.14	−0.36±0.31	−0.64±0.36	−1.71±0.42	−2.71±0.54
Control	15	4.46±0.13	+0.20±0.20	−0.27±0.27	−1.40±0.49	−1.93±0.65
Patients admitted <2 weeks after onset of neuropathy						
Exchange	9	4.43±0.17	−0.44±0.34*	−0.67±0.37	−1.56±0.38	−2.67±0.58
Control	11	4.45±0.16	+0.27±0.20*	−0.18±0.30	−1.27±0.42	−1.73±0.78

From Greenwood et al. (1984), with permission.
Mean ± SE.
+ = deterioration; − = improvement.
$P>0.05$ all comparisons: for comparison marked* $P=0.07$, two-tailed.

Table 8.13. Outcome 12 months after randomisation in the British randomised trial of plasma exchange

	Exchange ($n=14$)	Control ($n=15$)
Death	2	4
Full recovery at 1 year	4	4
Residual disability grade at 1 year:		
1	6	5
2	1	0
3	1	1
4	0	0
5	0	1
Relapsed within 1 year	1	1
Rank sum (Wilcoxon 2-sample test) of times to disability grade 1	$P = 0.30$	

From Greenwood et al. (1984), with permission.

trial were chosen with the intention of identifying a 1.0 grade change difference between the two groups after one month. As we shall see the difference between the two groups in a larger trial was less than this and yet was clinically worthwhile. Apart from the small size of the trial another problem was that patients received treatment rather late, at a time when much of the damage to myelin and axons had already been done.

In a Swedish controlled trial of 38 patients there were significant benefits to the patients who received plasma exchange: weakness stopped progressing sooner, improvement began earlier and one and two months after trial entry disability was less in the patients who received plasma exchange. For instance the mean fall in disability grades (with a slightly different scale from Table 8.6) was 2.1 in the exchange group and 0.6 in the control group ($P<0.05$). In this trial there were also non-significant differences in favour of the plasma exchange group for duration of hospital stay and time to virtually complete recovery (Osterman et al. 1984). In a third small randomised trial 13 patients who received prednisolone 100 mg daily for 10 days as well as plasma exchange fared no differently from 12 patients who were treated without prednisolone or plasma exchange (Mendell et al. 1985). The small size of this trial gave it a very low power of detecting the amount of benefit identified in the two large trials, one North American and one French.

The North American trial randomised 122 patients to plasma exchange and 123 to conventional treatment without plasma exchange (Guillain–Barré Syndrome Study Group 1985). Exchanges totalling 200–250 ml/kg body weight were performed over 7–14 days with continuous or intermittent flow machines. Most patients received 5% albumin as the replacement solution. This regime was similar to that used in the other controlled trials. The large sample sizes were based on different statistical calculations from those in the British trial: the trial had two major outcome criteria: (1) one third increase in the proportion of patients improving by one disability grade in the treated group after four weeks (grades similar to but specified more precisely than in Table 8.6) and (2) a reduction of the proportion failing to improve one grade after 6 months by one third. The major analysis was conducted following exclusion of 12 patients who did not complete plasma exchange according to protocol because they or their

Table 8.14. Outcome four weeks after randomisation in the North American randomised trial of plasma exchange

Group	Plasma exchange	Control	P
All patients			
Improved at least 1 grade	64/108 (59%)	47/120 (39%)	<0.01
Mean grade change	1.1	0.4	<0.001
Respirator patient subgroup			
Improved at least 1 grade	26/52 (50%)	18/52 (35%)	0.08
Mean grade change	0.8	0.1	<0.001

From Guillain–Barré Syndrome Study Group (1985), with permission.
Note: probability values are one-tailed.

Table 8.15. Outcome six months after randomisation in the North American randomised trial of plasma exchange

	Plasma exchange	Control	P
Percentage of patients failing to improve one grade			
All patients	3	13	<0.01
Ventilated patients	7	15	NS
Percentage of patients unable to walk unaided			
All patients	18	29	<0.05
Ventilated patients	26	42	NS

From Guillain–Barré Syndrome Study Group (1985).
Note: probability values are one-tailed. NS = not significant.

physicians did not want to continue. Follow-up information was not available on these patients after the time at which they had been withdrawn from the study. However, the major conclusions of the study were not altered if these patients were retained in the study at the same disability grade which they had reached at their last assessment (probably an ultra-conservative assumption). Probabilities quoted by the authors were based on the one-tailed probability that they were only looking for/expecting improvement with plasma exchange. Even if the more conventional two-tailed probabilities had been used most tests would still be significant. In particular the two major outcome criteria showed significant benefits to the plasma exchange group. Significantly more exchanged patients than control patients had improved one disability grade four weeks after randomisation (Table 8.14) and significantly less exchanged patients had failed to improve one disability grade six months after randomisation (Table 8.15). Among the many analyses it was notable that the median time to weaning from the ventilator in the 52 exchanged patients who needed ventilation was only 24 days compared with 48 days in exactly the same number of ventilated control patients ($P<0.01$ one-tailed). In addition when all the patients were analysed the median time to walk unaided was over a month less in the exchanged group (53 days) than the control group (85 days) ($P<0.001$ one tailed).

Analysis of subgroups within the North American trial revealed that there was a more striking difference between exchanged and control groups among those randomised within the first week of their neuropathic symptoms than among those randomised later (Guillain–Barré Syndrome Study Group 1985). This is

Table 8.16. Outcome in French randomised trial of plasma exchange (median time, days)

Time	Plasma exchange ($n = 109$)	Control ($n = 111$)
To begin weaning off ventilator	18**	31**
To walk with aid	30*	44*
To walk unaided	70*	111*

Date from French Cooperative Group on Plasma Exchange in Guillain–Barré Syndrome (1987).
$*P<0.01$; $**P<0.005$.

what would be expected if plasma exchange is removing a harmful substance present in the plasma in the early stages of the disease. One unexpected outcome from subgroup analysis was that there was a difference in outcome according to the type of plasma exchange machine. Patients exchanged with continuous flow machines fared better than those exchanged with intermittent flow machines. After four weeks 64% of patients who had been treated with continuous exchange machines had improved one disability grade compared with 51% on intermittent exchange machines and 39% having no exchange. The median time to walk unaided was 48 days with continuous exchange machines, 62 days with intermittent exchange machines, and 91 days without exchange: the comparison between patients who had been treated with continuous flow machines and those treated with intermittent flow machines was significant ($P<0.01$). Analyses of the volume of plasma exchanged and size of each individual exchange did not reveal significant correlations with outcome. Furthermore, it has not been possible to identify any subgroup of patients with GBS who are more likely to respond to plasma exchange. Although older patients and those with small distally evoked muscle action potentials are more likely to do badly this does not mean, as has sometimes been suggested (Greuner et al. 1987) that plasma exchange is ineffective. Antibody tests to nerve have not proved a useful method of predicting the value of plasma exchange (Rostami et al. 1989).

A French controlled trial of similar size provides even stronger evidence of a beneficial effect from plasma exchange (French Cooperative group in plasma exchange in Guillain–Barré syndrome 1987). The treatment protocol was similar to that used in the North American trial except that the plasma exchanges involved fewer but larger plasma exchanges, four exchanges on alternate days, each exchange being two plasma volumes. Other differences were that half of the exchanged group received fresh frozen plasma and half 5% albumin. Those who received 5% albumin also received 10 g gammaglobulin after each exchange. The neurological scoring system was different from the disability scales used in the other trial but the conclusion was the same: 44% of the 111 control patients and 67% of the 109 exchanged patients had improved after four weeks ($P<0.001$). The times to start weaning from the ventilator, to walk with assistance, and to walk unaided were all shorter in the patients who received plasma exchange (Table 8.16). There was no difference in outcome between the exchanged patients who received 5% albumin replacement and those who received fresh frozen plasma but fresh frozen plasma was associated with more adverse reactions including fever, rash and hepatitis. The differences in favour of plasma exchange were more substantial than in the North American trial. This might have been attributable to the different treatment regimes but the most striking difference was that patients in the French trial were exchanged earlier in the course of their disease (mean 6.6 days from onset in the French trial compared with 11.1 days in the North American trial). In a small trial from Finland 26 patients were alternately allocated to receive plasma exchange or not. Although there was no significant difference in hospital stay or time to recover, mean hand strength improved after plasma exchange and weakened in the control group (Farkkila et al. 1987).

If all the trials are considered together the evidence in favour of plasma exchange being beneficial is overwhelming (Fig. 8.1). Consequently even the sternest critics (Dyck and Kurtzke 1985) and a specially convened Consensus Conference (1986) have recommended that plasma exchange be used for patients who have severe GBS. Severe was defined as inability to walk. It is logical to treat patients as early as possible preferably during the first week, and there is no evidence to support the use of plasma exchange after the second week (which does not mean that it does not work then). These conclusions must not disguise the difficulties. The trials were conducted in centres with more than usual interest, experience and possibly expertise in treating GBS. Plasma exchange adds extra dangers to the care of severely ill patients who are at risk of infection and have unstable autonomic control of blood volume, blood pressure and

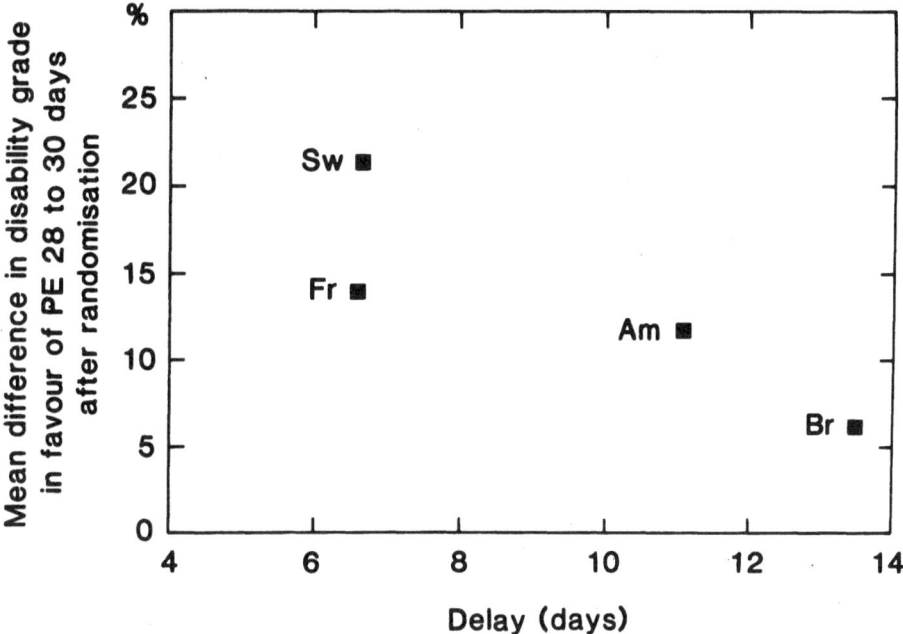

Fig. 8.1. Disability grade change differences between plasma exchange and control groups four weeks after randomisation for the Swedish (Sw) (Osterman et al. 1984), French (Fr) (French Cooperative group in plasma exchange in Guillain–Barré syndrome 1987), North American (Am) (Guillain–Barré Syndrome Study Group 1985), and British (Br) (Greenwood et al. 1984) trials. In order to make the comparison all disability scales have been converted to range from 0 to 100. The x-axis represents the mean delay from onset of neuropathy to the start of plasma exchange in each trial. A positive difference on the y-axis indicates that the plasma exchange group has improved more than the control group since randomisation.

pulse. Septicaemia occurred in 12.8% of the exchanged but only 4.5% of the control patients in the French trial ($P<0.05$) (French Cooperative group in plasma exchange in Guillain–Barré syndrome 1987) but infections requiring antibiotics occurred in 39% of both exchanged and control groups in the North American Trial (Guillain–Barré Syndrome Study Group 1985). Although autonomic circulatory instability is a theoretical objection to the use of plasma exchange, serious hypotension and cardiac arrhythmias were not significantly more frequent in the exchanged than the control group in either of the trials. Nevertheless it is well to remember the list of potential complications (Table 8.17). From a world wide survey it has been estimated that there have been three deaths per 10000 procedures (Consensus Conference 1986). The risks are substantially lower with 5% albumin as replacement fluid than with fresh frozen plasma.

Between approximately 10% and 25% of patients may show a limited relapse about two weeks (range one to six weeks) after plasma exchange which usually responds to a further plasma exchange (Osterman et al. 1986; Ropper et al. 1988).

Table 8.17.   Risks of plasma exchange

Risks	Cause
1. Discomfort, bacterial infection, haematoma, thrombosis, pneumothorax	Venous access
2. Fluid, electrolyte, protein depletion	Inappropriate replacement
3. Hypocalcaemia	Citrate
4. Allergic reactions	Fresh frozen plasma Complement activation
5. Haemorrhage	Heparin or citrate; platelet aggregation
6. Viral infection (hepatitis, CMV, HIV)	Fresh frozen plasma

Table 8.18.   Cost of plasma exchanging a GBS patient compared with savings in hospital care

	Plasma exchange cost[a] (£)	Saving based on reduction in median time	
		in intensive care unit stay[b] (£)	in hospital stay[c] (£)
Consumables	1500	1856	816
Transport (maximum)[d]	1400		
Staff, maintenance	1000	5400	2800
Total	3900	7296	3696
Nett cost		−3396 (saving)	+204

From Tharakan et al. (1989), with permission.
[a] Expense of 5 × 2.5 l plasma exchanges.
[b] Assuming reduction in intensive care unit stay is equivalent to reduction in time on ventilator.
[c] Assuming reduction in hospital stay is equivalent to reduction in time to walk with aid.
[d] Based on worst case of an ambulance journey from the south coast of England to London with a nurse and doctor in attendance

The introduction of a policy of plasma exchange for GBS was associated with shortening of the time on the ventilator and the time to walk with or without aid comparable with those anticipated from the controlled trials (Tharakan et al. 1989). The costs to the health service of providing plasma exchange were at least balanced by the savings in hospital and intensive care unit stay (Table 8.18).

Intravenous Immunoglobulin

Intravenous immunoglobulin has been found beneficial in some autoimmune disorders, particularly idiopathic thrombocytopenia. The mechanism of action is unclear: possibilities include blockade of the reticuloendothelial system or suppression of autoantibodies by naturally occurring anti-idiotypic antibodies in the normal gammaglobulin. High-dose intravenous gammaglobulin was reported beneficial in a series of patients with CIDP (van Doorn and Vermeulen 1988) and in five of six patients with GBS (Kleyweg et al. 1988). The results of a Dutch controlled trial comparing gammaglobulin and plasma exchange are awaited with interest. The dose of gammaglobulin used is large, 0.4 g/kg/day for five consecutive days, and expensive, but the preparative process removes the risk of transmitting infection and intravenous infusions are less troublesome and hazardous than plasma exchange.

Immunosuppressive Drugs

There have been few attempts to treat acute GBS with immunosuppressive drugs other than steroids. Any such drug would have to have a rapid action and be given early to have a chance of exerting a beneficial effect. Azathioprine was used in two patients with GBS and both improved within two weeks, before azathioprine itself would have been likely to exert any effect (Yuill et al. 1970). A few other cases of neuropathy have been treated with azathioprine or 6-mercaptopurine with possible beneficial effects but the course of the disease has been subacute and unlike our definition of GBS (Palmer 1965, 1966). Large intravenous doses of cyclophosphamide were given to a series of 15 patients of whom 11 would probably have fulfilled our criteria for GBS: three were treated in the first week and nine one to three weeks after onset (Rosen and Vastola 1976). The authors thought that disease progression stopped sooner than would have been expected from cases in the literature. However, they did not think that the ultimate outcome was different and three patients died. Dramatic improvement was briefly reported in three patients treated with low dose oral cyclophosphamide and in a fourth with subacute neuropathy (Ahuja et al. 1980). The present evidence is insufficient to determine whether cytotoxic agents are valuable in GBS. Their known hazards contraindicate their use outside a therapeutic trial and are sufficiently grave to make it difficult to mount such a trial except in a subgroup of patients known to have a poor prognosis.

Rehabilitation

The patients who do not recover receive scant mention in neurological texts although their persistent problems deserve the greatest consideration. During the progressive stage severely affected patients are festooned with the paraphernalia of intensive care but careful positioning to avoid pressure on vulnerable nerves and shortening of muscles are important and a full range of passive movements for all joints must be maintained with passive exercises. Severely paralysed patients should be given splints to maintain finger extension and ankle dorsiflexion. Swelling of the hands and feet will occur unless these movements are conducted assiduously: raising the hands and feet on pillows probably helps. Common sense dictates that rest is appropriate during the early stages of GBS when inflammation remains active. Exercise increases the metabolic demands of the peripheral nerves but is usually balanced by an increased blood flow. Where blood flow is limited by oedema and raised endoneurial pressure, exercise might worsen ischaemic damage to nerve fibres. The programme of passive movements must be maintained during the plateau phase. As strength begins to return a programme of assisted active movements should be tailored to the needs of the individual. A speech therapist may help restore communication. A cuffed speaking tracheostomy tube (Editorial 1987) may be used or, when the airway can be adequately protected, an uncuffed silver tube with a flap.

The initial programme will include restoration of trunk control and exercises to strengthen the muscles which have some function. Getting the patient out of bed and into a chair is an important milestone which has to be monitored carefully because of the danger of postural hypotension. The patient should be observed carefully and the blood pressure measured until it is clear that these precautions are not necessary. Once sitting is possible a wheelchair, wheeled by the patient or an attendant, opens new horizons. The move from sitting to standing is made more easily with a tilt table to which the patient can be strapped and gently raised to the standing position under careful observation. If the upper limbs remain paralysed the first steps often have to be taken with support such as that provided by a pulpit frame. If upper limb function has returned the parallel bars, crutches or a walking frame can be used to help walking. The ability to stand and walk depends particularly on the strength of quadriceps and the ability to lock the knees. For patients whose ankles remain weak out of proportion to proximal muscles light ankle–foot orthoses are helpful. Monitoring progress can be undertaken by scoring the strength of major muscle groups according to the 0–5 Medical Research Scale familiar to most neurologists. Much of the improvement occurs in the range 4–5 where the scale becomes insensitive. A hand-held myometer (Penny and Giles Ltd, Airfield Road, Christchurch, Dorset) can be used to provide measurements of the force generated by each muscle group in an isometric contraction. With care and training reproducible measurements can be obtained and used to monitor the improvement (Karni et al. 1984). When long intervals separate milestones in recovery, such as the delay from sitting to standing, improvement in readings of muscle strength may help persuade a patient that progress is being made. The strength measurements may give an early warning of a relapse and are of greater value in the management of patients with chronic idiopathic demyelinating polyradiculoneuropathy.

The patient with severe wasting has to rely on the slow process of axonal regeneration and may become very despondent. It is not always appreciated how long the recovery process can continue. I have seen a patient with weak hands and needing sticks to walk after one year but returning to work as a bricklayer by the end of the second year. Other patients have demonstrated to me improvement which has been continuing even longer after the acute illness. However, if significant deficit persists after a year has passed it is appropriate to look for ways of coping with disability rather than vain attempts to cure it. Suitable wheelchair, limb splints and aids must be provided by physiotherapists and occupational therapist. Home and work environments may need to be provided. Transport problems must be overcome; special adaptations can be made to cars to permit driving despite weak hands or feet. Detailed arrangements differ from district to district and country to country and can be provided by hospital therapy and social work services. The Guillain–Barré Syndrome Support Group exists to provide counselling and support to patients (Appendix).

Summary

The prognosis of GBS is variable but up to 13% die and a further 20% are left significantly disabled. Older age, requirement for ventilation, and small distally evoked muscle potentials increase the likelihood of persistent significant disability. Excellent intensive care is the mainstay of treatment to detect and combat respiratory failure and complications of ventilation, cardiac arrhythmias and an unstable circulation due to autonomic involvement. Adults should be treated with low-dose subcutaneous heparin because of the danger of pulmonary embolism. Steroid treatment has only been tested in small controlled trials not powerful enough to detect moderate beneficial effects and the outcome of a trial of high-dose intravenous methylprednisolone is awaited. Two large randomised, but not blind, trials have demonstrated that plasma exchange helps to shorten the duration of ventilation and the time to walk unaided by clinically and economically significant amounts. Plasma exchange is more effective if given early, during the first week, and there is no evidence that it helps after the second week. Plasma exchange should preferably be given with a continuous flow machine. Typical regimes are five 50 ml/kg exchanges carried out over 10–14 days. Albumin is safer than fresh frozen plasma as replacement fluid. Plasma exchange does not prevent some patients becoming severely disabled and better treatments must be sought.

References

Ahuja GK, Mohandas S, Virani V (1980) Cyclophosphamide in Landry–Guillain–Barré syndrome. Acta Neurol (Napoli) 35:186–190
Albers JW, Donofrio PD, McGonagle TK (1985) Sequential electrodiagnostic abnormalities in acute inflammatory demyelinating polyradiculoneuropathy. Muscle Nerve 6:504–509
Bredin CP (1976) Thrombotic complications in acute polyneuritis. Br Med J 1:837
Brettle RP, Gross M, Legg NJ, Lockwood M, Pallis C (1978) Treatment of acute polyneuropathy by

plasma exchange. Lancet ii:1100

Brumback RA (1980) Failure of oral versus parenteral corticosteroids in a case of acute inflammatory polyradiculoneuropathy (Guillain–Barré syndrome). Aust NZ J Med 10:224–226

Cole GF, Matthew DJ (1987) Prognosis in severe Guillain–Barré syndrome. Arch Dis Child 62:288–291

Consensus Conference (1986) The utility of therapeutic plasmapheresis for neurological disorders. Statement. JAMA 256:1333–1337

Cook JD, Tindall RAS, Walker J, Khan A, Rosenberg R (1980) Plasma exchange as a treatment of acute and chronic idiopathic autoimmune polyneuropathy: limited success. Neurology 30:361–362

Cornblath DR, Mellits, ED, Griffin JW et al. (1988) Motor conduction studies in Guillain–Barré syndrome; description prognostic value. Ann Neurol 23:354–359

Dalos NP, Borel C, Hanley DF (1988) Cardiovascular autonomic dysfunction in Guillain–Barré syndrome. Therapeutic implications of Swan–Ganz monitoring. Arch Neurol 45:115–117

de Jager AE, The TH, Smit Sibinga CT, Das PC (1981) Plasma exchange in five patients with acute Guillain–Barré syndrome. Int J Artif Organs 4:230–233

Dowling PC, Bosch VV, Cook SD (1980) Possible beneficial effect of high dose intravenous therapy in acute demyelinating disease. Neurology 30:33–36

Dureux JB, Gerard A, Roche G et al. (1980) Treatment of Guillain–Barré syndrome by plasma exchange: 6 cases. Nouv Presse Med 9:3696–3697

Durward WF, Burnett AK, Watkins R, Reid JM (1981) Plasma exchange in Guillain–Barré syndrome. Br Med J 283:794

Dyck PJ, Kurtzke J (1985) Plasmapheresis in Guillain–Barré Syndrome. Neurology 35:1105–1107

Editorial (1987) Speech with a cuffed tracheostomy tube. Lancet ii:432

Farkkila M, Kinnlinen E, Haapanen E, Inanainen M (1987) Guillain–Barré syndrome: quantitative measurement of plasma exchange therapy. Neurology 37:837–840

Fergusson RJ, Wright DJ, Willey RF, Crompton GK, Grant IWB (1981) Suxamethonium is dangerous in polyneuropathy. Br Med J 282:298–299

Ferner RE, Hughes RAC (1987) Modern management of Guillain–Barré syndrome. Br J Hosp Med 38:525–530

Fiorini G, Bigi G, Paracchini ML, Marinig C, Gibelli A (1982) Immunological monitoring of a patient with Guillain–Barré syndrome successfully treated with plasma exchange. Vox Sang 42:304–307

French Cooperative group in plasma exchange in Guillain–Barré syndrome (1987) Efficiency of plasma exchange in Guillain–Barré syndrome: role of replacement fluids. Ann Neurol 22:753–761

Frick E, Angstwurm H (1968) Zur Kortikosteroid-Behandlung der idiopathischen Polyneuritis. Münchener Med Wochenschr 110:1265–1271

Gerard A (1981) Successful plasmapheresis in the Miller-Fisher syndrome. Br Med J i:1627

Goodall JAD, Kosmidis JC, Geddes AM (1974) Effect of corticosteroids on course of Guillain–Barré syndrome. Lancet i:524–526

Greenwood RJ, Newsom Davis JM, Hughes RAC et al. (1984) Controlled trial of plasma exchange in acute inflammatory polyradiculoneuropathy. Lancet i:877–879

Greuner G, Bosch EP, Strauss RG, Klugman M, Kimura J (1987) Prediction of early beneficial response to plasma exchange in Guillain–Barré syndrome. Arch Neurol 44:295–298

Gross M, Legg NJ, Lockwood MC, Pallis C (1982) The treatment of inflammatory polyneuropathy by plasma exchange. J Neurol Neurosurg Psychiatry 45:675–679

Guillain–Barré Syndrome Study Group (1985) Plasmapheresis for acute Guillain–Barré syndrome. Neurology 35:1096–1104

Haass A, Trabert W, Gressnich N, Schimrigk K (1988) High dose steroid therapy in Guillain–Barré syndrome. J Neuroimmunol 20:305–308

Heller GG, de Jong RN (1963) Treatment of the Guillain–Barré Syndrome with corticotrophin and glucocorticoids. Arch Neurol 8:179–193

Hughes RAC, Newsom Davis JM, Perkin GD, Pierce JM (1978) Controlled trial of prednisolone in acute polyneuropathy. Lancet ii:750–753

Hughes RAC, Kadlubowski M, Hufschmidt A (1981) Treatment of acute inflammatory polyneuropathy. Ann Neurol 9 Suppl:125–133

Jackson RH, Miller H, Schapira K (1957) Polyradiculitis (Landry–Guillain–Barré syndrome). Treatment with cortisone and corticotrophin. Br Med J i:480–484

Karni Y, Mills KR, Archdeacon L, Wiles CM (1984) Clinical assessment and physiotherapy in Guillain–Barré syndrome. Physiotherapy 70:288–292

Kennard C, Newland AC, Ridley A (1982) Treatment of Guillain–Barré syndrome by plasma exchange. J Neurol Neurosurg Psychiatry 42:847–850

Kleyweg RP, Van der Meche FGA, Meulstee J (1988) Treatment of Guillain-Barré syndrome with high dose gammaglobulin. Neurology 38:1639-1642

Kleyweg RP, Van der Meche FGA, Loonen MCB, De Jonge J, Knip B (1989) The natural history of the Guillain-Barré syndrome in 18 children and 50 adults. J Neurol Neurosurg Psychiatry 52:853-856

Leese J (1976) Thrombotic complications in acute polyneuritis. Br Med J 38:585

Löffel NB, Rossi LN, Mumenthaler M (1977) The Landry-Guillain-Barré syndrome - complications, prognosis and natural history in 123 cases. J Neurol Sci 33:71-79

Mark B, Hurwitz BJ, Olanow CW, Fay JW (1980) Plasmapheresis in idiopathic inflammatory polyradiculoneuropathy. Neurology 30:361

Marshall J (1963) The Landry-Guillain-Barré syndrome. Brain 86:56-66

McKhann GM, Griffin JW, Cornblath DR et al. (1988) Plasmapheresis and Guillain-Barré syndrome: analysis of prognostic factors and the effect of plasmapheresis. Ann Neurol 23:347-353

McLeod JG (1981) Electrophysiological studies in the Guillain-Barré syndrome. Ann Neurol 9:20-27

Mendell JR, Kissel JT, Kennedy MS et al. (1985) Plasma exchange and prednisone in Guillain-Barré syndrome. A controlled randomised trial. Neurology 35:1551-1555

Nixon RA (1978) Quinine sulfate for pain in the Guillain-Barré syndromes. Ann Neurol 4:386-387

Osler LD, Sidell AD (1960) The Guillain-Barré syndrome: the need for exact diagnostic criteria. N Engl J Med 262:964-969

Osterman PO, Fagius J, Safwenberg J, Danersund A, Wallin B, Nordesjo LO (1982) Treatment of Guillain-Barré syndrome by plasmapheresis. Arch Neurol 39:148-154

Osterman PO, Lundemo G, Pirskanen R et al. (1984) Beneficial effects of plasma exchange in acute inflammatory polyradiculoneuropathy. Lancet ii:1296-1299

Osterman PO, Fagius J, Safwenberg J, Daniellson BG, Wikstrom B (1986) Early relapses after plasma exchange in acute inflammatory polyradiculoneuropathy. Lancet ii:1161

Palmer KNV (1965) Polyradiculoneuropathy (Guillain-Barré syndrome) treated with 6-mercaptopurine. Lancet i:733-734

Palmer KNV (1966) Polyneuropathy treated with cytotoxic drugs. Lancet 1:265

Peterman AF, Daly DD, Dion FR, Keith HM (1959) Infectious neuronitis (Guillain-Barré syndrome) in children. Neurology 9:533-539

Pleasure DE, Lovelace RE, Duvois RC (1969) The prognosis of acute polyradiculoneuritis. Neurology (Minneapolis) 18:1143-1148

Plum F (1953) Multiple symmetrical polyneuropathy treated with cortisone. Neurology 3:661-667

Ravn H (1967) The Guillain-Barré syndrome. A survey and clinical report of 123 cases. Acta Neurol Scand 43 suppl 30:1-164

Ropper AH, Shahani B, Huggins CE (1980) Improvement in 4 patients with acute Guillain-Barré syndrome after plasma exchange. Neurology 30:361

Ropper AH (1986) Severe acute Guillain-Barré syndrome. Neurology 36:429-431

Ropper AH, Shahani BT (1984) Diagnosis and management of acute areflexic paralysis with emphasis on Guillain-Barré syndrome. In: Asbury AK, Gilliatt RW (eds) Peripheral nerve disorders. A practical approach. Butterworths, London

Ropper AH, Albers JW, Addison R (1988) Limited relapse in Guillain-Barré syndrome after plasma exchange. Arch Neurol 45:314-315

Rosen AD, Vastola EF (1976) Clinical effects of cyclophosphamide in Guillain-Barré polyneuritis. J Neurol Sci 30:179-187

Rossi LN, Mumenthaler M, Lutsch J, Ludin HP (1976) Guillain-Barré syndrome in children with special reference to the natural history of 38 personal cases. Neuropadiatre 7:42-51

Rostami AM, Brown MJ, Pleasure DE (1989) Peptide 53-78 of myelin P_2 protein is a T-cell epitope for the induction of experimental allergic neuritis. Peripheral Nerve Study Group Abstracts 32

Rumpl E, Mayr U, Gerstenbr F, Hackl JM, Rosmanith P, Aichner F (1981) Treatment of Guillain-Barré syndrome by plasma exchange. J Neurol 225:207-217

Samantray SK, Johnson SC, Mathao KU, Pulimood BM (1977) Landry-Guillain-Barré syndrome. A study of 302 cases. Med J Aust 2:84-91

Schooneman F, Janot C, Streiff F et al. (1981) Plasma exchange in Guillain-Barré syndrome; ten cases. Plasma Therapy 2:117-121

Shy GM, McEachern D (1951) Further studies on the effect of cortisone and ACTH in neurological disorders. Brain 74:352-362

Stillman JS, Ganong WF (1952) The Guillain-Barré syndrome: report of a case treated with ACTH and cortisone. N Engl J Med 246:293-296

Swick HM, McQuillen MP (1976) The use of steroids in the treatment of idiopathic polyneuritis.

Neurology 26:205–212

Tharakan J, Ferner RE, Hughes RAC et al. (1989) Plasma exchange for Guillain–Barré syndrome. J R Soc Med 82:458–461

Tindall RSA (1982) The role of therapeutic apheresis in acute, relapsing and chronic inflammatory demyelinating polyneuropathy. In: Therapeutic apheresis and plasma perfusion. pp 205–217

Valbonesi M, Garelli S, Mosconi L, Zerbi D, Celano I (1980) Plasma exchange as therapy for Guillain–Barré syndrome with immune complexes. Vox Sang 41:74–78

van Doorn PA, Vermeulen M (1988) Improvement in chronic inflammatory demyelinating poly-radiculoneuropathy following γ-globulin infusion. Arch Neurol 44:897–898

Vedeler CA, Nyland H, Fagius J et al. (1982) The clinical effect and the effect on serum IgG antibodies to peripheral nervous tissue of plasma exchange in patients with Guillain–Barré syndrome. J Neurol 228:59–64

Wiederholt HM, Mulder DW, Lambert EH (1964) The Landry Guillain–Barré Strohl syndrome of polyradiculoneuropathy – historical review report on 97 patients and present concepts. Mayo Clin Proc 49:427–451

Winer JB, Hughes RAC (1988) Identification of patients at risk of arrhythmia in the Guillain–Barré syndrome. Q J Med 68:735–739

Winer JB, Hughes RAC, Greenwood R, Healy MH (1985) Prognosis in the Guillain–Barré syndrome. Lancet i:1202–1203

Winer JB, Hughes RAC, Osmond CA (1988) Prospective study of acute idiopathic neuropathy. I Clinical features and their prognostic value. J Neurol Neurosurg Psychiatry 51:605–612

Yuill GM, Swinburn WR, Liversedge LA (1970) Treatment of polyneuropathy with azathioprine. Lancet ii:854–856

Zerbi D, Celano I, Forlani G, Garelli S, Mosconi L, Valbonesi M (1981) Plasmapheresis in the treatment of four cases of Guillain–Barré syndrome (acute form). Ital J Neurol Sci 2:331–336

Chapter 9

Immunology of Guillain–Barré Syndrome

Cerebrospinal Fluid

The combination of increased CSF protein concentration and normal cell count was used by Guillain et al. (1916) as a distinguishing feature of the two cases in their original description. This combination remains the most helpful laboratory guide to the diagnosis of GBS but is not an invariable feature. In large series this "albuminocytological dissociation" is reported in about 80%–90% of patients. For instance, in a prospective study of 100 cases in south east England CSF total protein concentrations were abnormal in 80%; the mean concentration was 1.2 g/l and range 0.1–6.0 g/l. The CSF white cell count was normal in 90%: those patients having a pleocytosis had very few cells, median 8 per µl and range 6–103 per µl (Winer et al. 1988d). Single observations on individual patients suggest that the CSF protein concentration commonly remains normal during the first ten days of the illness, increases as the disease progresses, continues to rise during the plateau and early recovery phases and eventually returns to normal (Wiederholt et al. 1964). This time course was confirmed by serial lumbar punctures in a typical case (Link 1975). No clinical differences have been discovered between those patients with persistently normal CSF protein concentrations and those exhibiting the classical albuminocytological dissociation, although differences, especially in outcome, have been sought in almost every published series.

The proportion of cases with increased CSF cell counts has been 10% or less in most series but a mononuclear pleocytosis was identified at some time in 55% of one series of 24 patients who underwent serial lumbar punctures: the raised cell counts were encountered ten days to four months after the onset of the neuropathy (Link et al. 1979). A raised CSF cell count should always redouble concern as to whether the diagnosis of GBS is correct and alternatives such as poliomyelitis and borreliosis should be considered. It has become customary to exclude from treatment trials patients with CSF cell counts exceeding 50/µl. The majority of the increase in CSF protein concentration is due to albumin. The likely explanation is leakage of albumin because of disruption of the blood–nerve barrier in the intrathecal nerve roots which are known pathologically to be a preferential site of involvement. Less likely alternative explanations such as altered albumin metabolism or stagnation of CSF transport have been disproved by studies of transport of radioactively labelled albumin injected into the plasma or CSF (Okuyama 1975).

Electrophoresis or iso-electric focusing of the CSF proteins shows a characteristic transudate pattern suggesting that the increased CSF protein concentration is caused by breakdown of the blood–CSF barrier (Livrea et al. 1980; Kruger et al. 1981; Segurado et al. 1986; Amarenco et al. 1987; Mactier and Khanna 1987). The breakdown of the blood–CSF barrier also increases the CSF concentration of aminoacids (Corston et al. 1981). The release and breakdown of myelin proteins increases the CSF concentration of myelin basic protein (Cornblath et al. 1986). The concentration of IgG, IgA and IgM in the CSF usually increases *pari passu* with the increase in CSF albumin. However, in some cases, especially during the plateau or recovery phases, the increase is greater than would have been predicted from the increase in albumin suggesting intra-thecal immunoglobulin synthesis (Link 1975; Livrea et al. 1980; Segurado et al. 1986; Amarenco et al. 1987). This conclusion that IgG is synthesised within the theca assumes that the normal relationship between IgG and albumin transfer into the CSF is retained as the blood–CSF barrier is broken down, an assump-tion which is probably invalid because the inflammatory process disrupts normal cell transport mechanisms.

In several reports oligoclonal bands have been detected by agarose electro-phoresis or iso-electric focusing both in the serum and the CSF. With agarose electrophoresis Link et al. (1975) discovered oligoclonal bands in the CSF and serum of 21% of 24 patients and in the CSF alone in a further 21%. In six patients with CSF oligoclonal bands the CSF kappa/lambda ratios were low, consistent with over-production of immunoglobulin of lambda light chain type: three of these patients also had abnormally low serum kappa/lambda ratios and these patients also had serum oligoclonal bands. These findings differ from those in multiple sclerosis in which immunoglobulin oligoclonal bands are present in CSF but not serum and the oligoclonal bands have predominantly kappa light chains. The overall conclusion from more recent studies is that any increase in CSF immunoglobulin is usually due to a leak from the systemic circulation into the CSF. Increase in the immunoglobulins is not detected during the early stages and so probably represents a response to rather than the cause of the disease.

Diseases Associated with Guillain–Barré Syndrome

The prediction that myasthenia gravis had an autoimmune mechanism came from its association with other autoimmune disorders so that it is worth search-ing for such associations with GBS. Such reports are rare and some do not bear close scrutiny. Take for example the suggestion that GBS is associated with Addison's disease (Behar et al. 1986). Necrotising adrenalitis was demonstrated at post-mortem examination in three patients with GBS who had died with untreated infections in the pre-antibiotic era (Sabin and Aring 1941) and in two cases of acute ascending paralysis of obscure cause (Spaar and Orthner 1968). Another alleged association between GBS and Addison's disease is extremely tenuous: the patient had classical severe Addison's disease and weakness attri-buted to GBS but this was not documented as being due to neuropathy (Abbas et al. 1977).

Even some of the associations which I have accepted as genuine (Table 9.1) have odd features. A man recovered from idiopathic thrombocytopenia purpura and was being treated successfully for autoimmune haemolytic anaemia with

Table 9.1. Diseases associated with GBS

	No. of cases	Reference
Putative autoimmune diseases		
Myxoedema	3	Potz and Neundorfer (1975); Behar et al. (1986); Winer et al. (1988b)
Thyrotoxicosis	1	Abramsky et al. (1980)
Autoimmune haemolytic anaemia	1	Ala and Shearman (1965)
Idiopathic thrombocytopenic purpura	2	Ala and Shearman (1965); Gross (1980)
Myasthenia gravis	2	Winer et al. (1988b); Bourouresque et al. (1981)
Pernicious anaemia	2	Winer et al. (1988b)
Disorders due to or associated with vasculitis		
Polyarteritis nodosa	1	Abramsky et al. (1980)
Temporal arteritis	1	Corston (1980)
Systemic lupus erythematosus	5	Abramsky et al. (1980); Kennett et al. (1986); Moreila Filho et al. (1980); Korn-Lubetzki and Abramsky (1986)
Rheumatoid arthritis	2	Grant and Leopold (1954); Korn-Lubetzki and Abramsky (1986)
Disorders associated with immunosuppression		
Steroid treatment	3	Grant and Leopold (1954); Steiner et al. (1986); Drachman et al. (1970)
Hodgkin's disease	8	Klingon (1965); Lassmann et al. (1981); Cuttner and Meyer (1978); Julien et al. (1980)
HIV infection	++	See Chapter 12

prednisolone when he developed a GBS-like illness confined to the legs from which he recovered. Eighteen months later he developed a progressive spastic paraparesis. When he died of a pulmonary embolus four years later it was remarkable that no abnormality of the cord or nerve roots could be discovered and disappointing that the peripheral nerves were not examined (Ala and Shearman 1965). It is difficult to know whether the small number of reported associations with putative autoimmune disease are more than a chance occurrence. A better idea of the true frequency of autoimmune disorders comes from a quasi-population study in which two examples of pernicious anaemia and one each of myasthenia and myxoedema were discovered among 100 cases of GBS (Winer et al. 1988b).

Reports of the occurrence of GBS with colitis (Pradalier et al. 1982; Steiner et al. 1986) are interesting in view of a report that ulcerative colitis and Crohn's disease are more commonly associated with multiple sclerosis and autoimmune disease than would be expected by chance (Sadovnick et al. 1989). This might be because such patients are exposed to more gut organisms which might put them at greater risk of developing an immune response to a gut pathogen which cross-reacts with an autoantigen.

GBS-like pictures are not uncommonly encountered in association with vasculitic disorders and are sometimes published (Table 9.1). It is difficult to tell from clinical and even neurophysiological reports whether the neuropathy is due to vasculitis or an EAN-like mechanism. Some cases of neuropathy resembling GBS associated with SLE have been demonstrated by biopsy (Hughes et al. 1982) or autopsy (Rozen et al. 1980) to be caused by vasculitis (Chapter 12). Those listed in Table 9.1 have been claimed to have true (non-vasculitic) GBS.

There are several reports of glomerulonephritis in association with GBS. The literature has been confused by the inclusion as GBS of cases with CIDP which will be considered in Chapter 10. Bradford et al. (1918) first reported the occurrence of glomerulonephritis in their autopsy study but the association has not been noted in other post-mortem series. Rodriguez-Iturbe et al. (1973) encountered a patient with GBS who also had oedema, mild renal failure and hypertension associated with glomerulonephritis. This discovery prompted them to undertake renal biopsies on eight consecutive patients with GBS. All showed glomerulonephritis with varying degrees of membranous and proliferative changes. At follow-up there was a reduction in creatinine clearance and biopsies revealed persistent changes in some glomeruli. Bertinelli et al. (1989) reported a child who had typical GBS with the second of four episodes of proliferative glomerulonephritis. Looking at the problem the other way round, Murphy et al. (1986) discovered two cases of acute GBS in a series of 170 cases of membranous glomerulonephritis in which the onset of nephrotic syndrome and neuropathy coincided. In another report nephrotic syndrome was associated with lipoid nephrosis rather than glomerulonephritis (Froelich et al. 1980). I have not encountered clinically significant renal failure or nephrotic syndrome in the course of GBS, so that it must be rare, but systematic studies of large series of patients with GBS have not been undertaken. Such studies might be rewarding since the reported cases have subepithelial deposits visible by electron microscopy which are thought to represent immune complexes whose identity might provide a clue to both the glomerular and myelin damage.

Rare examples of Guillain–Barré syndrome have been reported as occurring in the course of multiple sclerosis (Forrester and Lascelles 1979, 2 cases; Sanders and Lee 1987). In a case of florid acute multiple sclerosis, inflammatory demyelinating lesions were found at post-mortem in the nerve roots (Lassmann et al. 1981). Conversely in a fatal case of GBS, multiple sclerosis plaques were found in the brain at post-mortem (Best 1985). Guillain–Barré syndrome has also rarely been reported in close temporal association with optic or retrobulbar neuritis. Some cases quoted as examples of this association cannot be accepted. In one case, the so-called GBS was atypical including flaccid tetraparesis with coma and subsequent optic atrophy (Mouren et al. 1971). Another consisted of CIDP with probable papilloedema (case 1 of Brock and Davidson 1947), and yet another had neuromyelitis optica (Nikoskelainen and Riekkinen 1972). However, four published cases of acute GBS and optic neuritis do seem to me acceptable (Behan et al. 1976; Dummolard et al. 1937; Martin 1961; Uncini et al. 1988). It must be stressed that these cases are exceptional, and, just as GBS normally spares the CNS, so optic neuritis and multiple sclerosis usually spare the PNS. The rarity of these coincidences is more striking than their occurrence.

There are several examples of GBS occurring in the context of treatment or diseases involving partial immunosuppression. The occurrence of GBS in association with HIV infection is well-recognised and might be relevant, although the timing is such that the GBS develops before overt immunodeficiency, albeit at the same stage as other autoimmune phenomena such as thrombocytopenia. The occurrence of cases of neoplastic disease with GBS is difficult to evaluate. Klingon (1965) reported a fatal case with carcinoma of the bronchus and I have seen cases of classical GBS with recovery in association with gastric carcinoma and bronchial carcinoma. The number of cases reported in association with Hodgkin's disease has been particularly notable. Eight cases had been reported by 1980 (Klingon 1965; Lisak et al. 1977; Cuttner and Meyer 1978; Julien et al. 1980). Several cases have been reported to start during steroid treatment either alone for miscellaneous medical conditions (Grant and Leopold 1954; Steiner et al. 1986) or during combined treatment with steroids and immunosuppressive agents in a renal transplant recipient (Drachman et al. 1970). These occurrences might be used as evidence against an autoimmune mechanism of GBS. However, it is possible to argue that partial immunosuppression has inhibited normal suppressor mechanisms and permitted an autoimmune process to escape from control.

HLA Associations

Most autoimmune diseases are associated with particular histocompatibility antigens which regulate immune responses and are coded for by genes on chromosome 6. Several attempts have been made to identify such an association with GBS. Most of these studies have been small and have not shown an association (Adams et al. 1977; Stewart et al. 1978). A Mexican study showed an increased frequency of a major histocompatibility class II antigen HLA-DR3, occurring in 17% of 38 GBS patients compared with 6% of controls (Gorodezky et al. 1983). This association was of potential interest because in young women HLA-DR3 is strongly associated with myasthenia gravis, which is the best defined of all human autoimmune diseases. However, no association with HLA-DR3 was found in a study of 97 GBS patients in south east England (Winer et al. 1988c) or in a study of 92 GBS patients in the USA (Kaslow et al. 1984). In the English study there was a weak association between HLA-DR2 and disease severity: eight of twelve severely paralysed patients had HLA-DR2 compared with 20 of 67 of the remainder ($P<0.05$) and 35% of the controls. The search for HLA associations with CIDP has been a little more rewarding and a possible association with Class I antigens has been identified (Chapter 10). This might be because the ability to suppress the autoimmune response is immunogenetically determined but the initial autoimmune response is not.

The only other immunogenetic research in GBS has been a study of a gene controlling protease inhibitors. The gene frequency of the alpha-1 antitrypsin allele M3 in GBS patients (0.1428) is twice that in control subjects (0.0640). Alpha-1 antitrypsin is a protease inhibitor and abnormalities might permit uninhibited proteolysis of myelin (McCombe et al. 1985). While this idea is attractive, the low frequency of the M3 gene means that this allele cannot have an important role. Similar small increases in the frequency of this gene were found in CIDP and also in MS.

Reports of familial occurrence of GBS are rare. Saunders and Rake (1965) reported a brother and sister who both developed GBS, five years apart, and could find no other familial cases in the literature. MacGregor (1965) reported a father with GBS and a daughter with an acute febrile illness with painful sensory neuropathy and normal CSF which, although diagnosed as GBS at that time, would probably not have fulfilled our criteria for the diagnosis. Despite advertising for familial examples in the British Guillain–Barré Syndrome Support Group Newsletter I have not identified properly documented examples of GBS occurring in first degree relatives.

Immunoglobulins, Complement and Immune Complexes

In the acute stage of GBS it is quite common for the erythrocyte sedimentation rate to be raised and for increased concentrations of immunoglobulins to be found in the serum. According to a brief report IgG or IgM concentrations were increased in more than half of the sera from 62 patients (Cook et al. 1970). In a study of 100 patients IgG concentrations were increased above the normal range in 17 patients, IgM in 21 patients and IgA in 23 patients (Winer et al. 1988a). Serum concentrations of complement components C3, C4 and C9 have been either normal or increased (Winer et al. 1988a; Tonnessen et al. 1982; Hughes 1979). Low concentrations of C3 or C4 may be a feature of vasculitic disorders in which immune complexes are formed and complement is consumed. Failure to find such low concentrations does not rule out a role for complement or complement-fixing antibodies in GBS. Raised concentrations of complement may indicate stimulation of complement synthesis following its utilisation. Lower than predicted concentrations of C3 were found in the CSF of patients with GBS (Amarenco et al. 1987). More sensitive indices would be the appearance of the split products of complement molecules which are released by complement activation. We were unable to identify C3a and C4a in the serum of patients with GBS (Sanders et al. 1988) but increased concentrations of C3a and C5a were identified in the CSF by Hartung et al. (1987). In addition C9 neoantigen has been shown to appear in the serum and CSF of patients with GBS indicating formation and release of the terminal membrane attack complex components of complement (Sanders et al. 1986; Koski et al. 1987). There are some reports of complement components, including C9, on peripheral nerve myelin in biopsies and autopsies but the issue remains unresolved (Chapter 4).

Immune complexes have been identified in the serum of various proportions of patients with GBS. In general the more sensitive the assay the greater the proportion of positive sera but also the greater the proportion of the positive results in the controls. Thus Cook and Dowling (1981) found immune complexes with an ultracentrifugation technique in 40% of patients with GBS but not in normal controls. Tachovsky et al. (1976) used an assay in which a Raji lymphoma cell line binds C3 present in immune complexes and obtained positive results in five of 11 GBS patients and four of 27 normal controls. Hughes (1979) reported positive results in four of nine GBS patients and not in any of 19 normal controls by a technique involving inhibition of K cell killing of Chang human liver cells. Goust et al. (1978) reported positive results in 15 of 16 GBS patients with a C1q binding assay. On the other hand with another version of the same test positive results were only found in three out of 100 patients (Winer et al.

1988a). The discrepancies between the proportion of positive results presumably reflect differences in technique. The presence of acute phase reactants and immune complexes may be an aftermath of the viral or other infection which so often precedes GBS or a consequence of any complicating chest infection. The fact that the immune complexes are an aftermath of a preceding viral infection does not mean that they are not relevant to the pathogenesis. If they are directly relevant they are producing the inflammation and demyelination by a mechanism which is unfamiliar in conventional immunology because the histological lesions of GBS do not involve classical vasculitis with inflammation of the vessel wall, occlusion and necrosis (Chapter 4).

Antibodies

Demyelinating Serum Factors

In early experiments a demyelinating effect of GBS serum on myelinated tissue cultures was reported. These observations were made possible by development of techniques to maintain fetal neural tissue in culture conditions which would induce demyelination. Pioneering experiments showed that serum from animals with EAE would demyelinate fetal mouse cerebellar cultures (Bornstein and Appel 1961). Subsequently sera from animals with EAN were shown to demyelinate dorsal root ganglion, but not cerebellar cultures (Yonezawa et al. 1968; Yonezawa et al. 1969). A complement-fixing antibody against a nerve component was considered to be involved since the demyelinating effect was lost after heat treatment and could be absorbed out with peripheral nerve tissue. Several attempts to demonstrate similar demyelinating factors have been made in GBS. Although no consistent demyelination was found in the early reports (Arnason et al. 1969; Winkler 1965), three studies have demonstrated demyelinating factors in the serum of GBS patients which resemble those reported in EAN (Cook et al. 1971; Dubois-Dalcq et al. 1971; Yonezawa et al. 1970). The subject has been reviewed by Seil (1977) and Cook and Dowling (1981). In the most comprehensive study, sera from almost all (18/19) patients in the acute stage of GBS showed demyelinating activity compared with two-thirds (8/12) in the convalescent stage (Cook et al. 1971). The effect was not specific for GBS, being found in 10% of sera from normal subjects and up to 40% of patients with a variety of other neurological diseases. The effect was abolished by heat treatment of the serum, and restored by addition of fresh complement, consistent with a complement-dependent antibody. The effect could be absorbed with peripheral nerve and was more frequently found in 19S (IgM) than 7S (IgG) ultracentrifuge fractions. Ultrastructural studies showed that the myelin had broken down and Schwann cells had beome vacuolated but the axons were preserved confirming a primary demyelinating effect (Hirano et al. 1971). Elegant as these experiments were, the high concentration of serum applied to the tissue culture, the inconsistency of the effect with different aliquots of the same serum applied to sister cultures, and the lack of specificity of the effect for GBS make it difficult to accept this test as demonstrating an important pathogenetic factor. The assay technique has been refined and myelinotoxic effects of GBS serum on CNS as well as PNS myelinated cultures were demonstrated as part of an investigation of demyelinating factors in MS (Bradbury et al. 1985). Similar myelinotoxic effects on CNS cultures were obtained with sera from non-demyelinating diseases such as stroke.

Many attempts have been made to demonstrate a demyelinating effect by injecting GBS serum into rat sciatic nerves. The difficulty with this model is that the insertion of the needle, needle movement, and injection of protein-rich fluid into the endoneurium inevitably damage nerve fibres and might induce primary demyelination. The endoneurium has a limited capacity and low protein content. Results have been assessed morphologically and electrophysiologically. In the most carefully performed experiments small volumes (10 µl) were injected with very fine bore (33 g) needles into rat sciatic nerves. Ten GBS sera were compared with 10 control sera (Low et al. 1982). The GBS sera did produce a small reduction in the muscle action potential evoked in a distal muscle by stimulation proximal to the site of injection during the hour following the injection. However, when the electrophysiological observations were delayed until seven days after the injection there was no difference between the groups. The sera used in these experiments had been stored at −70°C and complement was not added.

In experiments in which larger volumes of GBS serum were injected and complement was added, more impressive neurophysiological evidence of demyelination has been reported. However, there are differences of opinion as to the biochemical properties of the factor blocking conduction and whether the conduction block is specific for GBS. Sumner et al. (1982) reported that larger volumes of GBS serum injected into rat sciatic nerve do cause conduction block. Harrison et al. (1984) obtained a similar result but only with fresh serum. When they injected serum which had been frozen and then thawed, no more conduction block was obtained with GBS than control serum. When fresh serum was used, however, they consistently found that serum from patients with acute GBS would produce conduction block which was significantly greater than that produced by control serum. The conduction block was found to develop gradually within 24 hours of injection, and reach a maximum three to six days afterwards. The electrophysiological results were supported by corresponding pathological evidence of demyelination. It would be surprising if the serum factor producing conduction block is an antibody in view of its susceptibility to freezing and thawing. It might be an enzyme, or an acute phase reactant. It was absent from the serum of patients who had recovered from GBS. Neurophysiological observations on nerves injected with GBS serum have unfortunately not given clear answers.

Pathological observations do suggest that endoneurial injection of serum induces demyelination and that the demyelination reaches a maximum after about 7 days. Brown et al. (1987) used 40 µl of previously frozen serum with 10 µl of added complement and found similar demyelination and inflammation in the nerves injected with sera from GBS patients or normal controls ($n = 7$ each). Similarly Winer et al. (1988a) injected 40 µl of previously frozen serum, but without added complement, into rat sciatic nerves and then found similar numbers of demyelinated fibres one week later in nerves injected with sera from patients with severe acute GBS or normal subjects ($n = 5$ each). However, authors who have harvested the injected nerves earlier to identify the occurrence of demyelination have obtained different results. Feasby et al. (1982) injected

30 µl of serum and 10 µl of guinea-pig complement and harvested the nerves after one to four days. They reported significant demyelination with sera from 14 of 17 patients with GBS and only two of 30 subjects with no disease or with other neurological diseases. The demyelination produced was greater when the sera were from patients with earlier more severe disease. Further investigations on a small number of sera suggested that the demyelinating factor was not destroyed by freezing and thawing but was destroyed by heating to 56°C and not restored by adding back complement. Their observations suggested that the factor was not an antibody and might have been a heat-labile enzyme. Saida et al. (1982) harvested nerves two days after injection and found that stored serum from 11 of 26 patients produced demyelination. Demyelination was induced less commonly by control sera, i.e. two sera from 11 normal subjects and one serum from 29 patients with other neurological diseases. Brown et al. (1987) harvested nerves 24 hours after injection and found that vesicular demyelination without macrophage infiltration was induced by five of six GBS sera and none of seven control sera. The vagaries of the injection technique preclude pursuing this test of demyelinating activity much further.

Complement Fixing Antibodies to Myelin Antigens

Since complement-fixing antibodies to myelin antigens provide a theoretically plausible mechanism of inducing demyelination, studies of such antibodies in GBS are particularly important. Melnick (1963) examined sera from 38 patients with classical acute GBS, 56 patients with other types of neuropathy and from 1200 subjects who were either normal or had miscellaneous conditions. Homogenates of human nerve, which had been stored at −70°C, were used as antigen in a standard complement fixation test with the addition of three minimum haemolytic doses of complement (Fig. 9.1). A serum was regarded as positive when it fixed complement at a dilution of 1:8 or greater. Positive results were obtained in 50% of GBS patients, 31% of patients with hypersensitivity diseases but without neuropathy, 10% of patients with diabetic neuropathy, 2.8% of 178 normal subjects, and 6.3% of 969 patients with other diseases. In serial studies antibody titres were usually highest in the earliest samples and fell to become undetectable in 2–15 weeks. A similar proportion of GBS sera also reacted with a homogenate of spinal cord. This work has been confirmed by other groups although there is disagreement as to the proportion of GBS patients whose sera give positive reactions. Ryberg (1984) reported that 14 of 18 samples from patients examined in the first three weeks of their illness gave positive results in a similar complement fixation test with human nerve root homogenate as antigen. Positive results were less common later in the disease and only occurred in one of 12 patients with borreliosis and in one of 50 normal subjects. In most cases the antibody activity could be absorbed by staphylococcal protein A and also appeared in the CSF, both observations suggesting that the antibody is IgG rather than IgM, although this disagrees with the conclusions of Koski et al. (1985). Unfortunately other authors have not obtained positive results in as high a proportion of cases. With lyophilised myelin as antigen, three of 11 GBS and two of nine CIDP sera gave positive complement-fixation tests compared with none of eight normal sera (Latov et al. 1981). With human sciatic nerve homo-

CLASSICAL COMPLEMENT FIXATION TEST

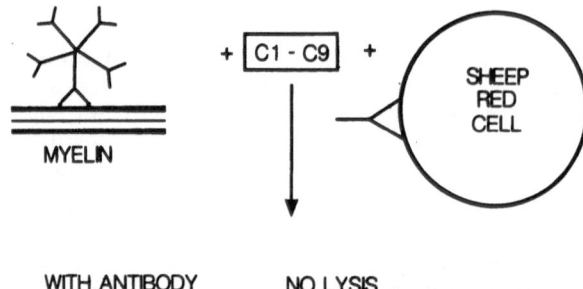

WITH ANTIBODY   NO LYSIS

Fig. 9.1.   Diagram of the classical complement fixation test which is a "back titration" to detect the consumption of complement which is added in excess. Immune complexes of aggregated immunoglobulin and other factors in the serum may also fix complement, rendering the serum anticomplementary and preventing detection of antibody.

genate as antigen we only obtained positive results with two of 17 GBS sera in our first series (Hughes et al. 1984), but three sera were anticomplementary which would have prevented detection of low titres of antibody. In a second larger series complement-fixing antibodies to human sciatic nerve homogenate were only found in seven out of 100 sera, compared with one of 100 subjects without neurological disease (Winer et al. 1988a). This low proportion of positive results was obtained despite the fact that the same test identifies positive results at very high dilutions, greater than 2^{10}, with sera from patients having IgM paraproteinaemic demyelinating neuropathy and antibodies to myelin-associated glycoprotein. The conclusion at present is that early in the disease some patients with GBS do have low titres of complement-fixing antibody to myelin antigens detected by conventional tests. The proportion of positive results depends on the sensitivity of the test. If the test system is too sensitive, a significant proportion of sera from patients with other inflammatory or neurological disorders also give positive results, albeit less frequently. Consequently this test has not been adopted as a diagnostic tool.

Fortunately the presence of complement-fixing antibodies to myelin antigens in GBS has been placed on a firmer footing by a series of papers from the laboratory of Dr Koski in the University of Maryland. She has employed a very sensitive variant of the complement-fixation test. This test avoids the problem of anticomplementary activity which prevents the detection of low titres of antibody in the classical complement fixation test. The test involves the fixation of C1 by antibody in the presence of its antigen and then the detection of the fixed C1 by indicator sheep red cells which are lysed in its presence (Fig. 9.2). Human nerve myelin is incubated with test serum overnight, and then washed and resuspended with a source of C1 (usually C2-depleted human serum). The

C1 FIXATION AND TRANSFER TEST

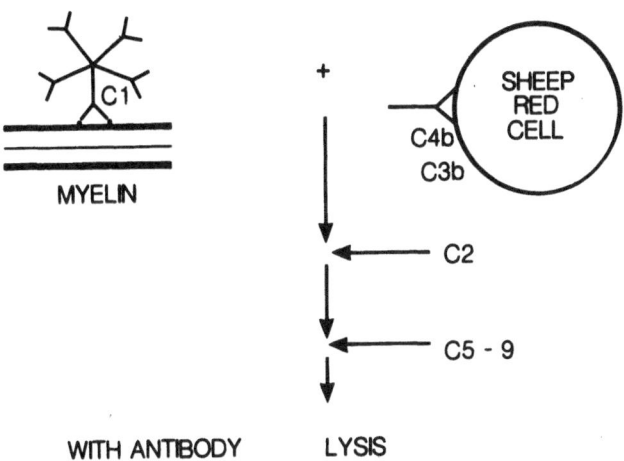

Fig. 9.2. Diagram to show the principle of the C1 fixation and transfer assay. Myelin and antibody are incubated together and then washed. C1 is added and the slurry is washed again. The fixed C1 is added to sensitised sheep red cells. Finally the remaining complement components are added. One molecule of C1 is sufficient to cause lysis of one cell.

myelin is then washed and the pellet consists of antigen–antibody and any C1 that has been fixed. The fixed C1 is assayed by adding the pellet to sheep erythrocytes which have been coated with anti-sheep erythrocyte antibody and C4b and C3b. Finally the whole mixture is incubated with all the other complement components except C1 so that the amount of lysis is proportional to the amount of C1 fixed. With this assay Koski et al. (1985) found complement-fixing antibodies to human myelin in 17 of 19 acute GBS sera. The titre of antibodies was highest in the samples collected earliest after onset and rapidly fell within three or four weeks to undetectable levels (Fig. 9.3). Similar antibodies were not found in other neurological diseases or in other inflammatory disorders (Koski et al. 1986).

Consistent with a pathogenic role for this antibody, activation of the terminal complement cascade has been demonstrated in GBS by showing the presence of C9 neoantigen in the serum. During complement activation membrane attack complexes are formed. Some are inserted into membranes and cause lysis. Excess C9 is incorporated into the inactive complex SC5b-9 consisting of S protein, C5b-8 and C9. During the formation of SC5b-9, C9 undergoes conformational changes which permit the recognition of C9 neoantigen. Raised

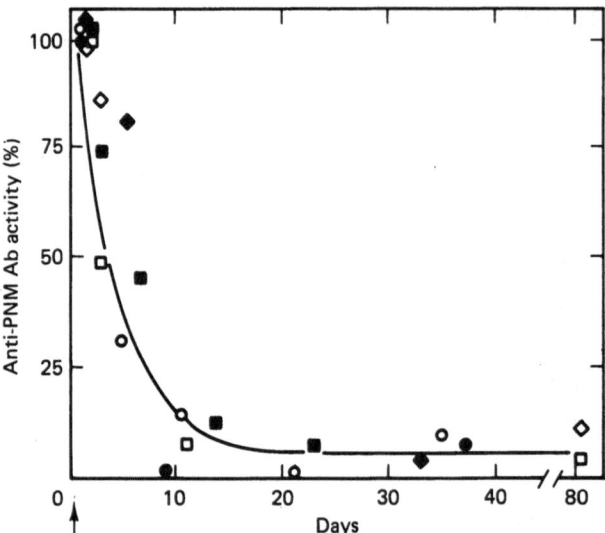

Fig. 9.3. Time course of disappearance of complement-fixing antibody to peripheral nerve myelin (Anti-PNM Ab) as detected by the C1 fixation and transfer assay in six patients with GBS. Each symbol represents a different patient. Antibody titres are expressed as a percentage of the initial titre obtained on admission. (From Koski et al. (1987), with permission.)

concentrations of SC5b-9 were detected with antibody to C9 neoantigen in the serum of all 19 patients with acute GBS, and six of seven patients with CIDP, but not in 10 healthy subjects (Koski et al. 1987).

Further definition of the antigen(s) with which antibodies in GBS serum react has been attempted. In experimental antimyelin antisera galactocerebroside is the major antigen with which complement-fixing antibodies react. Attempts to identify complement fixing antibodies to galactocerebroside have been negative except in rare cases (Hughes et al. 1984; Winer et al. 1988a). In Koski's C1 fixation and transfer assay human sciatic nerve myelin still retained antigenic activity with GBS serum after treatment with pronase suggesting that the antigen is not a protein. The myelin lipids were fractionated biochemically and incorporated into liposomes with the auxiliary lipids lecithin and cholesterol. Each fraction was tested with GBS sera in the C1 fixation and transfer assay and the major antigen reactivity was found to reside in a neutral glycolipid fraction. All of 10 GBS sera gave positive results with this fraction but none of six sera from disease controls or nine from normal subjects. In immunoblot studies, IgM from six of eight patients reacted with a minor glycolipid having a similar mobility to the Forssman lipid. The GBS sera also reacted with Forssman antigen incorporated into liposomes. Forssman antigen absorbed much of the antibody activity in some but not all of the sera (Koski et al. 1989). This evidence indicates that many patients with GBS have IgM antibodies which react or cross-react with the Forssman antigen. This antigen is widely distributed in many mammalian species, being present on the surface of sheep erythrocytes for instance. It consists of ceramide with five sugar residues: N-Ac-galactosaminyl-α(1–3)N-Ac-galactosaminyl-β-(1–3)galactopyranosyl-α-(1–3)galactopyranosyl-β-(1–4) glucopyranosyl-(1–1) ceramide (Siddiqui and Haromori 1971). It is generally regarded as being absent from human tissues. However, there are reports of its expression by neoplasms, and Koski et al. (1989) have shown that a monoclonal antibody to Forssman antigen reacts with a band in the neutral glycolipid fraction of human nerve myelin. It is going to take more work to resolve the pathogenetic importance of these antibodies to glycolipids since antibodies to protein and ganglioside antigens have also been identified in some patients with GBS.

Antibodies to Whole Nerve Preparations

Great immunological ingenuity has been brought to bear on trying to demonstrate antibodies to peripheral nerve in GBS. Nyland and Aarli (1978) used an antiglobulin consumption technique to measure the amount of globulin binding to lyophilised nerve. Positive results were obtained with 14 of 40 GBS and four of 40 control sera. Vedeler et al. (1982) used the mixed haemagglutination technique in which the test serum is applied to a section of human nerve. Binding of IgG is identified by adding an indicator system consisting of staphylococcal protein A attached to sheep erythrocytes by rabbit anti-sheep erythrocyte antibody. Fifteen sera from 42 patients with GBS, especially those with acute severe disease, gave positive results compared with none of 20 sera from normal subjects or 16 sera from patients with other diseases. In a subsequent study antibodies were reported in six of 11 patients with GBS and the presence of antibodies was thought to correlate with a response to plasma exchange (Vedeler et al. 1982). A more recent study confirmed the presence of these antibodies in 19 of 36 patients with GBS but there was no correlation between presence or titre of antibody and response to plasma exchange (Osterman et al. 1988). Antibodies to human nerve were identified in 11 of 16 GBS patients in a confirmatory report from another group who found positive results in only 3% of patients with other neuropathies (van Doorn et al. 1987). Thus the two groups using this mixed haemagglutination assay have identified antibodies to nerve tissue in about two-thirds of GBS sera and only a small percentage of control sera.

A more conventional technique for detecting binding of antibody to tissues is an indirect immunofluorescence method in which sections are incubated with test serum and washed and then any bound immunoglobulin is identified with an antibody bound directly or indirectly to fluorescein or peroxidase. Such techniques are fraught with difficulties, especially because immunoglobulin has the property of sticking non-specifically to tissues in general and myelin in particular by its Fc portion (Aarli et al. 1975). Tse et al. (1971) reported that four of six GBS patients had anti-myelin antibodies in their sera which could not be identified in a small number of control sera. Lisak et al. (1975) reported similar antibodies in sera from patients with a wide range of diseases but only at low dilutions. It was not clear from the earlier papers whether the antibody binding was to the myelin or axon. In our own experience binding to the axon is more common: 15.5% of 58 GBS sera gave this reaction compared with 3.4% of 58 controls without neurological disease (Winer et al. 1988a). From time to time these antibodies are rediscovered and increased titres of antiaxonal antibody have also been reported in Chagas disease (Khoury et al. 1979) and leprosy (Eustis-Turf et al. 1986). Binding to myelin was identified in 12% of the GBS and 1.7% of the control sera in our series (Winer et al. 1988a).

Antibodies to miscellaneous peripheral nerve proteins have been demonstrated in the sera of normal and diseased individuals with the immunoblot technique. This involves dissolving the tissue in a denaturing solution containing sodium dodecylsulphate and separating the proteins by electrophoresis in an acrylamide gel according to molecular size. The proteins are then transferred from the gel onto a nitrocellulose strip placed on its surface. The strip is probed with a test serum, washed and finally treated with a labelled anti-human immunoglobulin (class or light chain specific) to identify antibody binding. The results are difficult to interpret because binding to large numbers of bands may be found and the bands may be difficult to identify. Furthermore binding of the anti-Ig antibody to immunoglobulin heavy or light chains in the tissue extract must be distinguished. Finally, important antigenic epitopes may be lost when the proteins are dissolved and denatured in sodium dodecylsulphate. Nevertheless this technique has shown that normal individuals have antibodies to the 200 000 dalton neurofilament protein (Stefansson et al. 1985). Antibodies to other elements of the cytoskeleton have also been found in the sera of normal subjects and are more common following infections. IgM antibodies to intermediate filaments were demonstrated by immunofluorescence on cultured fibroblasts in three-quarters of post-infection sera and only 10% of controls (Toh et al. 1979). Antibodies to tubulin have also been demonstrated in normal sera and are present in higher titre after infectious mononucleosis (Mead et al. 1980; Guilbert et al. 1984). Antibodies to an unidentified 25–30 kd myelin protein were detected by an immunoblot technique as often in sera from normal blood donors as from patients with neuropathy (Cruz et al. 1987). Although GBS sera were included in this study no mention of a characteristic immunoblot pattern was made.

Antibodies to Glycolipids

The complement-fixing antibodies to neutral glycolipid already considered are only one example of several investigations of antibodies to glycolipid antigens. Following several small and conflicting reports of antibodies to cardiolipin (Harris et al. 1983; Colaco et al. 1984; Palosuo et al. 1985), we used an ELISA to screen a panel of 92 sera from patients with GBS and matched controls without neurological disease (Frampton et al. 1988). There was a highly significant increase in IgA antibodies to cardiolipin which were present in 23% of the GBS patients and only 4% of the controls (Fig. 9.4). There were also small increases in the percentage of sera having IgG (7.6% cp 3.7%) and IgM antibodies (18.5% cp 9.6%). It is possible that these anticardiolipin antibodies represent a response to the initial infection which triggered the GBS. While this does not preclude a role for these antibodies in the pathogenesis of GBS the presence and titre of antibodies could not be related to the course of GBS. Reactivity of anticardiolipin antibodies with myelin has not been reported. Cardiolipin is the antigen responsible for the Wasserman reaction in syphilis. Anticardiolipin antibodies are responsible for interfering with the partial thromboplastin clotting reaction and producing the "lupus anticoagulant effect", which is associated with a syndrome of recurrent abortions, thromboses and systemic lupus erythematosus-like illnesses (Harris et al. 1983).

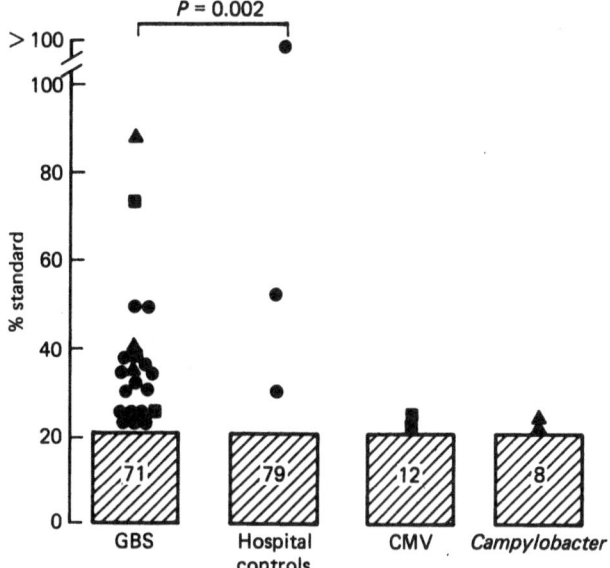

Fig. 9.4. IgA anticardiolipin antibodies in 92 GBS patients, 82 hospital controls without neurological disease, 14 patients with recent cytomegalovirus infection without neurological disease and 10 patients with recent *Campylobacter* infection without neurological disease. *Squares*, serological evidence of recent cytomegalovirus infection. *Triangles*, serological evidence of recent *Campylobacter* infection. (From Frampton et al. (1988), with permission.)

The failure to demonstrate complement-fixing antibodies in GBS to the major myelin glycolipid, galactocerebroside, has already been mentioned. Attempts to identify antibodies to galactocerebroside by ELISA have revealed values which are slightly higher in GBS than in normal control sera but no different from values obtained in controls with other neurological diseases (Hughes et al. 1984; Winer et al. 1988a; Rostami et al. 1987).

Using a sophisticated immunoblot technique one study reported antibodies to gangliosides in 20% of GBS sera but in none of the controls (Ilyas et al. 1988). The technique involves separating gangliosides on a thin-layer chromatograph or nitrocellulose strip, adding the test serum, and identifying antibody binding with a labelled antibody to human immunoglobulin. Five of 26 patients were found to have antibodies compared with none of 29 controls with or without neurological disease. The fine specificity of the antibodies differed from case to case being LM_1 (one case), GD1b (two cases), and both GD1a and GT1b (two cases). These differences indicate that the antibody specificity depends on the sugar sequences and not the ganglioside backbone. The significance of the presence of such antibodies to minor myelin constituents is not clear. Antibodies to gangliosides have been demonstrated by ELISA in motor neuron disease, lower motor neuronopathy and a wide range of autoimmune conditions. These antibodies are clearly not specific for demyelinating neuropathy. Nevertheless this remains an area of active research and the possibility remains that antibodies to an unidentified Schwann cell surface molecule which is a minor constituent of the membrane might have a major effect on its ability to maintain the integrity of the myelin.

Antibodies to Myelin Proteins

Because myelin P_2 protein was the first purified antigen demonstrated to induce EAN, much interest centred on identifying antibodies to this protein in GBS. The search has generally been unrewarding. Antibodies to P_2 were identified by ELISA in only one of 19 GBS sera in one series (Zweiman et al. 1983) and by binding of radioiodinated P_2 in only one of 16 GBS sera in our first series (Hughes et al. 1984). In a second series we used a sandwich ELISA in which P_2 is fixed to plastic by an antibody bridge and binding of the test antibody is then sought with a second antibody: even with this sensitive assay we could not identify any positive results among 20 GBS sera, despite selecting patients with both early and late severe disease (Winer et al. 1988a). Hemachudha et al. (1988) were also unable to identify antibodies to P_2 in any of eight sera from sporadic GBS cases or eight cases of GBS after rabies vaccine. There was no difference between the ELISA readings of 16 GBS patients and those of various controls with P_2, P_0 or myelin as antigen in another recent report (Geczy et al. 1985). In a more complicated assay (Luijten et al. 1985) reported the presence of B cells producing the anti-P_2 antibody in the blood of patients with GBS. The assay involves culturing lymphocytes in the presence of P_2 and showing that some of these lymphocytes will lyse sheep red cells coated with P_2: the sheep red cells and lymphocytes are incubated together in agar on a Petri dish in the presence of complement so that each lymphocyte secreting antibodies produces a plaque of lysis detectable by microscopy. In our attempt to reproduce this result we found that normal and GBS lymphocytes gave similar reactions with P_2 and the same effect could be produced if histone was used instead of P_2 as a control antigen (Winer et al. 1988a).

Antibodies to myelin basic protein are likewise not found in the sera of patients with GBS (Hughes et al. 1984; Hemachudha et al. 1988), although not surprisingly they are found in some patients with GBS following rabies vaccination (Hemachudha et al. 1987a,b).

Antibodies to Schwann Cells

Although it has been deduced from pathological evidence that the myelin sheath and not the Schwann cell is the primary target of attack in EAN and probably GBS, this is not firmly established. It is possible that a subtle immunological insult to the Schwann cell might first become evident as breakdown of the myelin sheath. Injection of a calcium ionophore into peripheral nerve induced just such a non-lethal insult to Schwann cells accompanied by myelin dissolution (Smith and Hall 1988). Accordingly consideration of antibodies to Schwann cells in GBS is appropriate. Cultured rat Schwann cells were incubated in the presence of test sera and lysis was assayed by ^{51}Cr release. Although GBS sera caused slightly more lysis than normal control sera, the results with GBS sera did not differ from those with sera from patients with axonal neuropathies, an all too familiar story (Lisak et al. 1984). Antibodies to human or rat Schwann cells could not be identified by ELISA in the sera from any of eight patients with sporadic GBS, or four patients with GBS following suckling mouse brain rabies vaccine, but were present in the sera of four patients who had GBS following Semple vaccine (Hemachudha et al. 1988).

Antibodies reacting with neuroblastoma cells in sera from a small number of GBS patients have also been reported (Rosenberg et al. 1975) but their significance is obscure and the work has not been reproduced or pursued.

Cellular Immunity

Circulating Cells

In the acute phase of GBS gross changes in the blood count cannot be detected in individual patients, but analyses of large groups show a slight reduction in lymphocytes (Winer et al. 1988a). This is due largely to a reduction in T lymphocytes and B cells are slightly increased in number (Nyland and Naess 1978). The occurrence of lymphocytotoxic antibodies in the serum of six of 13 patients with acute GBS compared with three of 93 controls has been briefly reported but not pursued (Seares et al. 1981). In early studies an abnormally large proportion of blood lymphocytes was shown to incorporate tritiated thymidine, indicating that they were actively dividing in the acute stage (Cook et al. 1971; Birnbaum 1973). We have recently shown that there is an increased percentage of T cells in the blood which bear the activation markers, DR, transferrin receptor or IL-2 receptor (Taylor and Hughes 1989) (Table 9.2). The increase in activated T cells is corroborated by the presence of higher concentrations of soluble IL-2 receptors, which are shed during cycles of T cell activation, in the serum of patients with GBS (Fig. 9.5). These studies of activated T cells do not indicate the nature of the antigen responsible for the activation but serve to confirm the activation of the immune system at the time of the development of what is usually an acute demyelinating process in GBS. There is no other way of demonstrating "inflammation" *in vivo*. Even nerve biopsies rarely include lymphocytic infiltration.

There is a minor change in the proportions of different T cell subsets in the acute stage of GBS. The details of the change have varied from study to study (Goust et al. 1978; Lisak et al. 1985; Hughes et al. 1983). In the largest study of 71 patients compared with age and sex-matched hospital controls there was no significant difference of total lymphocytes, CD3+ (T) lymphocytes or CD4+ (putative helper-inducer) lymphocytes but the CD8+ (putative cytotoxic-suppressor) lymphocytes were reduced in number compared with hospital controls, especially in the first week of the disease (Fig. 9.6). A similar reduction of CD8+ lymphocytes has been reported in the active stage of other inflammatory diseases including multiple sclerosis. Since the reduction is not specific for GBS it is unlikely to have an exotic explanation such as the sharing of antigenic epitopes between myelin or Schwann cells and CD8+ lymphocytes. It might be due to the migration of CD8+ lymphocytes into peripheral nerve tissue but this is unlikely: such information as is available suggests that the proportion of CD8+/CD4+ lymphocytes infiltrating into nervous tissue approximates that in the blood and that there is no differential influx of CD8+ cells (Cornblath et al. 1990). It seems more likely that the reduction in CD8+ cells is a non-specific accompaniment of an active immune response whose functional significance is uncertain.

Table 9.2. Clinical features and activated T cells in GBS patients and controls

Number	Sex	Age	Preceding infection		GBS		Infection at time	% T cells positive for:		
			Type	Day	Day	Severity		DR	TFR	IL-2R
1	F	75	0	0	3	5	chest	3.0	1.5	0
2	F	35	GEI	−7	3	5	chest	3.4	1.0	1.0
3	M	19	0	0	4	4	0	3.3	4.5	5.4
4	F	54	GEI	−14	6	5	septicaemia	9.4	5.9	6.0
5	M	45	GEI	−14	10	5	0	1.0	4.2	5.3
6	M	65	URTI	−14	11	5	chest	6.0	6.2	4.3
7	M	68	URTI	−28	12	5	0	0.5	0.5	0
8	M	61	URTI	−8	13	5	0	4.5	4.4	5.0
9	M	43	chicken pox	−28	24	2	0	3.7	1.8	2.7
10	F	33	sinusitis	−34	28	5	0	3.5	3.3	3.7
All patients, median 49					10.5	5		3.4	3.3	3.7
range 19–75					3–28	2–4		0.5–9.4	0.5–6.2	0–6.0
All controls, median 49								1.4	0	0
range 18–80								0–4.3	0–2.5	0–2.9

From Taylor and Hughes (1989), with permission.
Abbreviations: GE, gastroenteritis; URTI, upper respiratory tract infection. Preceding infection day: day of onset of preceding infection with respect to day of onset of GBS. GBS day: days since GBS onset at time of testing.

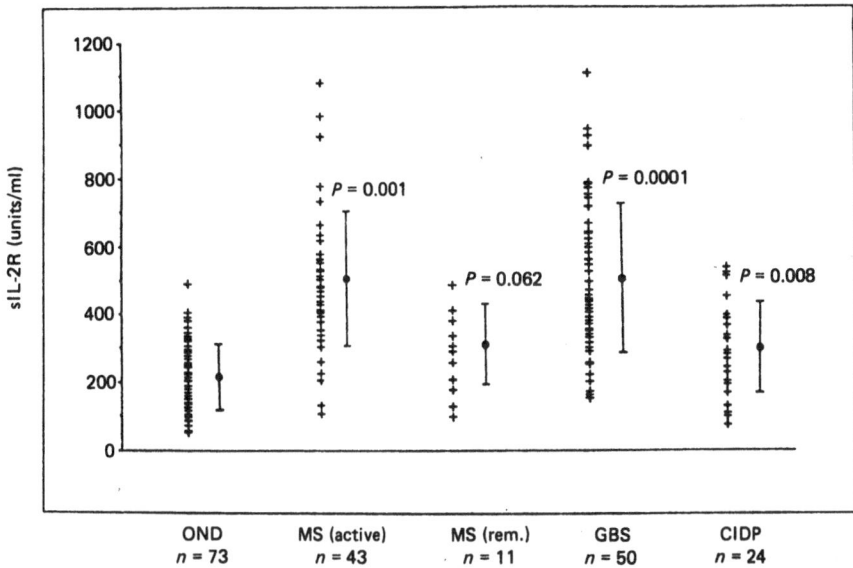

Fig. 9.5. Soluble IL-2 receptor concentrations in patients with GBS, CIDP, multiple sclerosis (MS) and normal control subjects. (From Hartung et al. (1990), with permission.)

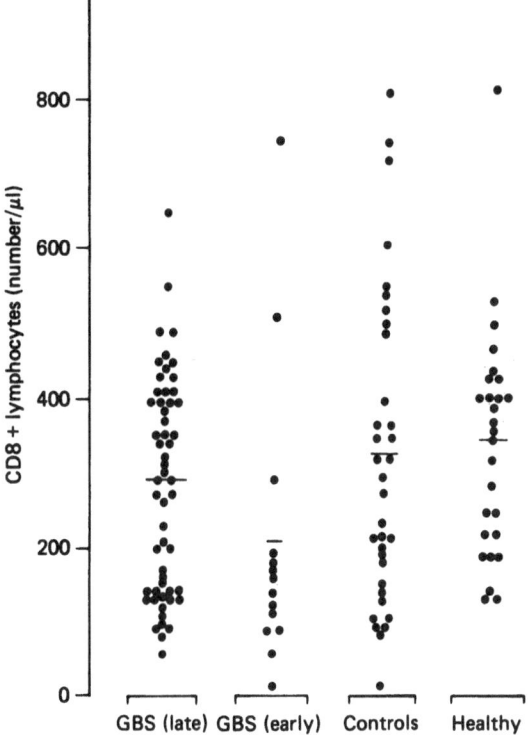

Fig. 9.6. CD8+ lymphocytes/μl blood in patients with late GBS (after the first week), early GBS (during the first week after onset), age, sex-matched hospital controls without neurological disease and healthy blood donors. (From Winer et al. (1988a), with permission.)

Lymphocyte Transformation Tests with Nerve Antigens

The simplest method of testing cell-mediated immunity in vitro is to incubate blood mononuclear cells with antigen in tissue culture and then to measure the amount of lymphocyte stimulation by adding tritiated thymidine, and counting the amount of radioactivity incorporated into the cells. The culture conditions, antigen concentration and incubation periods have to be optimised and there is plenty of room for variation in techniques from laboratory to laboratory. This may explain why there has been tremendous variation in the results of tests with this technique with both crude and purified antigens (Table 9.3). The reports divide into those by authors who have found positive responses in almost all patients and those who have found responses in only a small proportion. Having obtained negative results in earlier series of patients, we have been impressed by finding marked transformation to P_2 in the first few days of GBS in two patients which disappeared later in the disease and which were not found in CIDP or control patients. In this assay the technique was modified in line with modern practice and small numbers of responding cells and irradiated autologous mononuclear cells were used as antigen presenting cells (Taylor and Hughes, unpublished). This recipe appears to have identified a significant response to P_2 but only in a small proportion of patients.

Other In Vitro Tests of Cell-Mediated Immunity to Nerve Antigens

Several other techniques have been used to demonstrate cell-mediated immunity to nerve antigens in GBS. One of the earliest used the slowing of macrophage migration in an electrophoretic field in the presence of sensitised lymphocytes. Positive results were obtained with acid extracts of both human brain and nerve

Table 9.3. Lymphocyte transformation with nerve antigens in GBS: proportion of subjects with positive results

Reference	Antigens	GBS		Other neuropathy	OND or normal
		Acute	Convalescent		
Currie and Knowles (1971)	Human nerve acid extract	8/12		0/9	1/9
Behan et al. (1970)	Human nerve	9/9	0/9	–	–
Abramsky et al. (1975)	Bovine P_2	7/9	0/4	–	0/24
Zweiman et al. (1983)	Bovine P_2	0/10		–	0/6
Hughes et al. (1984)	Human or bovine P_2	0/11		–	0/14
Korn-Lubetzki and Abramsky (1986)		8/10[a]		1/44[a]	
Taylor and Hughes (unpublished)	Bovine P_2	2/10		0/10	0/10
	P_2 synthetic peptide 55–81	1/10		0/10	0/10

[a] these are the proportions taking a stimulation index >3.0 as the threshold for abnormality. Setting the threshold at >2.0 the proportions would be 18/30 and 2/44.

with all of 16 GBS patients, especially in the acute stage (Caspary et al. 1971). This test has not been adopted by other laboratories and the work has not been pursued. An ambitious early attempt to demonstrate a demyelinating effect of blood mononuclear cells involved adding buffy coat cells from patients with GBS to myelinated rodent tissue cultures: cells from seven of nine patients were reported to induce demyelination whereas cells from seven control patients did not (Arnason et al. 1969). This result was not reproduced in a subsequent study (Feasby et al. 1982) and its interpretation is difficult. Since T helper cells are now known to require the presentation of antigen in the context of their own MHC antigen it seems unlikely that this demyelination was mediated via a specific T cell response. It might have been due to the presence of activated lymphocytes or macrophages releasing lymphokines.

The most exploited test has been the macrophage migration inhibition test in which lymphocytes are incubated with antigens and their supernatant is added to guinea-pig macrophages. Sensitised lymphocytes secrete macrophage-inhibitor factor which slows the migration of macrophages. Macrophages are packed into a capillary tube and allowed to migrate out into medium containing supernatant from the lymphocytes being tested. The area of migration can be drawn or photographed and measured. The test is complex and involves the interaction of many factors, some inhibiting and some enhancing macrophage migration. It is a tribute to the enthusiasm of those who have employed this test that they have consistently found inhibition of macrophage migration by either whole nerve homogenates or P_2 protein (Table 9.4).

One laboratory has reported strikingly positive results indicating immunisation to P_2 specifically in GBS with a little used test of macrophage/monocyte procoagulant activity. The test is reputed to be more sensitive than lymphocyte transformation or macrophage-inhibiting factor assays and to correlate closely with delayed hypersensitivity. Guinea pigs with EAN have been shown to have specific immunity to peripheral nerve antigens with this test (Geczy et al. 1985). The test involves incubating blood mononuclear cells with antigen overnight, then adding the washed suspension to citrated plasma with calcium and simply measuring the time taken by the plasma to coagulate. The activated cells accelerate the coagulation of the plasma and results are expressed by comparison with a standard curve constructed from the results of adding a known amount of thromboplastin. Sheep myelin, myelin basic protein and P_2 proteins were used as antigens. Very low concentrations of P_2 (0.01 µg/ml) induced significantly positive results with this test in 13 of 16 patients with GBS. Two hundred fold greater concentrations of myelin basic protein and myelin were necessary to produce a positive result and even that was weaker. Cells from normal subjects and a wide range of controls with other diseases gave negative results (Geczy et al. 1985). These results appear very convincing but they stand alone. We have not succeeded in setting up the assay and no confirmatory reports have been published.

Summary

Summarising the rather messy literature concerning the immunology of GBS is not easy and remains susceptible to the bias of the reviewer. It is clear that no

Table 9.4. Macrophage migration inhibitory factor production with nerve antigens in GBS

Reference		Acute	Convalescent	Other neuropathy	Other neurological disease or normal	Comment
Rocklin et al. (1971)	Human nerve homogenate	4/7		0/13	0/5	
Behan et al. (1972)	Human nerve homogenate	3/3	1/6	–	0/40	
Sheremata et al. (1975)	Human nerve homogenate	19/34	–	3/33	0/33	
Castaigne et al. (1972)	Human nerve homogenate	10/13	7/18	3/19	11/60	Leucocyte not macrophage migration inhibition
Sheremata et al. (1975)	Rabbit P_2	8/12	–	0/5	0/12	P_2 contained myelin basic protein

single laboratory test has been adopted as defining the condition. The charact-eristically elevated CSF protein concentration and normal cell count remain the best laboratory guideline. Cerebrospinal fluid immunoglobulin concentrations are increased and oligoclonal bands may be present, partly due to leakage and partly due to intrathecal synthesis. Associations between GBS and other auto-immune diseases are rare but there is an association with glomerulonephritis, which is usually mild and has not been explained. GBS occurs more commonly than would be expected in situations of partial immunodeficiency. There is no striking HLA association. Immunoglobulin concentrations are somewhat in-creased in the serum. Immune complexes are found in some cases and sensitive tests display evidence of complement fixation and activation. Demyelinating factors are present in the serum but their antibody nature and importance in pathogenesis have not been demonstrated. Antibodies to whole nerve, and especially myelin, are present in some patients. The exact proportion is not clear: some reports claim positive results in almost all patients early in the disease and others positive results in only 10%–20% of patients. The antigen(s) which these antibodies recognise has not been unequivocally identified. The likelihood is that several antigens are involved and the identity of antigen varies according to the individual and the precipitating infection. Candidates are neutral glycolipids related to Forssman antigen, phospholipids related to cardiolipin, gangliosides, myelin P_2 protein and P_0 glycoprotein. Circulating T lymphocytes, especially CD8+ cells, are reduced in acute GBS and the pro-portion of activated cells is increased. Tests of cell-mediated immunity have demonstrated sensitisation to nerve antigens in almost all patients according to some reports but only 10%–20% in other studies. The most recent results with the most straightforward techniques suggest that a small proportion of patients show cell-mediated immunity to P_2 protein. The results are consistent with a heterogeneous pathogenesis for GBS, some patients having disease induced by antibodies to myelin glycolipids or gangliosides and others by a T cell driven response to myelin P_2 protein.

References

Aarli JA, Aparicio SR, Lumsden CE, Tonder O (1975) Binding of normal human IgG to myelin sheaths, glia and neurons. Immunology 28:171–185

Abbas DH, Schlagenhauff RE, Strong HE (1977) Polyradiculoneuropathy in Addison's disease. Neurology 27:494–495

Abramsky O, Webb C, Teitelbaum D, Arnon R (1975) Cell mediated immunity to neural antigens in idiopathic polyneuritis and myeloradiculitis. Neurology 25:1154–1159

Abramsky O, Korn-Lubetzky I, Teitelbaum D (1980) Association of autoimmune diseases and cellular immune response to the neuritogenic protein in Guillain–Barré syndrome. Ann Neurol 8:117

Adams D, Gibson JD, Thomas PK et al. (1977) HLA antigens in Guillain–Barré syndrome. Lancet ii:504–505

Ala FA, Shearman DSC (1965) A case of autoimmune hemolytic anaemia, thrombocytopenia, and the Guillain–Barré syndrome. Acta Haematol (Basel) 34:319–369

Amarenco P, Sauron B, Schuller E, Chain F, Castaigne P (1987) Serum and CSF humoral immunity in Guillain–Barré syndrome: clinical correlations. J Neurol Sci 80:129–142

Arnason BGW, Winkler GF, Hadler NM (1969) Cell-mediated demyelination of peripheral nerve in tissue culture. Lab Invest 21:1–10

Behan PO, Lamarche JB, Feldman RG, Sheremata WA (1970) Lymphocyte transformation in the Guillain–Barré syndrome Lancet i:421

Behan PO, Behan WMH, Feldman RG, Kies MW (1972) Cell mediated hypersensitivity to neural antigens. Occurrence in human patients and non-human primates with neurological diseases. Arch Neurol 27:145–152

Behan PO, Lessels S, Roche M (1976) Optic neuritis in the Landry–Guillain–Barré–Strohl syndrome. J Ophthal 60:58–59

Behar R, Penny R, Powell HC (1986) Guillain–Barré syndrome associated with Hashimoto's thyroiditis. J Neurol 233:233–236

Bertinelli A, Giani M, Rossi L et al. (1989) Ex novo cases of acute glomerulonephritis and Guillain–Barré syndrome: a case report. Clin Nephrol 31:269–273

Best PV (1985) Acute polyradiculoneuritis associated with demyelinated plaques in the central nervous system: report of a case. Acta Neuropathol 67:230–234

Birnbaum G (1973) Guillain–Barré syndrome. Increased lymphoproliferative potential. Arch Neurol 28:215–218

Bornstein MB, Appel SH (1961) The application of tissue culture to the study of EAE: I. Patterns of demyelination. J Neuropathol Exp Neurol 20:141–157

Bourouresque G, Delpuech F, Giudicelli R et al. (1981) Polyradiculoneuritis and myasthenia gravis. Nouv Presse Méd 10:253–254

Bradbury K, Aparicio SR, Sumner DW et al. (1985) Comparison of in vitro demyelination and cytotoxicity of humoral factors in MS and other neurological diseases. J Neurol Sci 70:167–181

Bradford JB, Bashford EF, Wilson JA (1918) Acute infective polyneuritis. Q J Med 12:88–126

Brock S, Davidson C (1947) Fatal cryptogenic neuropathy. Arch Neurol Psychiatry 58:559–569

Brown MJ, Rosen JL, Lisak RP (1987) Demyelination in vitro by Guillain–Barré syndrome and other human serum. Muscle Nerve 10:263–271

Caspary EA, Currie S, Walton JN, Field EJ (1971) Lymphocyte sensitisation to nervous tissues and muscle in patients with Guillain–Barré syndrome. J Neurol Neurosurg Psychiatry 34:179–181

Castaigne P, Berthaux P, Brunet P (1972) Polyradiculonévrites inflammatoires et immunité cellulaire: étude de la réponse immunitaire cellulaire envers des antigènes de nerf périphérique par le test de migration des leucocytes. Nouv Presse Méd 1:2445–2449

Colaco CB, Scadding GK, Newsom Davis J (1984) Anticardiolipin antibodies in neurological diseases. Lancet i:164

Cook SD, Dowling PC (1981) The role of autoantibody and immune complexes in the pathogenesis of Guillain–Barré syndrome. Ann Neurol 9 suppl:70–79

Cook SD, Dowling PC, Whitaker JN (1970) Serum immunoglobulins in the Guillain–Barré syndrome. Neurology 20:403

Cook SD, Dowling PC, Murray MR, Whittaker JN (1971) Circulating demyelinating factors in acute idiopathic polyneuropathy. Arch Neurol 24:136–144

Cornblath DR, Griffin JW, Tennekoon GI (1986) Immunoreactive myelin basic protein in cerebrospinal fluid of patients with peripheral neuropathies. Ann Neurol 20:370–372

Cornblath DR, Griffin DE, Welch D, Griffin JW, McArthur JC (1990) Quantitative analysis of endoneurial T-cells in human sural nerve biopsies. J Neuroimmunol 26:113–118

Corston RN (1980) Temporal arteritis in association with Guillain–Barré syndrome. Br Med J 280:292–293

Corston RN, McGale EHF, Stonier C, Aber GM, Hutchinson EC (1981) Abnormalities of CSF amino acids in patients with the Guillain–Barré syndrome. J Neurol Neurosurg Psychiatry 44:86–89

Cruz M, Ernerudh J, Olsson T, Hoseberg B, Link H (1987) Occurrence and isotype of antibodies against peripheral nerve myelin in serum from patients with peripheral neuropathy and healthy controls. J Neurol Neurosurg Psychiatry 51:820–825

Currie S, Knowles M (1971) Lymphocyte transformation in the Guillain–Barré syndrome. Brain 94:109–116

Cuttner J, Meyer R (1978) Guillain–Barré syndrome in a patient with Hodgkin's disease. Mt Sinai J Med (NY) 45:415–417

Drachman DA, Paterson PY, Berlin B, Rogusa J (1970) Immunosuppression and Guillain–Barré syndrome. Neurology 20:390

Dubois-Dalcq M, Buyse M, Buyse G, Gorce F (1971) The action of Guillain–Barré serum on myelin. A tissue culture and electron microscope analysis. J Neurol Sci 13:67–83

Dummolard A, Sarrouy J, Schonshoe R, Baredoux N (1937) Polyradiculonévrite avec hyperalbuminose du liquide cephalorachidien sans réaction cellulaire et névrite optique-evolution rapide, vers le guérison (Guillain–Barré). Rev Oto-Neuro-Ophthalmol (Paris) 15:26–28

Eustis-Turf EP, Benjamins JA, Lefford JL (1986) Characterization of the anti-neural antibodies in the sera of leprosy patients. J Neuroimmunol 10:313–330

Feasby TE, Hahn AF, Gilbert JJ (1982) Passive transfer studies in Guillain–Barré polyneuropathy. Neurology 32:1159–1167

Forrester C, Lascelles RG (1979) Association between polyneuritis and multiple sclerosis. J Neurol Neurosurg Psychiatry 42:864–866

Frampton G, Winer JB, Cameron JS, Hughes RAC (1988) Severe Guillain–Barré syndrome: an association with IgA anti-cardiolipin antibody in a series of 92 patients. J Neuroimmunol 19: 133–139

Froelich CJ, Searles RP, Davis LE, Goodwin JS (1980) A case of Guillain Barré syndrome with immunologic abnormalities. Ann Int Med 93:563–565

Geczy C, Raper R, Roberts IM, Meyer P, Bernard CCA (1985) Macrophage procoagulant activity as a measure of cell-mediated immunity to P_2 protein of peripheral nerves in the Guillain–Barré syndrome. J Neuroimmunol 9:179–191

Gorodezky C, Varela B, Castro-Escobar LE, Chavez-Negrete A, Escobar-Gutierrez A, Martinez-Mata J (1983) HLA-DR antigens in Mexican patients with Guillain–Barré syndrome. J Neuroimmunol 4:1–7

Goust JM, Chenais F, Carnes JE, Hames CG, Fudenberg HH, Hogan EL (1978) Abnormal T-cell populations and circulating immune complexes in the Guillain–Barré syndrome and MS. Neurology 28:421–428

Grant H, Leopold HN (1954) Guillain–Barré syndrome occurring during cortisone therapy. JAMA 155:252–253

Gross PT (1980) Acute idiopathic polyneuritis and idiopathic thrombocytopenic purpura. JAMA 243:256–257

Guilbert L, Dighiero G, Avramea S (1984) Naturally occurring antibodies against urine common antigens in neural humans. I. Detection, isolation and characterization. J Immunol 128:2779–2787

Guillain G, Barré JA, Strohl A (1916) Sur un syndrome de radiculonévrite avec hyperalbuminose du liquide cephalo-rachidien sans réaction cellulaire. Remarques sur les caractères cliniques et graphiques des reflexes tendineux. Bull Soc Méd Hôp Paris 40:1462–1470

Harris EN, Engelert H, Derve G, Hughes GRV, Gharavi A (1983) Antiphospholipid antibodies in acute Guillain–Barré syndrome. Lancet ii:1361–1362

Harrison BM, Hansen LA, Pollard JD, McLeod JG (1984) Demyelination induced by serum from patients with Guillain–Barré syndrome. Ann Neurol 15:163–170

Hartung H-P, Schwenke C, Bitter-Suermann D, Toyka KV (1987) Guillain–Barré syndrome: activated complement components C3a and C5a in CSF. Neurology 37:1006–1009

Hartung H-P, Hughes RAC, Taylor WA, Heininger K, Reiners K, Toyka KV (1990) T cell activation in the Guillain–Barré syndrome and in MS: elevated serum levels of soluble IL-2 receptors. Neurology 40:215–219

Hemachudha T, Griffin DE, Giffels JJ, Johnson RT, Moser AB, Phanupak P (1987a) Myelin basic protein as an encephalitogen in encephalomyelitis and polyneuritis following rabies vaccination. N Engl J Med 316:369–374

Hemachudha T, Phanuphak P, Johnson RT, Griffin DE, Ratanvongsiri J, Siripramsomsup W (1987b) Neurologic complications of semple-type rabies vaccine: clinical and immunologic studies. Neurology 37:550–556

Hemachudha T, Griffin DE, Chen WW, Johnson RT (1988) Immunologic studies of rabies vaccination-induced Guillain–Barré syndrome. Neurology 38:375–378

Hirano A, Cook SD, Whitaker JN, Dowling PC, Murray MR (1971) Fine structural aspects of demyelination in vitro the effects of Guillain–Barré serum. J Neuropathol Exp Neurol 30: 249–265

Hughes RAC (1979) Acute inflammatory polyneuropathy. In: Rose FC (ed) Clinical neuroimmunology. Blackwell, Oxford, pp 170–184

Hughes RAC, Cameron JS, Hall SM, Heaton J, Payan JA, Teoh R (1982) Multiple mononeuropathy as the initial presentation of systemic lupus erythematosus – nerve biopsy and response to plasma exchange. J Neurol 228:239–247

Hughes RAC, Aslan S, Gray IA (1983) Lymphocyte subpopulations and suppressor cell activity in acute polyradiculoneuritis (Guillain–Barré syndrome). Clin Exp Immunol 51:448–454

Hughes RAC, Gray IA, Gregson NA et al. (1984) Immune responses to myelin antigens in Guillain–Barré syndrome. J Neuroimmunol 6:303–312

Ilyas AA, Willison HS, Quarles RH, Jungalwala KB, Cornblath DR (1988) Serum antibodies to gangliosides in Guillain–Barré syndrome. Ann Neurol 23:440–447

Julien J, Vital CL, Aupy G, Lagueny A, Darriet D, Brechenmacher C (1980) Guillain–Barré syndrome and Hodgkin's disease – ultrastructural study of a peripheral nerve. J Neurol Sci 45:23–27

Kaslow RA, Sullivan-Bolyai JZ, Hafkin B et al. (1984) HLA antigens in Guillain–Barré syndrome. Neurology 34:240–242

Kennett RP, Morgan SH, Dudley C, Mackworth-Young C, Hull R, Hughes GR (1986) Acute polyradiculoneuropathy complicating systemic lupus erythematosus. Postgrad Med J 62:291–294

Khoury EL, Ritacco V, Cossio PM et al. (1979) Circulating antibodies to peripheral nerve in American trypanosomiasis (Chagas disease). Clin Exp Immunol 36:8–15

Klingon GH (1965) The Guillain–Barré syndrome associated with cancer. Cancer 18:157–163

Korn-Lubetzki I, Abramsky O (1986) Acute and chronic demyelinating inflammatory polyradiculo-neuropathy. Arch Neurol 43:604–608

Koski CL, Humphrey R, Shin ML (1985) Anti-peripheral myelin antibodies in patients with de-myelinating neuropathy: quantitative and kinetic determination of serum antibodies by com-plement component transfer and fixation. Proc Natl Acad Sci USA 82:905–909

Koski CL, Gratz E, Sutherland J, Mayer RF (1986) Clinical correlation with anti-peripheral-nerve myelin antibodies in Guillain–Barré syndrome. Ann Neurol 19:573–577

Koski CL, Sanders ME, Swoveland PT et al. (1987) Activation of terminal components of comple-ment on patients with Guillain–Barré syndrome and other demyelinating neuropathies. J Clin Invest 80:1492–1497

Koski CL, Chou DKH, Jungalwala FB (1989) Anti-peripheral nerve myelin antibodies in Guillain–Barré syndrome bind a neutral glycolipid of peripheral myelin and cross-react with Forssman antigen. J Clin Invest 84:280–287

Kruger H, Englert D, Pfughaupt KW (1981) Demonstration of oligoclonal immunoglobulin G in Guillain–Barré syndrome and lymphocytic meningoradiculitis by isoelectric focusing. J Neurol 226:15–24

Lassmann H, Budka H, Schnabeth G (1981) Inflammatory demyelinating polyradiculitis in a patient with multiple sclerosis. Arch Neurol 38:99–102

Latov N, Gross RB, Kastelman J et al. (1981) Complement fixing antiperipheral nerve myelin antibodies in patients with inflammatory polyneuritis and with polyneuropathy and parapro-teinemia. Neurology 31:1530–1534

Link H (1975) Demonstration of oligoclonal immunoglobulin in Guillain–Barré syndrome. Acta Neurol Scand 52:111–120

Link H, Wahren B, Norrby E (1979) Pleocytosis and immunoglobulin changes in CSF and herpes virus serology in patients with Guillain–Barré syndrome. J Clin Microbiol 9:305–316

Lisak RP, Zweiman B, Norman M (1975) Antimyelin antibodies in neurologic diseases: immuno-fluorescent demonstration. Arch Neurol 32:163–167

Lisak RP, Mitchell M, Zweiman B, Orrechio E, Asbury AK (1977) Guillain–Barré syndrome and Hodgkin's disease: three cases with immunological studies. Ann Neurol 1:72–78

Lisak RP, Kuchmy D, Armati-Gulson PJ, Brown MJ, Sumner AJ (1984) Serum mediated Schwann cell cytotoxicity in the Guillain–Barré syndrome. Neurology 34:1240–1243

Lisak RP, Zweiman B, Guerrero F, Moskovitz AR (1985) Circulating T cell subsets in Guillain–Barré syndrome. J Neuroimmunol 8:93–101

Livrea P, Zimatore GB, Simone IL et al. (1980) Isoelectric focusing and quantitative estimation of cerebrospinal fluid and serum IgG in idiopathic polyneuropathy. J Neurol 223:1–12

Low PA, Schmelzer J, Dyck PJ, Kelly JJ (1982) Endoneurial effects of sera from patients with acute inflammatory polyradiculoneuropathy: electrophysiological studies on normal and demyelinated rat nerves. Neurology 32:720–724

Luijten JAFM, De Jong WAC, Demel RA, Heijnen CJ, Ballieux RE (1985) Peripheral nerve P_2 basic protein and the Guillain–Barré syndrome. In vitro demonstration of P_2-specific antibody secreting cells. J Neurol Sci 66:209–216

MacGregor GA (1965) Familial Guillain–Barré syndrome. Lancet ii:1296

Mactier RA, Khanna R (1987) Guillain–Barré syndrome in kappa light chain myeloma. South Med J 80:1054–1055

Martin JJ (1961) Polyradiculonévrite avec dissociation albuminocytologique (du type Guillain–Barré) et atteinte optique (contribution anatomo-clinique). Psychiatry Neurol 142:265–292

McCombe PA, Clarke P, Frith JA et al. (1985) Alpha-1 antitrypsin phenotypes in demyelinating disease: an association between demyelinating disease and the allele PiM3. Ann Neurol 18:514–516

Mead GM, Cowin P, Whitehouse IMA (1980) Antitubulin antibody in healthy adults and patients with infectious mononucleosis and its relationship to smooth muscle antibody. Clin Exp Immunol 39:228–336

Melnick SC (1963) 38 cases of the Guillain–Barré syndrome; an immunological study. Br Med J 1:368–373

Moreila Filho PF, Cini Cinatus D, Freitas MRG, Nascimento JM, Porto FJS, Santos PC (1980) Guillain–Barré syndrome as a manifestation of SLE. Arq Neuropsiquiatr 38:165–170

Mouren P, Vital-Bernard P, Poinso Y et al. (1971) Névrite optique retrobulbaire au cours d'un syndrome de Guillain–Barré. Rev Oto-Neuro-Ophtalmol (Paris) 43:216–221

Murphy BF, Gonzales MF, Ebeling P, Fairley KF, Kincaid-Smith P (1986) Membranous glomerulonephritis and Landry–Guillain–Barré syndrome. Am J Kidney Dis 8:267–270

Nikoskelainen E, Riekkinen P (1972) Retrobulbar neuritis as an early symptom of Guillain–Barré syndrome: report of a case. Acta Opthalmol (Copenh) 50:111–115

Nyland H, Aarli JA (1978) Guillain–Barré syndrome; demonstration of antibodies to peripheral nerve tissue. Acta Neurol Scand 58:35–43

Nyland H, Naess A (1978) Lymphocyte subpopulations in blood and cerebrospinal fluid from patients with acute Guillain–Barré syndrome. Eur Neurol 17:247–252

Okuyama H (1975) Mechanism of increase of the cerebrospinal fluid (CSF) protein content in Guillain–Barré syndrome. Clin Neurol (Tokyo) 15:817–824

Osterman PO, Vedeler CA, Ryberg B, Fagius J, Nyland H (1988) Serum antibodies to peripheral nerve tissue in acute Guillain–Barré syndrome in relation to outcome of plasma exchange. J Neurol 233:285–289

Palosuo T, Vaarala O, Kinnunen E (1985) Cardiolipin antibodies in Guillain–Barré syndrome. Lancet ii:839

Potz G, Neundorfer B (1975) Polyradiculoneuritis and Hashimoto's thyroiditis. J Neurol 210:283–289

Pradalier A, Gettler V, Feyeux C, Dry J (1982) Guillain–Barré polyradiculopathy and segmental colitis. Nouv Presse Méd 11:3573

Rocklin RE, Sheremata WA, Feldman RG, Kies MW, David JR (1971) Cellular response in Guillain–Barré syndrome and multiple sclerosis. N Engl J Med 284:803–808

Rodriguez-Iturbe B, Garcia R, Rubio L, Zabala J, Maros G, Torres R (1973) Acute glomerulonephritis in the Guillain–Barré–Strohl syndrome. Report of 9 cases. Ann Intern Med 78:391–395

Rosenberg RN, Aung MH, Tindall RSA (1975) Idiopathic polyneuropathy associated with cytotoxic anti-neuroblastoma serum: IgG and IgM immunoglobulin studies. Neurology 25:1101–1110

Rostami AM, Burns JB, Eccleston PA, Manning MC, Lisak RP, Silberberg DH (1987) Search for antibodies to galactocerebroside in the serum and CSF in human demyelinating disorders. Ann Neurol 22:381–382

Rozen D, VanHaeverbeek M, Stenuit R, Ardichvili D (1980) Polyradiculitis, B. fragilis septicemia, peritonitis in a patient with SLE. Rev Méd Brux 1:395–398

Ryberg B (1984) Extra and intrathecal production of antinerve and antibrain antibodies in Guillain–Barré syndrome: evaluation by an antibody index. Neurology 34:1378–1381

Sabin AB, Aring CD (1941) Visceral lesions in infectious polyneuritis. Am J Pathol 17:469–481

Sadovnick AD, Paty DW, Yannakoulias G (1989) Concurrence of multiple sclerosis and inflammatory bowel disease. N Engl J Med 321:762–763

Saida T, Saida K, Lisak RP, Brown MJ, Silberberg DH, Asbury AK (1982) In vivo demyelinating activity of sera from patients with Guillain–Barré syndrome. Ann Neurol 11:69–75

Sanders EACM, Lee KD (1987) Acute Guillain–Barré syndrome in multiple sclerosis, J Neurol 234:128

Sanders EACM, Yewdall VMA, Hughes RAC, Cameron JS (1988) Absence of complement activation in demyelinating polyradiculoneuropathies. Ann Neurol 23:102–103

Sanders ME, Koski CL, Robbins D (1986) Activated terminal complement in CSF in Guillain–Barré syndrome and multiple sclerosis. J Immunol 136:4456–4459

Saunders M, Rake M (1965) Familial Guillain–Barré syndrome. Lancet ii:1106–1107

Seares RP, Davis LE, Hermanson S (1981) Lymphocytotoxic antibodies in Guillain–Barré syndrome. Lancet i:273

Segurado OG, Kruger H, Mertens HG (1986) Clinical significance of serum and CSF findings in the Guillain–Barré syndrome and related disorders. J Neurol 233:202–208

Seil FJ (1977) Tissue culture studies of demyelinating disease: a critical review. Ann Neurol 2:345–355

Sheremata WA, Colby S, Karkhanis Y, Eylar E (1975) Cellular hypersensitivity to basic myelin (P2) protein in the Guillain–Barré syndrome. Can J Neurol Sci 2:87–90

Siddiqui B, Haromori S-I (1971) A revised structure for the Forssman glycolipid hapten. J Biol Chem 246:5766

Smith KJ Hall SM (1988) Peripheral demyelination and remyelination initiated by the calcium-selective ionophore ionomycin: in vivo observations. J Neurol Sci 83:37–53

Spaar FW, Orthner H (1968) Adrenalitis inclusio necroticans in the Landry–Guillain–Barré syndrome. Dtsch Z Nervenheilkd 193:195–213

Stefansson K, Marton LS, Dieperink ME, Molinar GK, Schlaepfer WW, Helgason CM (1985) Circulating autoantibodies to the 200 000 dalton protein of neurofilaments in the serum of healthy individuals. Science 228:1117–1119

Steiner I, Wirguin I, Abramsky O (1986) Appearance of Guillain–Barré syndrome in patients during corticosteroid treatment. J Neurol 233:221–223

Stewart GJ, Pollard JD, McLeod JG, Wolnizer CM (1978) HLA antigens in the Landry–Guillain–Barré syndrome and chronic relapsing polyneuritis. Ann Neurol 4:285–289

Sumner AJ, Saida K, Saida T, Silberberg DH, Asbury AK (1982) Acute conduction block associated with experimental antiserum mediated demyelination of peripheral nerve. Ann Neurol 11: 469–477

Tachovsky T, Lisak RP, Koprowski H (1976) Circulating immune complexes in multiple sclerosis and other neurological disorders. Lancet ii:997–999

Taylor WA, Hughes RAC (1989) T lymphocyte activation antigens in Guillain–Barré syndrome and chronic idiopathic demyelinating polyradiculoneuropathy. J Neuroimmunol 24:33–39

Toh BH, Yildiz A, Sotelo J, Holborow EJ, Kanankoundi F (1979) Viral infections and IgM autoantibodies to cytoplasmic intermediate filaments. Clin Exp Immunol 37:76–82

Tonnessen TI, Nyland H, Aarli JA (1982) Complement factors and acute phase reactants in the Guillain–Barré syndrome. Eur Neurol 21:125–128

Tse KS, Arbesman CE, Tomasi TB, Tourville D (1971) Demonstration of antimyelin antibodies by immunofluorescence in Guillain–Barré syndrome. Clin Exp Immunol 8:881–887

Uncini A, Treviso M, Basciani M, Onofrj M, Gambi D (1988) Associated central and peripheral demyelination: an electrophysiological study. J Neurol 235:238–240

van Doorn PA, Brand A, Vermeulen M (1987) Clinical significance of antibodies against peripheral nerve tissue in inflammatory polyneuropathy. Neurology 37:1798–1802

Vedeler CA, Nyland H, Matre R (1982) Antibodies to peripheral nerve tissue in sera from patients with acute Guillain–Barré syndrome demonstrated by mixed haemagglutination technique. J Neuroimmunol 2:209–214

Wiederholt HM, Mulder DW, Lambert EH (1964) The Landry Guillain–Barré–Strohl syndrome of polyradiculoneuropathy – historical review: report on 97 patients and present concepts. Mayo Clin Proc 49:427–451

Winer JB, Gray IA, Gregson NA et al. (1988a) A prospective study of acute idiopathic neuropathy. III. Immunologic studies. J Neurol Neurosurg Psychiatry 51:619–625

Winer JB, Hughes RAC, Anderson MJ, Jones DM, Kangro H, Watkins RFP (1988b) A prospective study of acute idiopathic neuropathy. II. Antecedent events. J Neurol Neurosurg Psychiatry 51:613–618

Winer JB, Briggs D, Welsh K, Hughes RAC (1988c) HLA antigens in the Guillain–Barré syndrome. J Neuroimmunol 18:13–16

Winer JB, Hughes RAC, Osmond C (1988d) A prospective study of acute idiopathic neuropathy. I. Clinical features and their prognostic value. J Neurol Neurosurg Psychiatry 51:605–612

Winkler GF (1965) In vitro demyelination of peripheral nerve induced with sensitized cells. Ann NY Acad Sci 122:287–296

Yonezawa T, Ishihara Y, Matsuyama H (1968) Studies on experimental allergic neuritis. I. Demyelinating patterns studied in vitro. J Neuropathol Exp Neurol 27:453–463

Yonezawa T, Ishihara Y, Sato Y (1969) Demyelinating antibodies of EAE and peripheral neuritis represented by demyelinating pattern in vitro. J Neuropathol Exp Neurol 28:180–181

Yonezawa T, Robbins N, Ishipara Y, Iwanami H, Nakatani Y (1970) In vitro demyelination produced by sera from Landry–Guillain–Barré syndrome. Proc Int Congr Neurol 6:688–689

Zweiman B, Rostami A, Lisak RP, Moskovitz AR, Pleasure DE (1983) Immune reactions to P_2 protein in human inflammatory demyelinative neuropathies. Neurology 33:234–237

Chapter 10

Chronic Idiopathic Demyelinating Polyradiculoneuropathy

Introduction

The occurrence of chronic progressive or relapsing forms of polyradiculoneuropathy has been recognised since descriptions by Osler in 1892 and reports at the end of the nineteenth century (Targowla 1894). The precise definition of these disorders continues to exercise reviewers (Dyck and Arnason 1984; Hughes et al. 1987; Albers and Kelly 1989; Barohn et al. 1989).

The earlier literature concerning relapsing polyradiculoneuropathy was reviewed by Austin (1958). From 30 adequately documented early cases and two personal cases he deduced a typical picture of a young adult with a progressive motor polyradiculoneuropathy reaching its worst over an average of five months, recovering and then recurring again after an average of four years. The cranial nerves were occasionally involved, the sphincters rarely. Two-thirds were male. The clinical picture which Austin deduced is remarkably close to that described in subsequent larger series. The CSF protein concentration was usually increased. The peripheral nerves were considered thickened in a third, a factor which had given rise to confusion with hereditary forms of peripheral neuropathy and even led to the erroneous suggestion that CIDP is associated with neurofibromatosis. Austin recognised that relapses in nine of the 30 cases were related to withdrawal of corticosteroids. These cases might not have pursued a relapsing course in the absence of steroid treatment, a difficulty of interpretation which still pervades subclassification of CIDP. Austin had reported details of 20 attacks occurring during five years in one of two personal cases: most but not all the attacks occurred followed reduction or withdrawal of steroids, and steroid treatment was always followed by improvement. Austin proposed that interstitial oedema compromised nerve function and might be reduced by steroid treatment.

Following Austin's report several small series of cases with relapsing or progressive courses were published. Hinman and Magee (1967) reported two patients with progressive and two with relapsing CIDP. One of the relapses in a relapsing case was precipitated by tetanus toxoid immunisation. They estimated that cases with a prolonged progressive phase represented only 20% of cases of GBS. Ashworth and Smyth (1969) reported three patients with the chronic relapsing form of CIDP, each with a slowly progressive onset: in two cases relapses sometimes developed very rapidly, even overnight. Thomas et al. (1969) reported five cases, which had recurrent acute (1), subacute (1), chronic progressive (1), and chronic relapsing courses (2). Their report illustrated nerve biopsies from three cases in which teased nerve fibres showed segmental remyelination. There was only minor reduction in myelinated nerve fibre density. Perivascular lymphocytic infiltrates were only found in one case and were sparse even in that one. Nerve roots from one fatal case of chronic progressive CIDP showed perivascular collections of lymphocytes which were most numerous in the ventral roots but also present in the dorsal roots and dorsal root ganglia (Thomas et al. 1969). The inflammatory changes in this autopsy were comparable to those in GBS and EAN, supporting the notion that these conditions represent a spectrum. In two of their five cases steroids seemed beneficial.

Diagnostic Criteria

Dyck et al. (1975) next described a series of 53 patients with "chronic inflammatory polyradiculoneuropathy". They included all those patients whose deficit had shown no improvement or was still worsening after six months or who had a recurrent course. "For the most part the maximum neurological deficit did not develop over days to a few weeks, as is common in acute inflammatory polyradiculoneuropathy". This statement glossed over the need to define an, albeit arbitrary, boundary of the progressive phase to distinguish GBS from CIDP. Their criteria consisted of features which favoured the diagnosis rather than features required for the diagnosis. These features included absence of associated disease, occurrence of preceding infection or immunisation, progressive or recurrent course, symmetrical proximal and distal limb involvement, and an increased CSF protein concentration with a normal cell count.

Prineas and McLeod (1976) included cases with a relapsing and remitting course or an onset progressive phase lasting more than three weeks and evidence of demyelination. Their three week cut off point between "acute idiopathic polyneuritis" and "chronic relapsing polyneuritis", as they called it, is earlier than the four weeks which I propose. Evidence of demyelination could be either 5% or more of teased nerve fibres showing segmental demyelination or demyelinated axons observed by electron microscopy or a motor nerve conduction velocity of 30 m/s in the median and ulnar nerves, or less than 20 m/s in the lateral popliteal nerve.

In the most recent report from the same group the criteria were made slightly more precise, by defining a relapse as an increase in disability by one or more grades on a seven point scale (McCombe et al. 1987b). The histological criteria for demyelination were changed to the presence of 15%, or more, of teased fibres affected by demyelination or active demyelination, and inflammatory infiltrates on electron microscopy, or both. The neurophysiological criteria for demyelination permitted a wider range of abnormalities (see pp.216–219).

Most recently Barohn et al. (1989) have proposed diagnostic criteria which are based on and define more rigidly the criteria of Dyck et al. (1975). Their criteria set the duration of the progressive phase as at least two or more months (without stipulating lunar to calendar). They viewed CIDP as a syndrome with a common pathogenesis which may occur in association with systemic disorders. They argued against excluding cases which have associated systemic disease but sensibly include them in a separate subgroup.

Table 10.1. Diagnostic criteria proposed for CIDP

Inclusion criteria
Progressive weakness of two or more limbs due to polyradiculoneuropathy
Loss or diminution of tendon reflexes
Progression for more than four weeks or recurrence or relapse[a]
Fulfilment of neurophysiological criteria for demyelination[b]

Exclusion criteria
Intoxication by drugs or environmental agents
Family history of similar polyradiculoneuropathy
Neuropathy attributable to metabolic causes including diabetes mellitus, vitamin deficiency, liver or renal failure
Systemic vasculitis, neoplasm, paraproteinaemia

---

[a] Deterioration of at least one disability grade (Chapter 6) following improvement of at least one disability grade for at least one week.
[b] See Table 10.2.

Against this background I propose the diagnostic criteria detailed in Table 10.1. In essence these consist of clinical inclusion criteria (progressive weakness lasting more than four weeks, or with a recurrent or relapsing course, and reduced or absent tendon reflexes), neurophysiological evidence of demyelination and exclusion of other causes of neuropathy. The statement that weakness is due to polyradiculoneuropathy carries the implication that the weakness is proximal as well as distal and often proximally predominant. Useful neurophysiological criteria for demyelination have recently been proposed by Albers and Kelly (1989) and are reproduced in Table 10.2.

CIDP is distinguished from GBS by virtue of its clinical course. In GBS the nadir is reached within four weeks whereas in CIDP the initial progressive phase lasts longer, usually much longer. Although this four week cut-off point is necessarily arbitrary, it is justified by the clinical impression that the distribution of durations of the progressive phase of cases of acquired demyelinating polyradiculoneuropathy is bimodal. Most cases of acquired polyradiculoneuropathy have reached their nadir within four weeks (GBS) while the less common (CIDP) cases mostly progress for several months. In a study of cases of idiopathic neuropathy, Prineas (1970) found that those with shorter courses than

Table 10.2. Neurophysiological criteria for demyelination in CIDP

Evaluation should satisfy at least three of the following in motor nerves:

1. Conduction velocity less than 75% of the lower limit of normal (two or more nerves)[a]
2. Distal latency exceeding 130% of upper limit of normal (two or more nerves)[b]
3. Evidence of unequivocal temporal dispersion or conduction block on proximal stimulation consisting of a proximal-to-distal amplitude ratio less than 0.7 (one or more nerves)[b,c]
4. F-response latency exceeding 130% of upper limit of normal (one or more nerves)[a,b]

From Albers and Kelly (1989), with permission.
[a] Excluding isolated ulnar or peroneal nerve abnormalities at the elbow or knee respectively.
[b] Excluding isolated median nerve abnormality at the wrist.
[c] Excluding the presence of anomalous innervation (e.g. median to ulnar nerve crossover).

Table 10.3. Subcategories of CIDP

Subacute:	reaching a nadir/plateau in >4 and <12 weeks
Chronic progressive:	reaching a nadir/plateau in > 12 weeks
Chronic relapsing:	either onset or at least one relapse to have a progressive phase >4 weeks
Recurrent GBS:	each bout to have a progressive phase <4 weeks

three weeks completely recovered more commonly than those with longer courses. It is helpful to subdivide CIDP into subacute (progressive phase more than four but less than 12 weeks), chronic progressive (progressive phase more than 12 weeks) and chronic relapsing forms (Table 10.3). A final group of recurrent acute cases must be added in which each attack reaches its nadir within less than four weeks. This subgrouping does not cater for the possibility that, in an individual case of recurrent or relapsing neuropathy, some attacks reach their nadir in less than four weeks and others take longer. Particularly severe cases of GBS, often those with an explosive onset, quite commonly leave a persistent deficit. Patients who have pursued such a course should be diagnosed as having had GBS with an incomplete recovery and distinguished from CIDP.

The incidence and prevalence of CIDP are very difficult to determine. Clinical experience and the small number of large series available suggest that it is an uncommon condition. Its incidence is certainly less than GBS. In one series of patients there were 102 patients with GBS, and 28 with CIDP of whom six had recurrent GBS (Seitz 1966). Because CIDP is by definition a more chronic illness than GBS the prevalence of patients with CIDP attending a neurological service is likely to be disproportionately high compared with the incidence. If a neurological service encounters CIDP four times less than GBS and the annual incidence of GBS is 1–2 per 100000, the annual incidence of CIDP must be only about 0.25–0.5 per 100000 or 2.5–5 per million. However, this estimate is based on very uncertain assumptions. The frequency of recurrence in GBS has ranged up to 10% in different series but the most reliable data from population-based surveys suggest a figure closer to 3% (Table 10.4).

Table 10.4. Frequency of recurrence following GBS

Reference	Total cases n	Recurrences n	%	Duration of follow-up
Peterman et al. (1959) children	26	2	8	
Eiben and Gersony (1963)	48	0	0	
Wiederholt et al. (1964)	97	10	10	
McFarland et al. (1966)	62	0	0	7 months
Ravn (1967)	127	0	0	
Hewer et al. (1968)	45	0	0	
Kennedy et al. (1978)	40	1	2.5	
Löffel et al. (1977)	90	4	4.4	
Pleasure et al. (1969)	81	8	10	
Winer et al. (1988)	100	3	3	12 months

Preceding Events

Although previous infections, immunisations or other events are characteristic of GBS, opinion has been divided concerning the occurrence and frequency of such events before the onset or relapses of CIDP. Dyck et al. (1975) regarded such events occurring a few weeks or months before the onset of a chronic neuropathy as a feature favouring the diagnosis. Of their 53 patients 13 gave a history of an infection or immunisation within three months before the onset (Dyck and Arnason 1984). Since then, examples of CIDP associated with HIV (Chapter 12) and other infections have been reported. Faber and Baslov (1970) described a patient with CIDP and chronic hepatitis B infection: a nerve biopsy showed granular deposits of viral antigen in the vasa nervorum. The authors suggested that immune complexes circulating in the blood and becoming deposited in the nerves might be related to the pathogenesis of CIDP. The problem is that such immune complexes are not unexpected in such a viral infection and might not be causative. In a series of 25 cases (Dalakas and Engel 1981) there were said to be no examples of onset or relapse of CIDP following soon after specific illnesses. Prineas and McLeod (1976) reported upper respiratory tract infections in 11, varicella in one and other symptoms of viral infection in three of their 23 patients during the month before the onset. In a subsequent paper from the Australian group, which presumably included the cases of Prineas and McLeod (1976), 93 patients were studied and only 29 (32%) gave a history of preceding events within six weeks before onset: eight patients had had upper respiratory tract infections; four patients had had immunisations: two developed CIDP following tetanus toxoid immunisation and one had relapses following re-immunisations (Pollard and Selby 1978). Two patients who had previously had GBS experienced recurrences soon after immunisation with influenza vaccine (Seyal et al. 1978). An interesting patient developed a first episode of GBS after cytomegalovirus infection and a second after receiving a renal transplant from a cytomegalovirus positive donor (Donaghy et al. 1989). It is probable that both were precipitated by the viral infection or reinfection/reactivation.

The conflict in the literature concerning the frequency and significance of preceding events in CIDP may reflect the length of time since the onset of neuropathy when information is collected. At one extreme the occurrence of preceding events may have been forgotten by the time of referral to a North American centre for a fourth or fifth opinion. Alternatively when a patient is admitted from the onset of his neuropathy to a single centre the contemporary case records may have captured information about an otherwise not very remarkable respiratory infection. None of the studies has been case controlled.

The possibility that relapses of CIDP might be more likely during pregnancy and the puerperium has been considered. Several individual case reports testify to this association (Castaigne et al. 1966; Schott et al. 1968; Calderon-Gonzalez et al. 1970; Dalakas and Engel 1981; D'Ambrosio and De Angelis 1985). In one series of 51 women with the chronic relapsing form of CIDP the average relapse frequency was 0.33 per year (McCombe et al. 1987a). These 51 patients included 16 premenopausal women of whom nine had had a total of 30 pregnancies. Disease onset occurred during four of these pregnancies. Eight relapses were associated with pregnancy or the puerperium and five of these were in the third trimester or puerperium. The women who had children had 0.53 relapses per year during their pregnancy years and 0.17 relapses per year during their non-pregnancy years. The women who did not have children had only 0.13 relapses per year. The increase in risk of relapse in the pregnancy year was significant compared with either their non-pregnancy years or all years in the women who did not have children. It is possible that relapses were more likely to be reported during pregnancy. The present provisional conclusion that relapses are more likely during pregnancy contrasts with the reported reduction in relapse rate of multiple sclerosis during pregnancy balanced by an increase during puerperium. However, the data of McCombe et al. (1987a) do not distinguish relapses occurring during the puerperium from those in the last trimester of pregnancy.

Clinical Features

Men are affected more commonly by CIDP than women, which is also true of GBS. There were 57 men and 35 women in the largest series (McCombe et al. 1987b), 35 men and 25 women in the series of Barohn et al. (1989) and 35 men and 18 women in the series of Dyck et al. (1975). These three series give a male/female ratio of 1.6 which is significantly different from 1.0 ($P < 0.02$).

The onset of CIDP may occur at any age. Dyck et al. (1975) found that most patients had the onset of disease in their forties or fifties. The average age of onset in the series of McCombe et al. (1987b) was about 45 (range 8–77) years. The chronic relapsing form ($n = 60$) was more likely to start in young adults at an average age of 27 years (range 2–70). The chronic progressive form ($n = 32$) was more likely to start older, at an average age of 51 years (range 2–72). The difference in age of onset between these groups was significant ($P < 0.005$). The older age of onset in patients with chronic progressive compared with relapsing disease is reminiscent of multiple sclerosis.

CIDP usually presents with weakness and sensory symptoms, sometimes with weakness alone and rarely with sensory symptoms alone. Weakness is usually both proximal and distal, consistent with the notion that both roots and peripheral nerves are affected by the pathological process. The upper and lower limbs are usually affected together, the lower limbs more than the upper. Paraesthesiae in the extremities are common and may be painful. The occurrence of positive sensory symptoms argues in favour of a diagnosis of acquired neuropathy rather than hereditary motor and sensory neuropathy type I (Dyck et al. 1981). Aching pain in the muscles also occurs. Tendon reflexes are usually lost. Cranial nerve involvement is not uncommon though less frequent than in GBS (Waddy et al. 1989). Facial weakness developed in 15%, bulbar involvement in 6% and ophthalmoplegia in 4% of McCombe et al.'s (1987b) series (Table 10.5). Postural and intention tremor are occasionally noted, being present in 3% of McCombe et al.'s 92 patients. Dalakas and Engel (1981) reported tremor in six of their 25 patients. In a subsequent report on eight patients, tremor was confined to the upper limbs, coarse, irregular and not associated with cerebellar or other CNS deficits (Dalakas et al. 1984). It developed as the

Table 10.5. Percentage of patients showing particular clinical features in CIDP

	Dyck et al. (1985) ($n = 53$)	McCombe et al. (1987b) ($n = 92$)	Barohn et al. (1989) ($n = 60$)
Weakness	85	94	100
Sensory loss	83	72	86
Loss of tendon reflexes	94		95
Facial weakness	4	15	13
Ophthalmoplegia		4	3
Papilloedema	7	1	2
Tremor			3

disease became more severe, often in the second or third attack, and improved when immunosuppressive treatment controlled the underlying disease. Papilloedema is a rare complication. Autonomic dysfunction, an important feature of GBS, was not reported in McCombe et al.'s (1987b) series. Thickening of the nerves is not commonly recorded, being present in only 11% of Dyck et al.'s (1975) series and not mentioned by McCombe et al. (1987b). Although symptoms and signs of CNS demyelination are not usually prominent, occasional patients have extensor plantar responses and evoked potential and magnetic resonance imaging studies sometimes provide evidence of brain or spinal cord involvement (see below).

In most cases of CIDP the brunt of the disease falls on the limbs but four patients with recurrent ophthalmoplegia resembling Miller Fisher syndrome have been reported (Kaplan et al. 1985; Schapira and Thomas 1986; Barohn et al. 1989). Rarely ophthalmoplegia may be the presenting symptom of CIDP (Donaghy and Earl 1985).

The severity of CIDP is extremely variable. The course is, by definition, more insidious than in GBS and autonomic complications, sudden death and respiratory failure are extremely rare. However, progression to the point of being bedbound with profound weakness of the arms is not uncommon. Despite the severity of the weakness, early in the disease muscle atrophy is relatively slight, in keeping with demyelination being the major pathogenetic mechanism and the axon remaining intact. Sensory impairment is usually mild and predominantly affects modalities mediated by the large myelinated nerve fibres. Sometimes sensory loss is profound and disabling. It is a matter of dispute whether a purely sensory form of CIDP occurs. Of 92 cases in McCombe et al.'s (1987b) series 6% presented with sensory neuropathy and the same percentage of the patients of Dyck et al. (1975) had an essentially pure sensory polyradiculoneuropathy (Dyck and Arnason 1984). Such cases would not be included in my definition of CIDP nor that of Barohn et al. (1989).

The usual course of the untreated disease is gradual progression over weeks or months, interspersed in the more common chronic relapsing form, with spontaneous remissions. About one-third of patients have the chronic progressive form, which includes some patients with subacute disease which plateaus and then disappears spontaneously. Patients with recurrent GBS form only a small percentage of CIDP patients. The largest series is the most helpful for evaluating the prognosis of CIDP (McCombe et al. 1987b). The mean (SD) follow-up from disease onset to final review or death was 10.6 (9.1) years and range 1 to 41 years. Most patients had recovered virtually completely and had no or only minor symptoms but 2% were unable to walk and 6% had died of their disease. In the series of Barohn et al. (1989) the mortality was 5%. The prognosis in the series of Dyck et al. (1975) was worse. After a mean follow-up of 7.4 (5.3) years from onset 60% were managing to work with or without symptoms, 25% required a wheelchair and 11% had died of their disease. The worse outcome in the Mayo Clinic series may have been because more severe cases had been referred, or because the patients were collected earlier and modern treatment is more effective.

Differential Diagnosis

There are two phases in the diagnosis of CIDP; first, establishing that the disease is a demyelinating polyradiculoneuropathy and second, excluding other causes (Table 10.6). The suspicion of polyradiculoneuropathy will be aroused by the occurrence of proximal as well as distal weakness from the early stages. Predominantly distal weakness is more characteristic of axonal polyneuropathy. Demyelination rather than axonal degeneration is also suggested by relatively severe weakness in relation to the amount of wasting. However, the diagnosis depends largely on neurophysiological tests showing that conduction is slowed beyond the range which is expected in axonal neuropathy (Table 10.2).

In practice borderline results are not uncommon so that all the causes of chronic axonal neuropathy may have to be considered (Table 10.7). Nerve

Table 10.6.   Causes of chronic demyelinating neuropathy

Hereditary
Hereditary motor and sensory neuropathy Type I
Hereditary motor and sensory neuropathy Type III
Refsum's disease
Metachromatic leucodystrophy
Globoid cell leucodystrophy
Cockayne syndrome

Paraproteinaemic
Monoclonal gammopathy of undetermined significance
Solitary osteosclerotic myeloma
Multiple myeloma
Waldenstrom's macroglobulinaemia

Toxic
Drugs: amiodarone, perhexilene

Chronic idiopathic demyelinating polyradiculoneuropathy
Subacute
Chronic progressive
Chronic relapsing
Recurrent GBS

Table 10.7. Causes of chronic axonal neuropathy

Metabolic
Diabetes mellitus
Acromegaly
Hypoglycaemia (insulinoma)
Myxoedema
Uraemia

Deficiency
Vitamin B_{12}
Folic acid
Thiamine, pyridoxine, nicotinamide
Vitamin E

Toxic
Alcohol
Chemicals: Acrylamide, arsenic, carbon disulphide, hexacarbons, lead, organ phosphates, thallium
Drugs: dapsone, disulfiram, gold, isoniazid, metronidazole, nitrofurantoin, platinum, thalidomide, vincristine

Leprosy

Paraneoplastic
Carcinoma
Lymphoma
Polycythaemia rubra vera

Paraproteinaemic
Monoclonal gammopathy of undetermined significance
Multiple myeloma
Waldenstrom's macroglobulinaemia

Connective tissue diseases
Rheumatoid arthritis
Systemic lupus erythematosus
Vasculitis

Hereditary
Giant axonal neuropathy
Hereditary motor and sensory neuropathy Type II
Hereditary sensory and autonomic neuropathies
Neuropathy associated with hereditary ataxia

Miscellaneous
Adult coeliac disease
Chronic obstructive airways disease
Fibrosing alveolitis
Primary amyloidosis
Primary biliary cirrhosis
Sarcoidosis

Undetermined

From Hughes et al. (1987), with permission.

biopsy may confirm the presence of demyelination but not always (see below). Recently Julien et al. (1989) have suggested on the basis of three personal cases that there exists a chronic relapsing idiopathic axonal polyneuropathy which resembles CIDP in its clinical course but is distinguished by profound wasting, axonopathy on nerve biopsy and an unsatisfactory outcome. Post-mortem examination in one of their cases showed no inflammation or demyelination in the nerve roots.

Most of the other causes of demyelinating neuropathy are rare (Table 10.6). However, hereditary motor and sensory neuropathy type I (HMSN I) is fairly common. Suspicion of this diagnosis is raised by the presence of claw toes and pes cavus or scoliosis in any patient with demyelinating neuropathy. Even if these are absent family members should be questioned and preferably examined for the stigmata of HMSN I (Dyck et al. 1981). HMSN I is autosomal dominant and very variable: its expression ranges from the classical picture of inverted champagne bottle legs, through mild neuropathy, to pes cavus and claw toes, or even no clinical manifestations. The other hereditary causes of demyelinating neuropathy in Table 10.7 are autosomal recessive and rare. HMSN III is a heterogeneous but usually severe, form of demyelinating peripheral neuropathy with an onset in infancy. Pathologically there is evidence of failure to form myelin properly with unmyelinated axons surrounded by prominent onion bulbs. Refsum's disease presents as a distal sensory and motor polyneuropathy in the second and third decades. It is usually progressive but may relapse and remit. It can usually be recognised by associated abnormalities including cerebellar ataxia, atypical pigmentary retinal degeneration, cataract, deafness and icthyosis. It is due to failure to metabolise phytanic acid which accumulates in the peripheral and central nervous system and can be measured in the serum. Metachromatic leucodystrophy, due to aryl sulphatase deficiency, occurs most commonly in a late infantile form but juvenile and adult onset forms also exist. In each form cerebral involvement is prominent and the diagnosis would be unlikely in a patient with demyelinating neuropathy alone. In globoid cell leucodystrophy, dementia, spasticity, and optic atrophy present in infancy and are due to deposition of galactocerebroside in the white matter. Onset slightly later, up to the age of 6, has been reported. Peripheral nerve involvement can be demonstrated by nerve conduction tests but is not prominent. Similarly in the Cockayne syndrome, peripheral neuropathy forms a minor part of a multisystem disorder causing growth retardation, progeria, ataxia, deafness and mental retardation (Harding and Thomas 1984).

Careful search needs to be made for a paraprotein in every case of suspected CIDP. Most cases can be discovered by screening the serum for a paraprotein band by conventional electrophoresis of serum on cellulose nitrate strips. However, in a few patients a small serum paraprotein band may be missed by this technique but detected by immunoelectrophoresis. Some plasma cell dyscrasias do not secrete heavy chains but produce light chains which can be detected as Bence-Jones protein by electrophoresis of concentrated urine. Some solitary myelomas associated with neuropathy do not secrete detectable paraproteins and yet are associated with neuropathy. Consequently a radiographic skeletal survey or radioisotope bone scan should be undertaken (Chapter 12).

With the exception of amiodarone and perhexiline the neuropathy produced by most drugs is axonal. Careful enquiry should always be made after drugs, alcohol, and other toxins which might complicate the picture.

A final possible differential diagnosis is vasculitis. The clinical differentiation between an asymmetrical form of CIDP and confluent multiple mononeuropathy due to vasculitis may be difficult. Vasculitis may sometimes present as neuropathy without symptoms or signs of involvement of other organs (Chapter 12). Usually nerve conduction studies in vasculitis reveal multifocal nerve damage with reduction of sensory and motor action potentials evoked by stimulating affected nerves: nerve conduction is usually only slightly slower than normal. Occasional patients fulfilling clinical and neurophysiological criteria for CIDP are found to have vasculitis on nerve biopsy. There is also a group of rare patients with multifocal conduction block and otherwise normal nerve conduction velocities who give the appearance of a progressive form of multiple mononeuropathy. In a series of five such cases, three underwent nerve biopsy which showed features interpreted as a demyelinating neuropathy without vasculitis (Lewis et al. 1982). In another patient multifocal conduction block was confined to motor nerves and a sural nerve biopsy was normal (van den Bergh et al. 1989). It is likely that these patients represent a variant of CIDP.

Evidence for CNS Involvement in CIDP

Both GBS (Chapter 6) and CIDP may rarely be associated with multiple sclerosis (MS). Two patients from a series of 40 with CIDP had clear cut episodes of optic neuritis but not other features of MS (Lewis et al. 1982). Rubin et al. (1987) reported two patients with progressive neurological signs indicating affection of both the peripheral and central nervous system. Nerve conduction was severely slowed consistent with demyelination. In one of the patients the CSF protein concentration was markedly increased, nerve biopsy showed depletion of myelinated nerve fibres and an MRI brain scan showed multiple white matter abnormalities. A patient presented with features of MS and also absent tendon reflexes: nerve conduction studies and biopsy demonstrated a demyelinating neuropathy (Ro et al. 1983). While assembling their series of 60 patients with CIDP Barohn et al. (1989) encountered three patients who had clear cut relapses, remissions and CNS signs suggestive of MS. All three had MRI scan white matter abnormalities. They therefore studied 13 more CIDP patients with MRI scans and found similar white matter abnormalities in another three (Mendell et al. 1987). However, these three patients were all in their seventies, an age when MRI white matter abnormalities are common and usually due to vascular disease. If these three patients are excluded, there were three patients (5%) with CNS involvement suggestive of MS in that series.

Thomas et al. (1987) reported six patients with relapsing and remitting or chronic progressive neurological illness and clinical features of both PNS and CNS involvement. All six had severely reduced motor nerve conduction velocity and increased distal motor latencies consistent with demyelinating neuropathy. Abnormally prolonged latencies of the P100 visual evoked potential were found in all six except for one in whom the results were borderline. When recordings could be obtained, the central component of somatosensory and brainstem-evoked potentials was also abnormal in a way which suggested delay in CNS conduction pathways in most instances. Central motor conduction, examined by percutaneous stimulation of the scalp, was prolonged in the four cases in which it was measured. MRI scans, undertaken in five cases, showed changes similar to those reported in MS which were probably all significant since the ages of the patients ranged from about 38 to 50 years when the scans were undertaken. The CSF protein was increased in all cases except one to more than 1.00 g/l, i.e. more than is usual in MS. Nerve biopsies showed a reduction of myelinated but not unmyelinated fibre density, onion bulb formation and regeneration clusters. Many surviving fibres had abnormally thin myelin sheaths indicating remyelination and teased fibres showed segmental demyelination. Active demyelination was rare and inflammation was absent. The same laboratory has recently reported a consecutive series of 28 patients: five had minor clinical evidence of CNS involvement, 14 had lesions in the brain demonstrated by MRI scanning, and one-third (of 18 tested) had abnormalities of central motor conduction detected by recordings following magnetic stimulation of the motor cortex (Ormerod et al. 1990).

Sometimes hypertrophic neuropathy has been discovered unexpectedly at autopsy in a case of MS. In a recently reported example a man died of multiple sclerosis which had pursued a relapsing and remitting course over 47 years with typical signs including brisk tendon reflexes. At autopsy multiple sclerosis was confirmed but the spinal roots were grossly thickened and showed florid onion bulb changes (Rosenberg and Bourdette 1983). Cases 1 and 3 of Schoene et al. (1977) and some earlier cases (Dinkler 1904; Schob 1907; Ninfo et al. 1967; Jellinger 1969) were similar to that reported by Rosenberg and Bourdette (1983). All had autopsy evidence of CNS plaques and grossly hypertrophic spinal roots. In most cases the hypertrophic changes were confined to the spinal roots but peripheral nerves were not always studied. The hypertrophic changes were associated with onion bulb formation but not with inflammatory changes. Although it is possible that these cases represent a form of CIDP associated with MS, the alternative that they represent a coincidence of HMSN type I and MS is more likely. Hypertrophic changes in the nerve roots of patients with HMSN type I may be very marked. The hypertrophic changes and onion bulb formation illustrated, for instance, by Rosenberg and Bourdette (1983) were much more marked than is usual in CIDP.

A conservative conclusion from these reports would be that CNS and PNS demyelination may occur together for several reasons. First, multiple sclerosis may occur coincidentally with CIDP or with hereditary demyelinating neuropathy. The association between multiple sclerosis and CIDP has not been noticed frequently enough to establish a significant association between the two. Second, the demyelinating lesions of multiple sclerosis may "spill over" into the PNS and the lesions of CIDP may "spill over" into the CNS, causing subclinical involvement. Third, there might be a separate disorder producing combined CNS and PNS demyelination. The third possibility would have to be proved by demonstrating separate biological markers for such a combined disorder, or by demonstrating differences in clinical features, prognosis, or response to treatment.

Neurophysiology

Each centre has adopted slightly different criteria for categorisation of a neuropathy as demyelinating, and individual centres have modified their criteria over the years. The strictest and clearest criteria were recently promulgated in a review and have been reproduced in Table 10.2 (see p. 207).

In a large series of CIDP patients from the Mayo Clinic, the authors usually found slowed motor and sensory nerve conduction, but considered that normal studies did not exclude the diagnosis (Dyck et al. 1985). They noted that there might be a disparity between the severity of weakness or sensory loss and the degree of slowing of nerve conduction, a point which has been made in subsequent papers.

In the first paper from the Sydney group, a motor nerve conduction velocity of 30 m/s or less in the median or ulnar nerves or less than 20 m/s in the common peroneal nerve was considered to be one criterion for demyelination (Prineas and McLeod 1976). In that study of 23 patients all except one had "severe" slowing of motor nerve conduction. Sensory and lateral popliteal nerve action potentials were absent in almost all cases with the apparatus available at the time. The distal motor latencies were on average significantly prolonged compared with a control group. There was little relationship between conduction velocity and clinical severity of neuropathy in different patients. Sometimes patients with mild weakness at onset or during recovery had markedly slowed nerve conduction. In their most recent paper the Sydney group have expanded their neurophysiological criteria for demyelination. The revised criteria were motor nerve conduction velocity less than 40 m/s in the median or ulnar nerves, and less than 30 m/s in the peroneal nerve; terminal latency more than 7 ms in the median or ulnar nerves or 10 ms in the peroneal nerve; or conduction block or dispersion of the compound action potential on proximal compared with distal stimulation (McCombe et al. 1987b).

In the most recent series from Ohio neurophysiological abnormalities were not regarded as a mandatory criterion for CIDP (Barohn et al. 1989). However, motor nerve conduction velocities were considered to indicate demyelination if they were slowed to less than 70% of the normal lower limit which was taken as two standard deviations below the normal mean. Forty (67%) of the 60 patients in that series had at least one nerve which exhibited slowing within the demyelinating range, two (4%) patients had normal motor nerve conduction velocities and the remainder showed lesser degrees of slowing. Conduction block, defined as a proximally evoked muscle action potential less than 50% of the distally evoked potential, was present in 13% of patients and dispersion of the proximally evoked potential in 11% of patients. Fibrillation was detected in 72%. There was little relationship between slowing of motor nerve conduction velocity and detection of demyelination in sural nerve biopsies. This discrepancy might be due to the multifocal nature of the disease, especially since nerve conduction was being measured in a motor nerve and the morphological observations were being made on a different sensory nerve. In a study of miscellaneous neuropathies there was a rough correlation between the finding of demyelinated fibres in the superficial peroneal nerve and the presence of abnormal sural nerve conduction (Bolton et al. 1979).

Although distal sensory nerve action potentials may be very small or absent in CIDP, sensory conduction can still be evaluated by recording potentials evoked over the scalp. Using this technique Parry and Aminoff (1987) showed that afferent nerve conduction velocity was slowed somewhere along its pathway following stimulation of 11 out of 15 nerves. They considered that afferent nerve conduction was less slowed than motor in four nerves, more slowed in one nerve and equivalently slowed in ten nerves.

In hereditary demyelinating neuropathy conduction is uniformly slowed in all nerves whereas multifocal slowing is characteristic of CIDP. This helpful diagnostic point was stressed in an elegant study comparing 40 CIDP patients with 18 patients who had autosomal dominant demyelinating peripheral neuropathy (Lewis et al. 1982). The multifocality was shown in several ways. The median and ulnar forearm segment motor conduction velocities were always similar in the familial cases but often differed in CIDP. The distal motor latencies and forearm nerve conduction velocities were always correspondingly abnormal in the familial cases but often showed discrepancies in CIDP (Fig. 10.1). The dispersion of the evoked muscle action potential following proximal compared with distal stimulation was greater in CIDP patients than in familial neur-

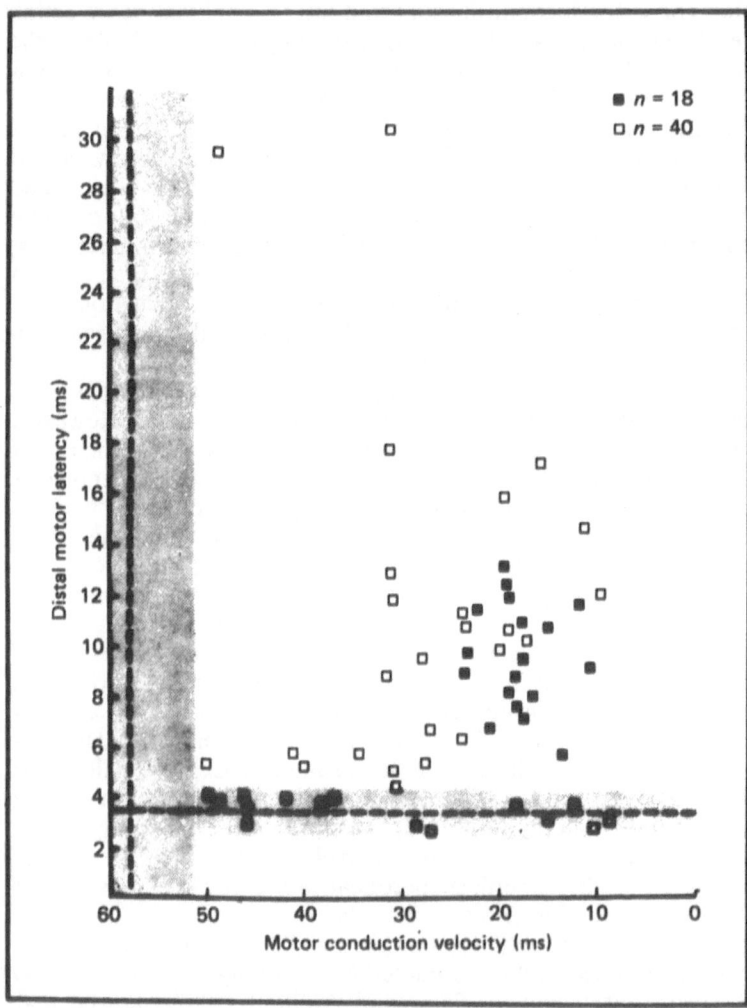

Fig. 10.1. Scattergram of the median nerve distal motor latency and forearm maximum motor conduction velocity in hereditary demyelinating neuropathy (*black squares*) and CIDP (*open squares*). (From Lewis and Sumner (1982), with permission.)

opathy patients. Twelve patients (30%) in the CIDP group but none of those in the familial group had complete or partial conduction block. In several CIDP patients conduction block was almost the only neurophysiological abnormality. Five of these patients were reported in more detail subsequently (Lewis et al. 1982), and nerve biopsy evidence was produced to support the idea that the underlying pathology was indeed demyelination. However, it was never possible to study the precise region of the neurophysiologically detected conduction block. Somewhat similar cases with multifocal conduction block confined to motor nerve have since been reported (Pestronk et al. 1988).

The results of nerve conduction studies on 30 patients with CIDP were briefly reported in a helpful review by Albers (1987). All patients showed abnormalities of motor conduction and 87% had abnormal sensory conduction. Unequivocal evidence of demyelination was identified in all but two patients: in these two demyelination was present in one nerve, and conduction was absent or slightly slowed in at least one other. Nerve conduction was more slowed than is usual in GBS. As in GBS median nerve sensory action potentials were more impaired than sural nerve action potentials. Electromyography showed fibrillation in almost all patients, and was more frequent in distal than proximal muscles. Most patients had an increased proportion of polyphasic motor unit action potentials and all had abnormally large motor unit action potentials. No correlation between nerve conduction or EMG parameters and outcome has yet been discovered.

A study seeking evidence of CNS involvement from neurophysiological tests showed abnormal brainstem and visual evoked responses in four of six patients with CIDP (Gigli et al. 1989). Six patients selected for having evidence of CNS involvement had prolonged visual evoked potential latencies (Thomas et al. 1987). Six of 18 cases had delayed central motor conduction (Ormerod et al. 1990). These results taken with other evidence suggest that CIDP is less specific for the peripheral nervous system than GBS.

Cerebrospinal Fluid

An increased CSF protein concentration with a normal cell count is as characteristic of CIDP as of GBS. In the first large series from the Mayo Clinic the CSF protein concentration was greater than the upper limit of normal in 40 of 44 cases and the cell count was increased in only five of 44 (Dyck et al. 1985). The average CSF protein concentration was 1376 mg/l and the highest value 6000 mg/l. The highest cell count was 75/µl. The majority of the protein was albumin and electrophoresis showed that gammaglobulin was increased above the normal range in only six of 42 cases. The CSF protein concentration was increased in 20 of the 23 patients in the Sydney series but the cell count was only increased in three and only marginally in those (highest count 16/µl). In seven of the 23 the gammaglobulin content was abnormal. In the Ohio series (Barohn et al, 1989) the CSF protein concentration was increased in 56 of 59 patients (95%): the mean value was 1340 mg/l and the highest value 7860 mg/l. Only two patients (3%) had CSF pleocytosis. In a study of 15 patients (Dalakas and Engel 1980) the CSF:serum ratios of IgG and albumin were normal and intrathecal

IgG synthesis was not detected. However, in the same study a monoclonal band was detected in the CSF in the gammaglobulin region by agarose gel electrophoresis. This band remained constant despite successful treatment in individual patients. This unusual band was not present in the serum: its presence has not been confirmed or pursued. In subsequent studies this band has not been a consistent finding, and, on higher resolution electrophoresis with immunofixation, some patients have oligoclonal bands and others do not (M. Dalakas, personal communication). The CSF is not diagnostic in CIDP. Since the CSF bathes the nerve roots where the pathology mainly lies, it might repay further study.

Pathology

There have been few autopsy studies of CIDP. Those that have been published often only contain fragmentary information (Table 10.8). The CNS has usually been normal apart from anterior horn cell chromatolysis and dorsal column pallor, which are the expected consequences of damage to the motor axons and dorsal root ganglion cells. However, in one patient who had shown signs of myelopathy in life, there were inflammatory cells in the spinal cord and degeneration of the lateral and dorsal columns (Case 1 of Dyck et al. (1975)). In one other patient there was very severe lymphocytic infiltration of the meninges extending into the brainstem (Borit and Altrocchi 1971). The pathological changes were so marked that the authors commented on the resemblance to lymphomatous meningeal infiltration and compared their case to Marek's disease in chickens (Chapter 11). Some infiltration of the peripheral nervous system by mononuclear cells or lymphocytes was reported in most cases. There was none in the case of Matthews et al. (1970) in which large doses of steroids had been given. In general the severity of lymphocytic infiltration was greater in those cases which had not received steroids during the terminal phase of their illness. It is difficult to discern from these reports whether the brunt of the pathology has fallen on the myelin sheath or the axon, although the usual implication was that segmental demyelination was present and axons were preserved. Onion bulb formations were prominent in the peripheral nerves of only one of the eight cases.

The light and electron microscopic appearances of a nerve biopsy from a patient with chronic relapsing CIDP were clearly described by Prineas (1970). There was a sparse inflammatory cell infiltrate in the endoneurium, and active macrophage-mediated demyelination, such as had recently been described in GBS and EAN. Differences from GBS were that only macrophage-associated stripping and phagocytosis of myelin sheaths were noted and extracellular vesicular demyelination had not occurred. An additional finding was the presence of layers of supernumerary Schwann cells around remyelinated axons forming onion bulbs. Since then several individual case reports and small series have demonstrated similar changes, although inflammatory changes have been sparse or absent (Bonnaud et al. 1974).

The light microscopic appearances of a large series of biopsies from patients with CIDP were first described by Dyck et al. (1975) and are representative of what has been reported since. A total of 26 sural nerve biopsies were studied.

Table 10.8. Post-mortem examinations of CIDP

Reference	Duration (years)	Type	Steroid treatment	CNS	Spinal roots			Peripheral or cranial nerves		
					Myelin loss	Axon loss	Mononuclear cell infiltration	Myelin loss	Axon loss	Mononuclear cell infiltration
Thomas et al. (1969)	0.4	CP	0	Normal	++	0	+			
Torvik and Lundar (1977)	0.6	CP	0	Dorsal column pallor, AHC chromatolysis	+			0–++	0	0–+++
Sibley (1972)	0.9	CR	+		++	++	0	++	++	0
Dyck et al. (1975) case 6	1.8	CP	+	AHC chromatolysis	+		+			
Goto et al. (1969)	2.3	CR	0	AHC chromatolysis, dorsal column degeneration	++	0–++	++	++	0–++	++
Dyck et al. (1975) case 47	3.2	CP	+	Lymphocyte infiltration of brainstem and dorsal and lateral columns of spinal cord	++	++	++	0–++	0–+	+
case 11	3.5	CR	+			0				0

(continued, p. 222)

Table 10.8. (*cont.*)

Reference	Duration (years)	Type	Steroid treatment	CNS	Spinal roots			Peripheral or cranial nerves		
					Myelin loss	Axon loss	Mononuclear cell infiltration	Myelin loss	Axon loss	Mononuclear cell infiltration
Matthews et al. (1970) case 1	3.5	CR	+	Dorsal column pallor	+	0	0	+	0	0
Dyck et al. (1975) case 51	15	CP with myelopathy	0	Lateral and dorsal column degeneration mononuclear cell spinal cord infiltration				Onion bulbs		++
Borit and Altrocchi (1971)	31	CR	0	AHC loss AHC chromatolysis dorsal column degeneration; marked lymphocytic meningeal infiltration extending into brainstem	0–+++	0–+++	+++	0–+++	0–+++	+++

Abbreviations: CP, chronic progressive; CR, chronic relapsing; AHC, anterior horn cell. Pathological changes are scored 0, none; +, minor; ++, moderate; +++, severe. Steroid treatment refers to the last months of the illness.

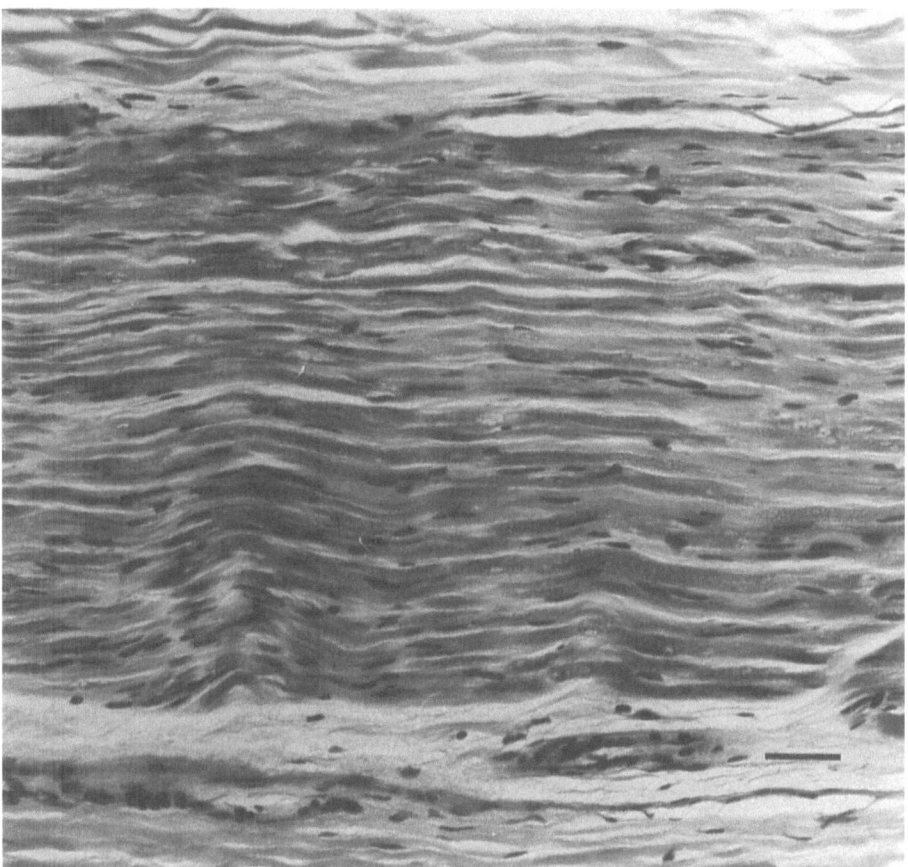

Fig. 10.2. Sural nerve biopsy from a patient with CIDP. Note the absence of inflammatory cells. Haematoxylin and eosin. Bar = 100 μm.

Inflammatory cell infiltration was infrequent and slight. Moderate perivascular infiltrates were encountered in the epineurium of two cases and diffuse endoneurial mononuclear cell infiltration was only present in six cases. In most cases, as in my own material, endoneurial infiltration is absent (Fig. 10.2). Endoneurial oedema was noted in five cases. Morphometric studies showed a reduction in the density of packing of myelinated nerve fibres compared with control nerves (Fig. 10.3). Onion bulb formation was noted in four cases (Figs. 10.4 and 10.5). In teased fibre studies the commonest finding was an increased percentage of fibres showing linear rows of myelin ovoids due to axonal degeneration. The average percentage of fibres showing this change was 25% compared with only 9% of fibres showing segmental demyelination and 6% of fibres showing segmental remyelination. The contradiction between pathological findings of predominant axonal degeneration and neurophysiological and clinical features pointing to a demyelinating neuropathy may be explained by the fact that the brunt of the pathology is in the proximal nerve trunks and spinal roots, as suggested by the limited autopsy data available. In a recent case

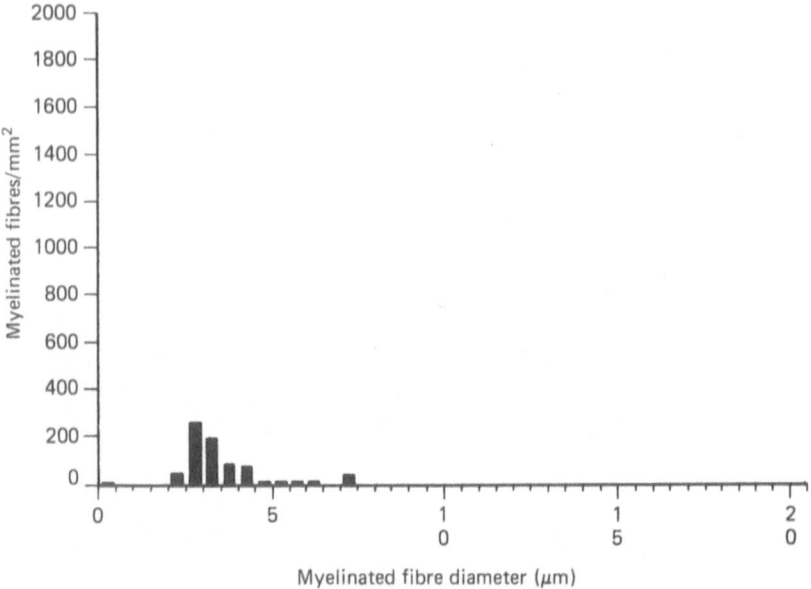

Fig. 10.3. Reduction in density of myelinated nerve fibres in the sural nerve of patient with CIDP.

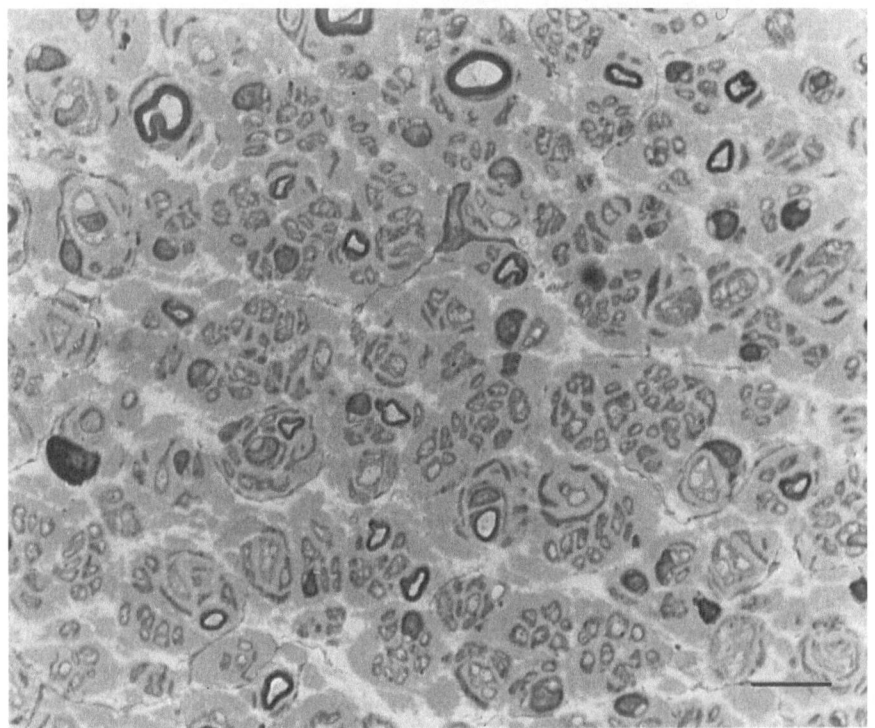

Fig. 10.4. Onion bulb formation in a nerve biopsy from a patient with CIDP. Resin-embedded section stained with thionin and acridine orange. Bar = 10 μm.

of monophasic chronic progressive CIDP and multifocal conduction block, the severity of changes varied markedly from one fascicle to another in a fashion reminiscent of vasculitis (Nukada et al. 1989). However, no vasculitis was demonstrated on semi-serial sections. Although the authors argued in favour of an ischaemic explanation for this multifocal distribution, such a pattern could be explained by damage from inflammatory reactions which had resolved by the time the biopsy was performed. This patient had not been treated. In some other cases where the biopsy or autopsy had shown little or no cellular infiltration steroid treatment may have suppressed cell infiltration. In a recent series of three patients with chronic relapsing CIDP all had histological evidence of onion bulb formation: in one a supraclavicular mass was biopsied and found to consist of a hypertrophic nerve trunk containing onion bulbs and marked perivascular lymphocytic infiltration (De Mello et al. 1989). In other histological studies macroscopic hypertrophy of nerve trunks has not been a feature. Prineas and

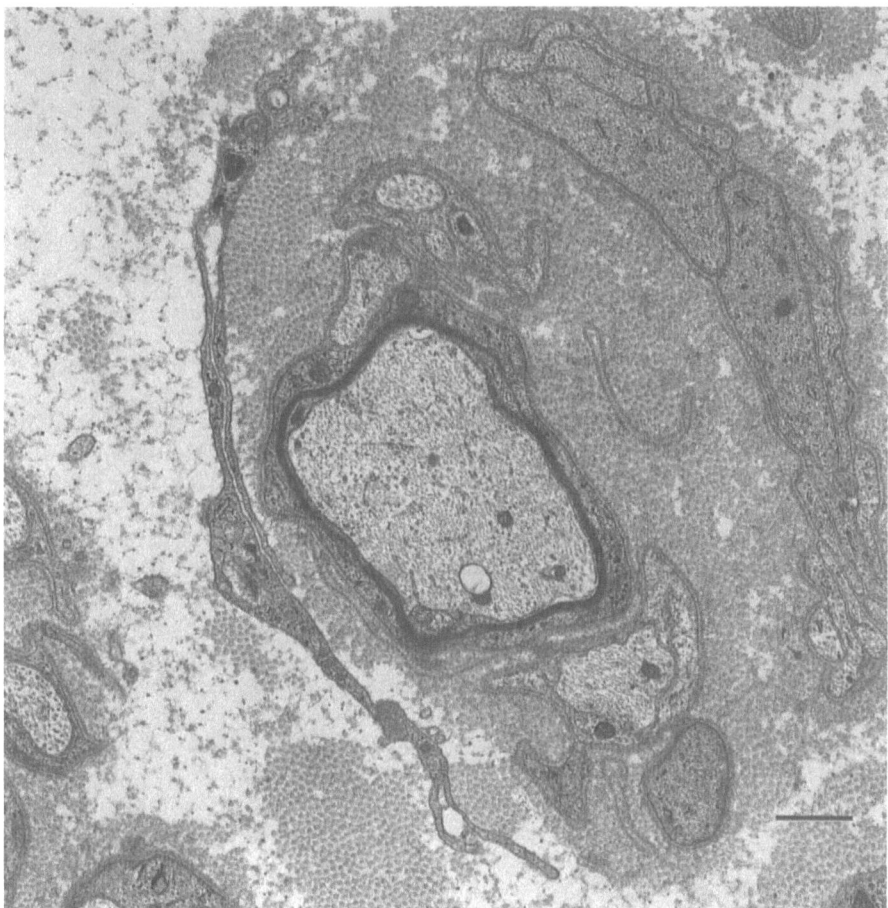

Fig. 10.5a. Onion bulb formation from a patient with CIDP. Onion bulb formation is uncommon in patients with CIDP and when found is usually sparse and multifocal. Bar = 1 μm

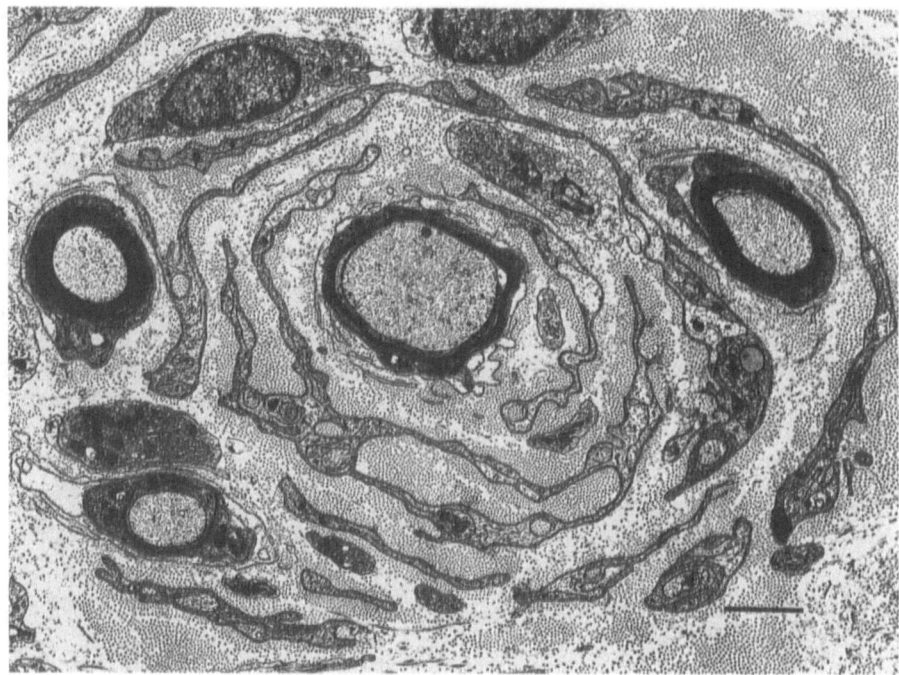

b

Fig. 10.5b. Onion bulb formation from a patient with hereditary motor and sensory neuropathy type I. Onion bulbs were uniformly abundant in this specimen. Electron micrograph. Bar = 2 μm.

McLeod (1976) found that the transverse fasicular area of the sural nerve did not differ from control values.

The electron microscopic appearances of 23 biopsies from patients with CIDP were described and beautifully illustrated by Prineas and McLeod (1976). In seven of these biopsies the myelinated fibre density was reduced below the normal range. In three of these there was preferential loss of large myelinated fibres causing loss of the normal bimodal fibre diameter distribution. Nine biopsies showed endoneurial oedema. Ten biopsies contained onion bulb formations. Inflammatory cell infiltrates were not reported in any of the biopsies. On electron microscopy seven biopsies were normal but 16 showed an expanded endoneurial space containing amorphous granular material and larger stellate mucopolysaccharide-like granules. A striking feature was masses of 11 nm diameter filaments associated with elongated probably fibroblastic cytoplasmic processes (Fig. 10.6). Similar masses of filaments had previously been reported in CIDP and familial neuropathy (Thomas and Lascelles 1969; Prineas et al. 1971). Fourteen nerves showed stacks of Schwann cell processes in a configuration which suggested loss of unmyelinated nerve fibres. Six nerves showed active demyelination. These tended to be biopsies from patients with a shorter history and the most florid case was biopsied only five weeks after the onset. Demyelinated axons were usually diffusely distributed among fascicles, but often especially in a subperineurial situation. In one case only a single fascicle was affected. Myelin breakdown only occurred in association with cells (Fig. 10.7) and closely resembled the changes already reported by Prineas in GBS and Lampert in EAN (Chapters 3 and 4). Mononuclear cells usually interpreted as macrophages invaded the Schwann cell basal lamina and could be seen to have inserted processes between myelin lamellae and ingested myelin debris into their cytoplasm (Fig. 10.8). Occasional cells within the basal lamina were identified by the authors as lymphocytes although from the illustration provided it is impossible to distinguish such cells confidently from macrophages. Recently demyelinated axons often appeared shrunken with a wavy axolemma and abnormal dense packing of neurofilaments and microtubules. Remyelinated axons initially had very thin myelinated sheaths, and were often surrounded by supernumerary Schwann cell processes in one or more layers interspersed with collagen fibrils eventually forming onion bulbs. Two cases demonstrating similar demyelinated axons and onion bulb formation but lacking inflammatory cell infiltration have been reported recently (Rizzuto et al. 1982). The conclusions of Prineas and McLeod (1976) were also supported by the most recent report of Barohn et al. (1989). These authors studied biopsies on 56 patients: 48% of biopsies showed predominant demyelination and remyelination, 21% axonopathy and 12.5% a mixed picture. Only six biopsies (11%) showed signs of inflammation which usually consisted of small clusters of inflammatory cells. An-

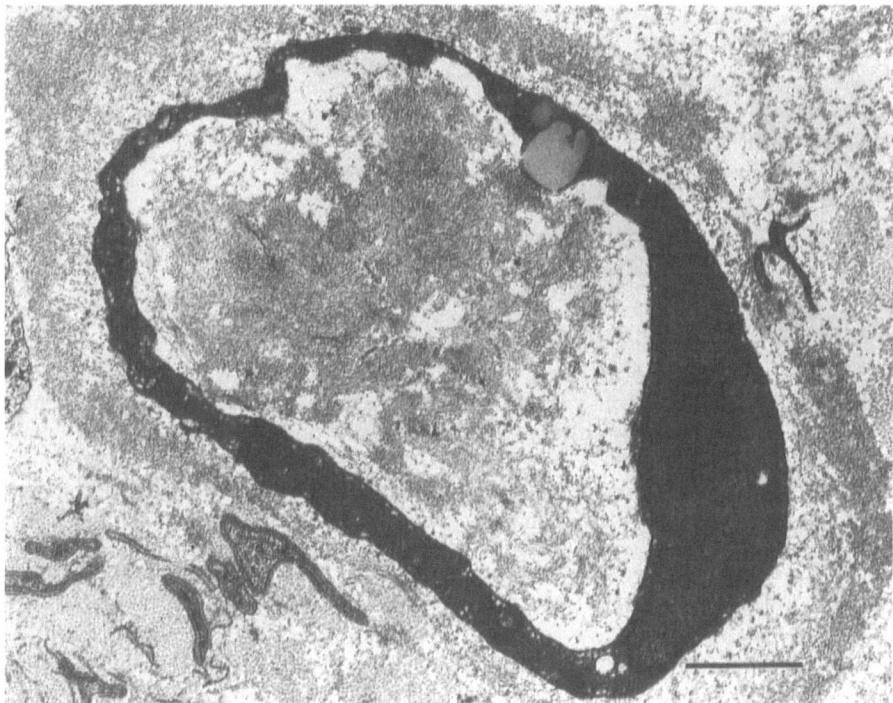

Fig. 10.6a.   Fibroblast-like cell surrounded by and enclosing fibrils in the sural nerve of a patient who had CIDP associated with a monoclonal IgG paraprotein. Electron micrograph. Bar = 20 μm.

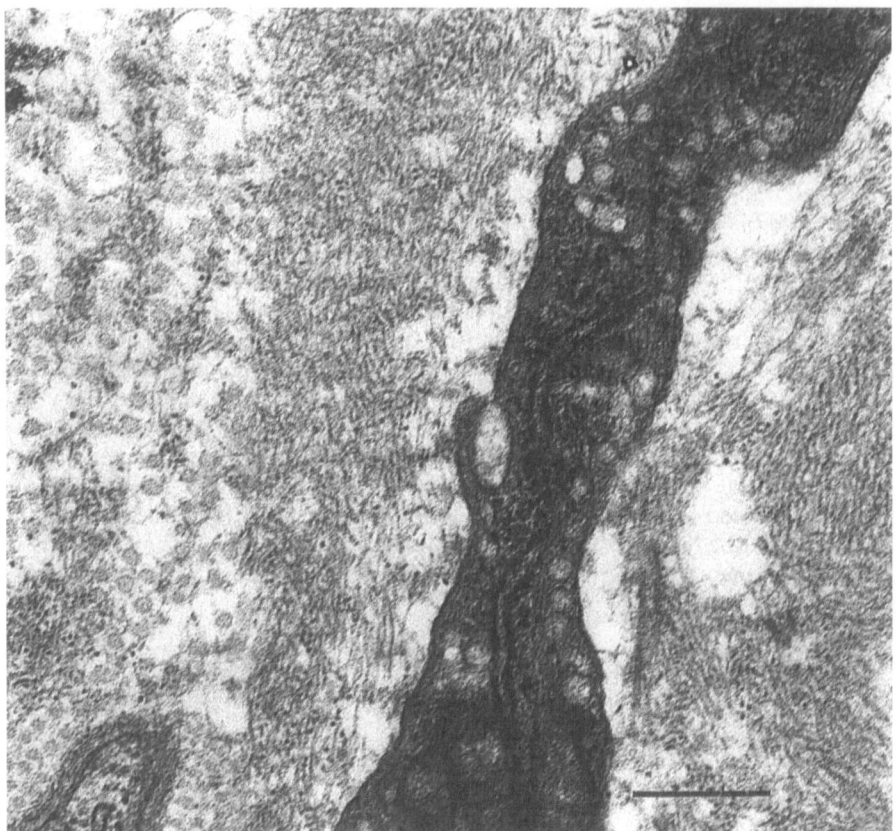

b

Fig. 10.6b. Enlargement of an area from **a** to show the fibrils in greater detail. The fibrils are a rare non-specific finding in chronic neuropathy of several causes and differ from amyloid fibrils. Bar = 0.5 μm.

other recent study of biopsies in 14 patients also identified demyelination and onion bulbs in seven, but cellular infiltration in only four, patients (Krendel et al. 1989). Inflammatory changes were found in one of three cases by Poewe et al. (1981).

Modern immunohistochemical techniques should enable us to distinguish inflammatory from non-inflammatory neuropathies with greater ease. In the first place monoclonal antibodies will permit unequivocal identification of lymphocytes. In six patients with CIDP small numbers of T cells were identified in frozen sections of biopsied nerve fascicles but were far outnumbered by macrophages (Pollard et al. 1986). Major histocompatability class II antigen was upregulated on the endoneurial cells of CIDP patients compared with normal subjects. The extent of the staining strongly suggested that Schwann cells, as well as invading or resident macrophages, were expressing this class II antigen. Furthermore, the staining included onion bulb layers which are for the most part Schwann cell processes. In both the study of Pollard et al. (1986) and another by Mancardi et al. (1988) this up-regulation of MHC class II antigen

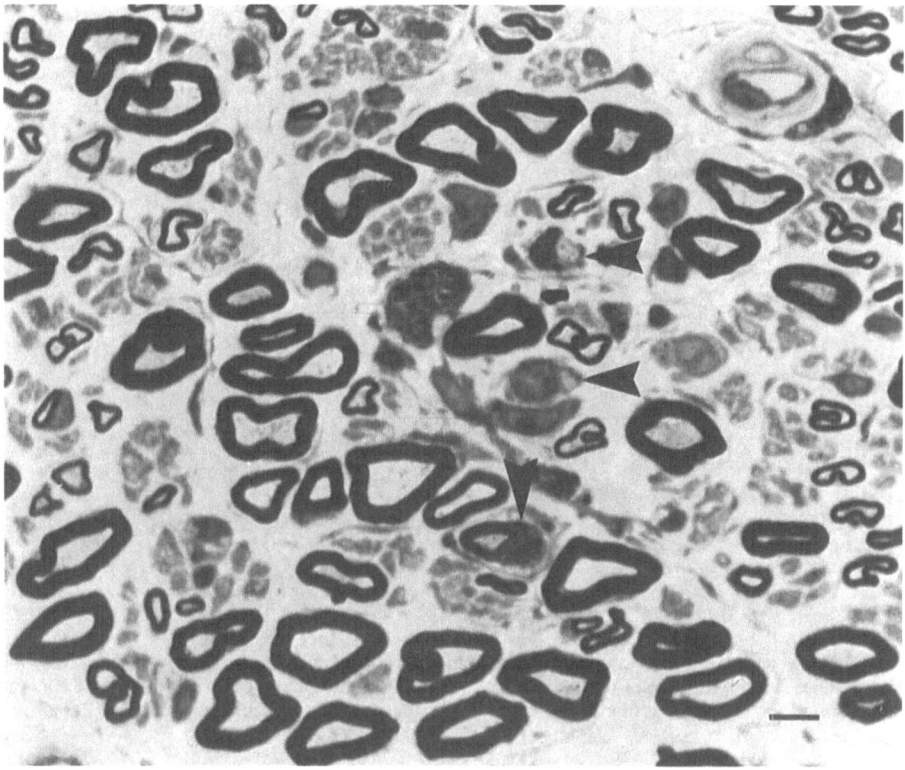

Fig. 10.7. Macrophage-associated demyelination in CIDP. The fibre marked with a *vertical arrow-head* has been invaded by a macrophage. Note demyelinated axons (*horizontal arrowheads*). See also Fig. 10.8. Resin embedded section stained with thionin and acridine orange. Bar = 10 μm.

on endoneurial cells was not specific for CIDP but also occurred in hereditary and diabetic neuropathy.

The overall conclusion is that the pathological evidence gives support for the demyelinating component of the pathology and pathogenesis of CIDP but demonstrates a surprising amount of axonal loss and degeneration and little sign of any inflammatory component. There is a need for more post-mortem studies.

Immunology

Immunogenetics

There is some evidence that immunogenetic factors play a part in CIDP. Since most patients do not give a family history of CIDP, the disease is clearly not explicable on the basis of a single gene. A possible example of CIDP occurring in two siblings has been published (Gabreels-Festen et al. 1986): two sisters

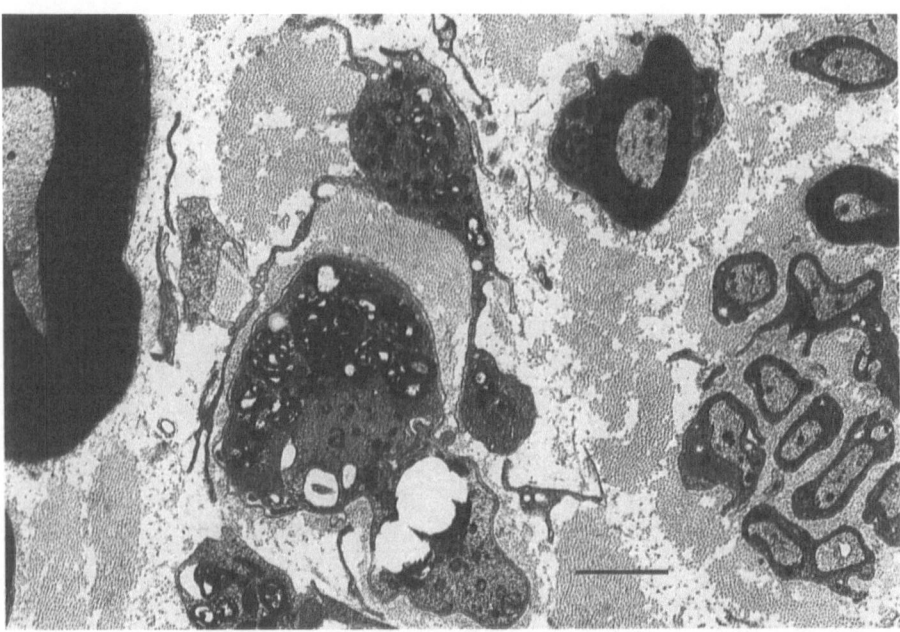

Fig. 10.8. Macrophage-associated demyelination in CIDP from the same biopsy as Fig. 10.7. A macrophage laden with myelin debris surrounds a demyelinated axon (a) within its basal lamina tube. Two other macrophages lie in the endoneurial space. Electron micrograph. Bar = 2.5 μm.

had mild progressive demyelinating neuropathy, scoliosis and, in one case, pes cavus. Nerve biopsy in both cases showed foci of endoneurial inflammation which persuaded the authors to diagnose CIDP rather than a mild form of HMSN type I with incomplete expression in the parents. Treatment was not given.

It is too simple to discard the diagnosis of CIDP in every patient with a family history. To do so would be to deny out of hand an immunogenetic basis for CIDP and prevent exploring a potentially treatable component of hereditary neuropathy. Dyck et al. (1982b) reported seven patients with slowly progressive demyelinating neuropathy, skeletal abnormalities including pes cavus and hammer toes, and similarly affected relatives, who responded to steroid treatment. Mitchell et al. (1987) reported a similar family. It is of course possible that patients with a hereditary demyelinating neuropathy may coincidentally develop a superimposed inflammatory or autoimmune reaction. Patients with a hereditary tendency to break down their own myelin will be releasing myelin antigens into the circulation to a greater extent than normal subjects and will be at greater risk of immunising themselves against their own myelin sheaths.

The search for an HLA association in CIDP has been slightly more rewarding in CIDP than in GBS in which no significant association was discovered. In two small early studies (Stewart et al. 1978; Adams et al. 1979) there were just significantly more patients who had HLA-B8. This class I antigen was also more common in a more recent larger study (Vaughan et al. 1990) (Table 10.9). A slight increase in HLA-DR3 which is in linkage dysequilibrium with HLA-B8 was found in the first two studies but not confirmed in the third (Table 10.10).

Table 10.9. Percentage of Class I HLA phenotypes in CIDP

HLA type	Adams et al. (1979)		Stewart et al. (1978)		Vaughan et al. (1990)		Mean	
	CIDP	Control	CIDP	Control	CIDP	Control	CIDP	Control
n	14	561	16	322	31	346	58	
A1	21	33	38	37	45	34	38	34
AW30, 31	8	5	48	4**	13	5	18**	5**
B7	21	24	25	26	39*	21*	31	24
B8	50*	25*	50*	26*	42	25	46**	25**
Cw7	–	–	–	–	84**	53**	84**	53**

*$P<0.05$; **$P<0.01$.

Table 10.10. Percentage of Class II HLA phenotypes in CIDP

HLA type	Adams et al. (1979)		Stewart et al. (1978)		Vaughan et al. (1990)		Mean	
	CIDP	Control	CIDP	Control	CIDP	Control	CIDP	Control
n	14*	99*	13*	59*	31	60		
DR2	7	21	–	–	32	36	24	27
DW2	7	18	39	21	–	–	22	19
DR3	37*	21*	–	–	32	33	33	26
DW3	37*	17*	46**	14**	–	–	40*	16*

*$P<0.05$; **$P<0.01$.

The most striking difference was the increased frequency of HLA-CW7 in the CIDP patients (84%) compared with the normal subjects (53%). At least two hypotheses would explain this association. Class I molecules bind endogenous peptides in their clefts and in a virally infected cell viral antigens may replace endogenous peptide. The viral antigen in the HLA-CW7 cleft might trigger an autoimmune response. Alternatively since class I molecules are recognised by at least one set of suppressor lymphocytes, HLA-CW7 or a linked gene product may be less capable of suppressing an autoimmune response to myelin antigens. This possibility is more likely because the frequency of HLA-CW7 is decreased rather than increased in GBS.

The only other immunogenetic evidence so far published is a report of an increased frequency of one of the alleles of the gene for α-1 antitrypsin. This enzyme is a protease inhibitor which behaves as an acute phase reactant and is present on the surface of macrophages. Macrophages and proteases released by them play important parts in experimental demyelination. There are more than 30 alleles of α-1 antitrypsin which can be identified by isoelectric focusing. The frequency of the M3 allele was increased to 29% in 52 CIDP patients compared with 11.5% in normal subjects. Similar increases in frequency were noted in GBS and multiple sclerosis. These differences in frequency suggest a role for this protease inhibitor system in permitting the generation of both central and peripheral nervous system demyelinating disease.

Table 10.11.   Association between CIDP and putative autoimmune diseases

	No. of cases	Reference
Addison's disease	1	Abbas et al. (1977)
Autoimmune haemolytic anaemia	1	Prineas and McLeod (1976) case 22
Coeliac disease	3	Buge et al. (1979)
Coagulopathy due to Factor VIII antibody	1	Berger et al. (1983)
Chronic active hepatitis	1	Barohn et al. (1989)
Hashimoto's disease	1	Korn-Lubetzki and Abramsky (1986)
Thyrotoxicosis	4	McCombe et al. (1987b)
Pernicious anaemia	1	Gross (1987)
Sicca syndrome	1	Gross (1987)
Systemic lupus erythematosus	4	Goldberg and Chitanondh (1959) Korn-Lubetzki and Abramsky (1986) Rechthand et al. (1984) Sahenk et al. (1977)

Association with Putative Autoimmune Diseases

Clues to the aetiology of CIDP come, as in myasthenia gravis, from the company it keeps. There are several reports of its co-occurrence with a wide variety of putative autoimmune diseases (Table 10.11). The occurrence of CNS lesions resembling multiple sclerosis and of episodes of optic neuritis in association with CIDP has already been accepted as more than a coincidence (see above). Distinguishing between inflammatory demyelinating polyradiculoneuropathy and vasculitic multiple mononeuropathy can be a difficult diagnostic exercise. Claims that polyarteritis nodosa is associated with CIDP lack histological proof (Korn-Lubetzki and Abramsky 1986). Although SLE usually produces neuropathy by way of vasculitis (Hughes et al. 1982), in some cases an inflammatory demyelinating neuropathy without vasculitis has been demonstrated.

The rare occurrence of membranous glomerulonephritis in CIDP, with deposits of IgG and C3 on the glomerular basement membrane, has given rise to speculation that immune complex deposition might be important in causing the neuropathy (Behan et al. 1973; Peters et al. 1973; Haslitt 1987; Witte and Burke 1987).

Considering the relative rarity of CIDP the reports of its association with organ-specific autoimmune diseases probably do represent more than a coincidence. The probable association with autoimmune diseases in general provides support for the idea that CIDP is itself autoimmune.

Antibodies

No formal studies of immunoglobulin concentrations in CIDP have been published. However, 11 of 23 patients with chronic acquired polyneuropathy of both axonal and demyelinating types had elevated serum IgM concentrations (Whitaker et al. 1973). The patients with high serum IgM concentrations had slower nerve conduction velocities than those with normal concentrations. This suggests that patients with CIDP might have increased IgM concentrations more commonly than controls.

Attempts to discover antibodies to myelin antigens in CIDP have been disappointing. In a search for myelinotoxic factors, sera from three CIDP patients were added to myelinated dorsal root ganglion cultures. One of the three produced demyelination compared with positive results from 26 of 31 GBS and one of 11 control subjects (Cook et al. 1970). This line of research has not been pursued in CIDP as much as in GBS. Endoneurial injection of CIDP serum into rat sciatic nerve gave more demyelination than control serum in one study (Dyck et al. 1982a) but not in another (Server et al. 1979). In the most serious attempt to identify demyelinating factors in CIDP, Toyka et al. (1982) undertook serial injections of immunoglobulin from CIDP or control patients into marmosets. Immunoglobulin from five of six patients produced slowing of motor nerve conduction in the sciatic nerve which was not observed with immunoglobulin from normal subjects or from patients with non-inflammatory neuropathy. The slowing was observed about five days after the start of the injections and lasted for the duration of the experiment (30 days). An immune response to the foreign immunoglobulin injected was prevented with isophosphamide and human immunoglobulin was shown to have reached the endoneurium by immunohistochemistry. Unfortunately the authors were unable to identify a morphological correlate of the slowed nerve conduction and this work has not been pursued.

If antibody were playing an important part in CIDP, deposition of antibody and complement would be expected in the peripheral nerve. According to Dalakas and Engel (1980), deposits of IgM, C3 and occasionally IgG can be found in the walls of endoneurial blood vessels, and linear deposits of IgM can be found along the surviving myelin sheaths in nerve biopsies. Unfortunately this work has not been confirmed and we have been unable to identify greater deposition of immunoglobulin, C3, C1q or C3d than in biopsies from control subjects (Leibowitz and Hughes, unpublished). This contrasts with the demyelinating neuropathy associated with anti-MAG antibodies and IgM_k paraproteinaemia in which deposition of IgM and C3d on myelin sheaths is prominent.

Limited attempts have been made to identify antibodies to nerve antigens with conventional antibody assays. One of four sera from patients with CIDP was shown to contain antibodies to nerve tissue with an antiglobulin consumption test (Nyland and Aarli 1978). Toyka et al. (1982) reported that serum from a patient with CIDP who had responded to plasma exchange reacted with the axon–myelin junction. This may have represented no more than the axonal antibodies present in a proportion of normal human sera. We have been unable to identify antibodies to myelin in the sera of CIDP patients by the indirect immunofluorescent technique (Leibowitz and Hughes, unpublished). Osuntokun et al. (1966) also could not identify antibodies to human peripheral nerve myelin by the immunofluorescent technique in any of ten sera from patients with CIDP: complement fixation tests to nerve extracts were also negative with the same sera. None of the sera from 11 patients with CIDP contained detectable levels of complement-fixing antibody to human or rabbit myelin and only one contained antibodies to galactocerebroside (Hughes et al. 1984). Galactocerebroside is an antigen of particular relevance since a chronic form of inflammatory demyelinating neuropathy can be induced in rabbits by injecting galactocerebroside. This has been proposed as a model for CIDP (Saida et al. 1981) but evidence of an immune response to galactocerebroside in any human disease is lacking. In another study two of nine patients with CIDP had demonstrable complement-fixing antibodies to human myelin: the titres were much lower than those given by sera from patients with IgM paraproteinaemia and neuropathy and did not alter with the passage of time (Latov et al. 1981). ELISA tests for antibody to P_2 myelin protein were weakly positive in one of 11 patients with CIDP tested by Hughes et al. (1984) and negative with all of the 11 sera studied by Zweiman et al. (1983).

Since the likeliest explanation of the beneficial effect from plasma exchange in CIDP (see below) is removal of a demyelinating factor or antibody, further studies along these lines would be appropriate and should be rewarding.

Cell-Mediated Immunity

The results of the few studies of cell-mediated immunity in CIDP have been conflicting. In an immunofluorescent study the percentage of T cells in the blood bearing the activation marker DR was shown to be slightly but significantly higher than that of healthy control subjects (Taylor and Hughes 1989). The percentage of activated T cells was small and less than in a parallel study of GBS. Complementary evidence of T cell activation in CIDP comes from a study demonstrating increased concentrations of soluble IL2 receptors in the serum (Hartung et al. 1990). The low percentage of activated cells may account for the difficulty investigators have had in identifying cell-mediated immunity against specific antigens. In fact such investigations have given conflicting results. Castaigne et al. (1972) found positive results in all of five cases using a leucocyte migration inhibition test and crude nerve homogenate as antigen. Sheremata et al. (1974) obtained negative results with a macrophage migration inhibition test and P_2 protein as antigen in three patients. Abramsky et al. (1980) reported positive lymphocyte transformation tests with human P_2 protein in all of four CIDP patients, whereas neither Zweiman et al. (1983) in a study of 11 patients nor Taylor and Hughes (unpublished information) in a study of 10 patients have been able to confirm this finding. These negative results have to be interpreted cautiously since activated T cells relevant to an ongoing autoimmune process may not be easy to detect in the blood. In our study of EAN we found specific responsiveness of blood lymphocytes to P_2 only in the onset phase of EAN, whereas cells rescued from the cauda equina could still be shown to respond to P_2 during the early convalescent stage (Taylor and Hughes 1988).

Treatment

Steroids

In Austin's (1958) case 20 bouts were documented to respond to ACTH or cortisone and to relapse when placebo was substituted. Since then it has been accepted that steroids are worth trying in CIDP. Thomas et al. (1969) described the clinical course of five cases and concluded that steroids had been beneficial in two but not in the other three. In Dyck et al.'s (1975) series 38 patients were treated with steroids: 39% were considered to have improved, 47% were unchanged and 14% worsened. Typical doses were prednisolone 60–80 mg daily or 120 mg on alternate days tapered over three to six months. These large doses were associated with the usual complications including gastric haemorrhage and perforation, and activation of tuberculosis and other infections. In the large Australian series 65% of 76 patients given steroids were considered to have improved (McCombe et al. 1987b). Of 59 patients in the Ohio series 95% showed an initial response to "immunosuppressive treatment" started with prednisolone 100 mg daily for two to four weeks and then 100 mg on alternate days (Barohn et al. 1989). Azathioprine 2.5 mg/kg was added if the response was poor or delayed. Improvement was seen within a mean (SD) 1.9 (3.6) months and a plateau was reached in 6.6 (5.4) months. Spontaneous remissions occurred in 40% of the Ohio series of patients at various times during their course and left the patient without any neurological deficit at all in 30%. The remissions of relapsing CIDP resemble the time course of multiple sclerosis. Remissions may also occur spontaneously in patients who have been gradually worsening for many months or years.

The clinical experience that steroids induce at least short-term benefit is supported by a controlled trial. Dyck et al. (1982b) randomised 40 patients to receive prednisone 120 mg on alternate days for the first week, 100 mg on alternate days for the second week and then a dose tapered to nothing over the next ten weeks. The trial had an open design. Five patients were withdrawn because the initial diagnosis proved incorrect. Five prednisone-treated and two control patients did not complete the study. The analysis was confined to the 28 patients who completed the trial. Unfortunately exclusion of the withdrawn patients can be construed as having biased the trial in favour of a beneficial result: one of the patients withdrawn from the steroid group died of cardiac arrhythmia possibly related to hyperglycaemia and another could not return for follow-up because of such severe disease that he was respirator bound at another institution. These comments should not disguise the fact that this trial represents the best evaluation of steroid efficacy in CIDP and further evaluations seem unlikely to be undertaken. The 14 evaluated prednisone-treated patients improved a median of 10 points on a neurological disability score while the 14 control patients deteriorated 1.5 points ($P < 0.02$). This score involves scoring abnormalities on the neurological examination and the changes ranged widely between 23 points and improving 79 points in the prednisone-treated group, and between deteriorating 44 points and improving 65 points in the control group. There were also significant improvements in some more easily measured data including hand grip strength and median nerve motor conduction velocity.

Although there is general agreement that steroids are beneficial in CIDP, a wide range of dosage schedules has been used and no work has been addressed to the question of which is best. Reports have not yet appeared of the use of high dose intravenous methylprednisolone pulse treatments which are becoming popular in multiple sclerosis (Thompson et al. 1989). Dalakas and Engel (1981) on the basis of experience treating 25 patients recommend using large doses for longer periods: they give details of a regime starting with 100 mg daily for three to four weeks, and then tapered to 100 mg on alternate days over twelve weeks. This dose is continued for another four to twelve weeks before tapering to zero over the next two or two and a half years. Such high dose regimes, even with alternate day dosing, carry a high risk of Cushingoid side-effects. The authors of this regime acknowledge the occurrence of osteoporosis, cataracts, diabetes mellitus, hypertension, and avascular necrosis of the femoral head. A severely Cushingoid facies, obesity and myopathy must also be common. Some of the other published regimes have already been mentioned. I prefer to start with prednisolone 60 mg daily and continue with this dose for four weeks. The dose is then decreased to an alternate day regime, so that by the end of twelve weeks the patient is taking 60 mg on alternate days. Further tapering is continued at 5 mg every week depending on the response. If the disease appears to escape from control I prefer to try adding azathioprine and plasma exchange rather than continue with very high doses of prednisolone. In order to minimise side-effects antacids and potassium supplements are often recommended. I prefer to prescribe these if they become necessary.

Immunosuppressive Drugs

Numerous different cytotoxic drugs have been tried in CIDP with anecdotal reports of success. The purine analogue 6-mercaptopurine was reported as sometimes beneficial 25 years ago (Palmer 1965, 1966) but has not been much used since. Apparent benefit from azathioprine, which is metabolised by the liver into 6-mercaptopurine, was reported in several cases (Yuill et al. 1970; Walker 1979; Pentland 1980). Because of its lack of toxicity compared with other cytotoxic drugs and its widespread use in autoimmune disease and transplant medicine, azathioprine has been more widely adopted in CIDP than any other immunosuppressive drugs. Its use was advocated by Dalakas and Engel (1981) if steroids alone were insufficient: they recommended using 3 mg/kg and then gradually increasing the dose to induce a slight reduction of the white cell count. We have drifted into using azathioprine if steroids fail to control CIDP without formal proof that it is effective. The only controlled trial, once again undertaken by Dyck and colleagues (1985), failed to show any benefit from adding azathioprine 2 mg/kg to prednisone: 13 patients received prednisone alone, 14 patients had the combined treatment. The trial measured many outcomes, including overall neurological disability score, and none showed benefit to the azathioprine group. Since the sample size was small its statistical power to detect a small effect was low. Furthermore the dose used was only 2 mg/kg. The authors stated that two of their patients subsequently appeared to respond to azathioprine 3 mg/kg. This trial result has not deterred groups who specialise in caring for patients with CIDP from using azathioprine. McCombe et al. (1987b) considered that azathioprine helped four of the five patients in whom it was tried. Barohn et al. (1989) used 2.5 mg/kg. I usually introduce the drug gradually to reduce gastrointestinal intolerance and increase the dose to 2.5 mg/kg with further increments only if control of the disease is not achieved. A very small proportion of patients cannot tolerate azathioprine because of allergic rashes, abnormal liver function or gastrointestinal intolerance. In a trial of azathioprine 2.5 mg/kg in multiple sclerosis it proved possible to keep about 80% of patients on the drug for three years (British and Dutch Multiple Sclerosis Azathioprine Trial Group 1988). Side-effects were more frequent in the treated than the control group (Table 10.12) but promptly disappeared when the drug was stopped. There is a theoretical risk of developing cancer because of long-

Table 10.12. Side-effects reported by 174 patients taking azathioprine 2.5 mg/kg or placebo in a double-masked controlled trial in multiple sclerosis

	Year 1		Years 2–4	
	Azathioprine	Placebo	Azathioprine	Placebo
Leucopenia (<3000/μl)	36	1	6	2
Anaemia	6	0	2	1
Thrombocytopenia	1	1	1	1
Other haematological abnormality	8	0	4	2
Abnormal liver function	15	4	9	4
Anorexia	29	10	8	7
Vomiting	21	7	4	3
Abdominal pain	19	5	1	3
Diarrhoea	10	5	4	4
Other gastrointestinal abnormality	18	5	4	6
Rash	13	2	8	8
Fever	4	0	1	1
Arthralgia	4	4	3	0
Myalgia	7	3	5	2
Other allergic conditions	17	5	11	5

From British and Dutch Multiple Sclerosis Azathioprine Trial Group (1988), with permission.

Table 10.13. Risk of cancer in patients taking azathioprine or cyclophosphamide for medical conditions other than transplant rejection

Type of cancer	Azathioprine		Cyclophosphamide	
	Observed	Expected	Observed	Expected
Non-Hodgkin's lymphoma	5	0.38	1	0.16
All skin:	4	2.58	1	1.00
Squamous cell	2	0.42	1	0.17
Basal cell	2	1.89	0	0.73
Bladder	1	1.14	5	0.47
Others	30	24.51	16	9.33
Total	40	28.61	23	10.96

From Kinlen (1985), with permission.

term immunosuppression, but in practice this risk is small and confined to non-Hodgkin's lymphoma and skin cancer (Table 10.13). Perhaps the main problem with azathioprine is the need to have blood tests. I recommend weekly liver function tests and blood counts including platelet counts for eight weeks, then four weekly for a year and thereafter six weekly. An increased red cell mean corpuscular volume is common and, provided the serum B_{12} and folate are normal, does not require treatment. I reduce the dose of azathioprine if the total leucocyte count is reduced below 4000/μl, or in the presence of anaemia or thrombocytopenia or rising liver enzyme concentrations. Reduced doses are rarely necessary if 2.5 mg/kg is taken as the usual top dose.

Cyclophosphamide, an alkylating agent, is the next most used cytotoxic drug in CIDP. It has not been subjected to any controlled trials. Benefit was reported in nine patients in early studies (Rosen and Vastola 1976; Prineas and McLeod 1976; Fowler et al. 1979; Dalakas and Engel 1981). More recently McCombe et al. (1987b) reported improvement following its use in four of five patients. Although it can be used in large intravenous doses as have been used in neoplastic disease and multiple sclerosis, oral doses of 2 mg/kg have usually been used in CIDP. It is a less pleasant drug to take than azathioprine, being more likely to induce gastrointestinal disturbances. Higher doses may cause alopecia, but this is not usually a problem with 2 mg/kg. Particular hazards of cyclophosphamide are haemorrhagic cystitis and bladder cancer (Table 10.13).

The latest fashion in immunosuppressive drugs is cyclosporin A and it has been used with success in a few cases of CIDP. In a case reported by Kolkin et al. (1987) the neuropathy improved only at the expense of unacceptable nephrotoxicity. In two cases reported in another connection the neuropathy abated during cyclosporin treatment (Gross and Thomas 1981). In an abstract report cyclosporin A appeared beneficial in four of seven patients (Tindall et al. 1989). The most recent report is very enthusiastic (Hodgkinson et al. 1990). Cyclosporin A appeared beneficial in all of three cases of CIDP alone and five cases of CIDP associated with paraprotein. Three patients had a very good response, two recovering completely. The dose was started at 5–15 mg/kg/day and reduced to a maintenance dose of 2–6 mg/kg/day.

A multitude of other agents have been tried in cases of CIDP refractory to treatment with steroids. Benefit has been claimed from intravenous injections of the interferon inducer polyinosinic-polycytidylic acid poly-L-lysine in two of four cases (Dalakas and Engel 1981; Engel et al. 1983), total lymphoid irradiation in another (Rosenberg et al. 1985), and chloroquine in another (De Vivo and Engel 1970). The problem with such reports is that it is impossible to know how many patients have been given these toxic treatments and not benefited or even been damaged. On the other hand it is easy to understand the frustration and desperation which severely disabled patients with CIDP and their physicians share.

Plasma Exchange

Several early series and abstracts reported the successful use of plasma exchange in progressive (Levy et al. 1979; Cook et al. 1980; Server et al. 1980) and relapsing CIDP (Fowler et al. 1979; Toyka et al. 1982; Lastavica et al. 1989). Gross and Thomas (1981) published the results of treatment in six cases. They reported rapid, substantial but temporary improvement in two of three relapsing cases, but only slight improvement in a third relapsing case and three progressive cases. Most subsequent anecdotal reports and small series have concluded that plasma exchange is of temporary benefit in CIDP (Maas et al. 1981; Osterman et al. 1982; Feasby et al. 1983; Donofrio et al. 1985). In a study of five patients two responded to plasma exchange, and in these the nerve biopsies showed more demyelinated nerve fibres and less axonal degeneration than in the three who did not respond (Pollard et al. 1983). The numbers in this comparison are too small to permit firm conclusions and reliable predictors of response to plasma exchange in CIDP have not been determined.

The value of plasma exchange in CIDP has been confirmed by the most elegant trial of treatment yet undertaken in CIDP. Dyck èt al. (1982, 1986) randomised 34 patients. After withdrawals for incorrect diagnosis and one case of myocarditis possibly related to a subclavian catheter, 15 patients were assigned to receive true plasma exchange and 14 to sham exchange. On average 47 ml of plasma per kg was exchanged per session and each patient received twice weekly plasma exchanges for three weeks. After the three-week treatment period there was significantly more reduction of overall neurological disability in the true exchange group ($P = 0.025$). Five patients who had received plasma exchange had improved more than the largest improvement achieved in the sham exchange group. There were also significant improvements on several measures of nerve conduction and the changes in nerve conduction showed a significant correlation with improvement in overall neurological disability score. In clinical practice plasma exchange seems to permit one to gain and maintain control of the disease in CIDP when steroids and azathioprine are insufficiently effective. Occasional patients become hooked on plasma exchange in the same way that patients with renal failure become dependent on dialysis. Feasby et al. (1983) report using plasma exchange as maintenance treatment for up to 30 months. Although this may seem undesirable, it could be preferable to the long-term effects of continued prednisolone and azathioprine treatment. To some extent the use of plasma exchange has to be determined by what is available. In North America where plasma exchange services are more readily available, plasma exchange is likely to be more widely used than in the United Kingdom. At present I am constrained by the limited availability of the service to use plasma exchange when the combination of prednisolone and azathioprine is insufficiently effective or when a rapid treatment effect is considered mandatory.

Immunoglobulin

Increasing interest is being aroused by reports of benefit from intravenous immunoglobulin. This is quite different from plasma exchange, in which plasma is usually removed and replaced with 5% albumin without added globulin. Following up the lead that intravenous gammaglobulin is beneficial in idiopathic thrombocytopenia, Vermeulen et al. (1985) tried a similar regime in CIDP. They first gave fresh frozen plasma 0.1 l/kg within 5–7 days to 17 patients of whom 13 improved. Improvement began within 3–7 days and was sometimes as dramatic as a change from being unable to walk at all to being able to walk unaided. Four patients had persistent stable benefit for at least four months after one course. Nine patients noted deterioration after an average of three weeks, only to respond again to a further infusion. The occurrence of allergic reactions and risks of HIV or other viral infection make this an unacceptable form of treatment for general use. Ten patients were treated with intravenous gammaglobulin 0.4 g/kg/day on five consecutive days. One of the patients had not responded to fresh frozen plasma and did not respond to gammaglobulin either. Eight of the other nine patients responded to gammaglobulin in the same way that they had responded to fresh frozen plasma. An Italian study also claimed benefit from high-dose intravenous gammaglobulin in all of four cases of CIDP (Curro Dossi and Tezzon 1987). Benefit was noted in a single case by

Albala et al. (1987). Most recently Faed et al. (1989) have reported benefit from treatment in all of nine patients. They used 0.4 g/kg daily infusions on three consecutive days, topped up with a further two days of infusions seven to ten days later if the response was incomplete. Benefit was seen in 2–7 days and lasted at least four weeks. Six patients achieved sustained remissions. The improvements reported were dramatic and unlikely to have been due to a placebo effect. In fact the responses were so dramatic and rapid that it ought to be easy to test them with a controlled trial. Possible mechanisms for the effect include antigen specific anti-idiotypic suppression of autoantibodies against myelin or Schwann cell constituents, or various non-specific mechanisms including Fc receptor blockade or solubilisation of immune complexes.

Summary

CIDP is defined as a chronic progressive polyradiculoneuropathy involving weakness, which is usually proximal and distal, associated with neurophysiological evidence of demyelination and lacking any systemic cause. It is an uncommon disease but more common in men than women. There are subacute, progressive, relapsing and recurrent acute forms. The disease may start at any age, the progressive form usually at the age of 40–50, and the relapsing form significantly younger, at about the age of 20–30. Preceding infections or immunisations have been reported in about a third of cases, which is less than in GBS. Several relapses have been reported following tetanus toxoid immunisation. Relapses are more likely to occur during a pregnancy year but it is not clear whether this relates to the last trimester or the puerperium. Although CIDP predominantly affects the limbs, facial weakness occurs in 15%, bulbar involvement in 6% and ophthalmoplegia in 4%. Respiratory failure and autonomic dysfunction are distinctly uncommon. The main differential diagnoses are hereditary and paraproteinaemic neuropathies. About 5% of cases have associated CNS involvement. The CSF protein concentration is usually increased without a pleocytosis. Limited information from autopsies suggests that there may be inflammation in the nerve roots, sometimes spilling into the CNS. In sural nerve biopsies, despite clear neurophysiological evidence that demyelination is the major pathogenetic mechanism, nerve fibre loss and axonal degeneration are usually more apparent than demyelinated or remyelinated fibres. Supernumerary Schwann cells are common and onion bulb formation may be seen, but endoneurial inflammation is rare. Major histocompatibility class II antigen expression is up-regulated on endoneurial cells, probably including Schwann cells, but this also happens in hereditary and metabolic neuropathies.

There is a possible association with putative autoimmune diseases. The frequency of the major histocompatibility class I antigen, HLA-CW7, is increased in CIDP, possibly reflecting an inherited impairment of the ability to suppress autoimmune responses to myelin antigens. There is an increased frequency of one allele of the gene of α-1 antitrypsin, a protease inhibitor which might control proteolysis involved in macrophage-mediated demyelination. Attempts to identify antibodies or cell-mediated immunity to myelin antigens have rarely been successful and have certainly not produced any helpful diagnostic test.

Benefit from steroids and plasma exchange has been demonstrated by controlled trials. Clinical experience strongly suggests benefit from immunosuppressive drugs including azathioprine, cyclophosphamide and cyclosporin A. Preliminary evidence suggests that intravenous gammaglobulin, as used for idiopathic thrombocytopenia, is also helpful. A conventional treatment plan would be to start with prednisolone, add azathioprine if too high or prolonged steroid dosage becomes necessary, and to use plasma exchange if that combination is ineffective. Where plasma exchange is readily available, it may be preferable to introduce plasma exchange earlier, even as the first treatment.

References

Abbas DH, Schlagenhauff RE, Strong HE (1977) Polyradiculoneuropathy in Addison's disease. Neurology 27:494–495

Abramsky O, Korn-Lubetzki I, Teitelbaum D (1980) Association of autoimmune diseases and cellular immune response to the neuritogenic protein in Guillain–Barré syndrome. Ann Neurol 8:117

Adams D, Festenstein H, Gibson JD et al. (1979) HLA antigens in chronic relapsing idiopathic inflammatory neuropathy. J Neurol Neurosurg Psychiatry 42:184–186

Albala M, McNamaram ME, Sokol M, Wyshock E (1987) Improvement of neurologic function in CIDP following intravenous γ-globulin infusion. Arch Neurol (Chicago) 44:248–249

Albers JW (1987) Inflammatory demyelinating polyradiculoneuropathy. In: Brown WF, Bolton CF (eds) Clinical electromyography, Butterworths, Boston, pp 211–244

Albers JW, Kelly JJ (1989) Acquired inflammatory demyelinating polyneuropathies: clinical and electrodiagnostic features. Muscle Nerve 12:435–451

Ashworth B, Smyth GE (1969) Relapsing motor polyneuropathy. Acta Neurol Scand 45:342–350

Austin JH (1958) Recurrent polyneuropathies and their corticosteroid treatment. Brain 81:157–192

Barohn RJ, Kissel JT, Warmolts JR, Mendell JR (1989) Chronic inflammatory demyelinating polyradiculoneuropathy. Clinical characteristics, course, and recommendations for diagnostic criteria. Arch Neurol 46:878–884

Behan PO, Lowenstein LM, Stilmant M, Sax DS (1973) Landry–Guillain–Barré syndrome and immune complex nephrosis. Lancet i:850

Berger JR, Rosenfeld WE, Sheremata WA et al. (1983) Chronic inflammatory polyradiculoneuropathy complicated by factor VIII antibody. Neurology 33:1224–1226

Bolton CF, Gilbert JJ, Girvin JP, Hahn A (1979) Nerve and muscle biopsy: electrophysiology and morphology and polyneuropathy. Neurology 29:354–362

Bonnaud E, Vital C, Cohere G, Castaing R, Loiseau P (1974) Recurrent and relapsing polyneuritis. Four cases with ultrastructural studies of the peripheral nerve. Pathol Eur 9:109–118

Borit A, Altrocchi PH (1971) Recurrent polyneuropathy and neurolymphomatosis. Arch Neurol 24:40–47

British and Dutch Multiple Sclerosis Azathioprine Trial Group (1988) Double-masked trial of azathioprine in multiple sclerosis. Lancet i:179–183

Buge A, Escourolle R, Rancurel G et al. (1979) Chronic inflammatory neuromyopathies in adults treated for gluten-sensitive enteropathy. A report on three cases with microvascular nerve and muscle lesions. Rev Neurol (Paris) 135:719–731

Calderon-Gonzalez R, Gonzalez-Cantu N, Rissi-Hernandez H (1970) Recurrent polyneuropathy with pregnancy and oral contraceptives. N Engl J Med 282:1307–1308

Castaigne P, Brunet P, Nouailhat F (1966) Enquête clinique sur les polyradiculonévrites inflammatoire en France. Rev Neurol (Paris) 115:849–872

Castaigne P, Berthaux P, Brunet P (1972) Polyradiculonévrites inflammatoires et immunite cellulaire: étude de la réponse immunitaire cellulaire envers des antigènes de nerf périphérique par le test de migration des leucocytes. Nouv Press Med 1:2445–2449

Cook JD, Tindall RAS, Walker J, Khan A, Rosenberg R (1980) Plasma exchange as a treatment of acute and chronic idiopathic autoimmune polyneuropathy: limited success. Neurology 30:361–362

Cook SD, Dowling PC, Whitaker JN (1970) Serum immunoglobulins in the Guillain–Barré syndrome. Neurology 20:403

Curro Dossi B, Tezzon F (1987) High dose intravenous gammaglobulin for chronic inflammatory demyelinating polyneuropathy. Ital J Neurol Sci 8:321–326

Dalakas MC, Engel WK (1980) Immunoglobulin and complement deposits in nerves of patients with chronic relapsing polyneuropathy. Arch Neurol 37: 637–640

Dalakas MC, Engel WK (1981) Chronic relapsing (dysimmune) polyneuropathy: pathogenesis and treatment. Ann Neurol 9 Suppl:134–145

Dalakas MC, Teravainen H, Engel WK (1984) Tremor as a feature of chronic relapsing and dys-gammaglobulinaemic polyneuropathies. Incidence and management. Arch Neurol 41:711–714

D'Ambrosio G, De Angelis G (1985) Syndrome de Guillain–Barré au cours de la grossesse. Rev Neurol (Paris) 141:33–36

De Mello AR, De Freitas MRG, Chimelli L (1989) Chronic recurrent Guillain–Barré syndrome. Report of 3 cases. Arq Neuropsiquiatr 47:84–90

De Vivo DC, Engel WK (1970) Remarkable recovery of steroid – responsive recurrent polyneuropathy. J Neurol Neurosurg Psychiatry 33:62–69

Dinkler F (1904) Zur Kasuistik der multiplen Herdsklerose des Gehirns und Ruckenmarks. Dtsch Z Nervenheilk 26:233–247

Donaghy M, Earl CJ (1985) Ocular palsy preceding chronic relapsing polyneuropathy by several weeks. Ann Neurol 17:49–50

Donaghy M, Gray JA, Squier W et al. (1989) Recurrent Guillain–Barré syndrome after multiple exposures to cytomegalovirus. Am J Med 87:339–341

Donofrio PD, Tandan RUP, Albers JW (1985) Plasma exchange in chronic inflammatory demyelinating polyradiculoneuropathy. Muscle Nerve 8:321–327

Dyck PJ, Arnason BGW (1984) Chronic inflammatory demyelinating polyradiculoneuropathy. In: Dyck PJ, Thomas PK, Lambert EH, Bunge R (eds) Peripheral neuropathy. WB Saunders, Philadelphia, pp 2101–2114

Dyck PJ, Lais AC, Ohta M, Bastron JA, Okazaki H, Groover RV (1975) Chronic inflammatory polyradiculoneuropathy. Mayo Clin Proc 50:621–651

Dyck PJ, Oviatt KF, Lambert EH (1981) Intensive evaluation of unclassified neuropathies yields improved diagnosis. Ann Neurol 10:222–226

Dyck PJ, Lais AC, Hansen SM et al. (1982a) Technique assessment of demyelination from endoneurial injection. Exp Neurol 77:359–377

Dyck PJ, Swansen CJ, Low PA, Bartleson JD, Lambert EH (1982b) Prednisone responsive hereditary motor and sensory neuropathy. Mayo Clin Proc 57:239–246

Dyck PJ, O'Brien PC, Oviatt KF et al. (1982) Prednisone improves chronic inflammatory demyelinating polyradiculoneuropathy more than no treatment. Ann Neurol 11:136–141

Dyck PJ, O'Brien P, Swanson C, Low P, Daube J (1985) Combined azathioprine and prednisone in chronic inflammatory demyelinating polyneuropathy. Neurology 35:1173–1176

Dyck PJ, Daube J, O'Brien P et al. (1986) Plasma exchange in chronic inflammatory demyelinating polyradiculoneuropathy. N Engl J Med 314:461–465

Eiben RM, Gersony WM (1963) Recognition, prognosis and treatment of the Guillain–Barré syndrome (acute idiopathic polyneuritis). Med Clin North Am 47:1371–1380

Engel WK, Askanas V, Levy HB, McFarlin DE (1983) Polyinosinic-polcytidylic acid poly-L-lysine (polyICLC): A new "anti-dysimmune" effect remarkably beneficial in neuropathies evokes a new concept of pathogenesis. Neurology 33 Suppl 2:83

Faber B, Baslov JT (1970) Immunofluorescent studies of renal biopsies in acute polyradiculoneuritis. Acta Pathol Microbiol Scand 78:655–656

Faed JM, Day B, Pollock M, Taylor PK, Nukada H, Hammond-Tooke GD (1989) High-dose intravenous human immunoglobulin in CIDP. Neurology 39:422–425

Feasby TE, Hahn AF, Brown WF (1983) Long-term plasmapheresis in chronic progressive demyelinating polyneuropathy. Ann Neurol 14:122

Fowler H, Vulpe M, Marks G, Egolg C, Dau PC (1979) Recovery from chronic progressive polyneuropathy after treatment with plasma exchange and cyclophosphamide. Lancet ii:1193

Gabreels-Festen AAWM, Hageman ATM, Gabreels FJM et al. (1986) Chronic inflammatory demyelinating polyneuropathy in two siblings. J Neurol Neurosurg Psychiatry 49:152–156

Gigli GL, Carlesimon A, Valente M, Mazza S, Di Trapani G (1989) Evoked potentials suggest cranial nerves and CNS involvement in chronic relapsing polyradiculoneuropathy. Eur Neurol 29:145–149

Goldberg M, Chitanondh H (1959) Polyneuritis with albuminocytologic dissociation in the spinal fluid in SLE: report of a case with a review of the pertinent literature. Am J Med 27:342–350

Goto Y, Hamaguchi K, Hirai S, Matsuyama H (1969) Chronic polyneuritis with repeated remissions and exacerbations. Report of a case with autopsy findings. Clin Neurol (Japan) 9:239–247

Gross M (1987) Chronic relapsing inflammatory polyneuropathy complicating Sicca syndrome. J Neurol Neurosurg Psychiatry 50:939–940

Gross MLP, Thomas PK (1981) The treatment of chronic relapsing and chronic progressive idiopathic inflammatory polyneuropathy by plasma exchange. J Neurol Sci 52:69–78

Harding AE, Thomas PK (1984) Genetically determined neuropathies. In: Asbury AK, Gilliatt RW (eds) Peripheral nerve disorders. Butterworths, London, pp 204–242

Hartung H-P, Hughes RAC, Taylor WA, Heininger K, Reiners K, Toyka KV (1990) T cell activation in the Guillain–Barré syndrome and in MS: elevated serum levels of soluble IL-2 receptors. Neurology 40:215–219

Haslitt J (1987) Membranous glomerulopathy associated with Landry–Guillain–Barré syndrome. Am J Kidney Dis 9:445

Hewer RL, Hilton PJ, Smith AC, Spalding JMK (1968) Acute polyneuritis requiring artificial respiration. Q J Med 37:479–491

Hinman RC, Magee KR (1967) Guillain–Barré syndrome with progressive onset and persistent elevation of CSF protein. Ann Intern Med 67:1007–1012

Hodgkinson SJ, Pollard JD, McLeod JG (1990) Cyclosporin A in the treatment of chronic demyelinating polyradiculoneuropathy. J Neurol Neurosurg Psychiatry 53:327–330

Hughes RAC, Cameron JS, Hall SM, Heaton J, Payan JA, Teoh R (1982) Multiple mononeuropathy as the initial presentation of systemic lupus erythematosus – nerve biopsy and response to plasma exchange. J Neurol 228:239–247

Hughes RAC, Gray IA, Gregson NA et al. (1984) Immune responses to myelin antigens in Guillain–Barré syndrome. J Neuroimmunol 6:303–312

Hughes RAC, Sanders EACM, Winer JB (1987) Guillain–Barré syndrome and chronic idiopathic demyelinating polyradiculoneuropathy. Prog Clin Neurosci 1:143–155

Jellinger K (1969) Einige morphologische Aspekte der multiplen Sklerose. Wien Z Nervenkeilk Suppl 2:12–37

Julien J, Vital C, Lagueny A, Ferrer X, Brechenmacher C (1989) Chronic relapsing idiopathic polyneuropathy with primary axonal lesions. J Neurol Neurosurg Psychiatry 52:871–875

Kaplan JG, Schaumburg HH, Sumner A (1985) Relapsing ophthalmoparesis – sensory neuropathy syndrome. Neurology 35:396

Kennedy RH, Danielson MA, Mulder DW, Kurland LT (1978) Guillain–Barré syndrome: A 42-year epidemiologic and clinical study. Mayo Clin Proc 53:93–99

Kinlen LJ (1985) Incidence of cancer in rheumatoid arthritis and other disorders after immunosuppressive treatment. Am J Med 78 Suppl A:44–49

Kolkin S, Nahman NS, Mendell JR (1987) Chronic nephrotoxicity complicating cyclosporin treatment of chronic inflammatory demyelinating polyradiculoneuropathy. Neurology 37:147–148

Korn-Lubetzki I, Abramsky O (1986) Acute and chronic demyelinating inflammatory polyradiculoneuropathy. Arch Neurol 43:604–608

Krendel DA, Parks HP, Anthony DC, StClair MB, Graham DG (1989) Sural nerve biopsy in chronic inflammatory demyelinating polyradiculoneuropathy. Muscle Nerve 12:257–264

Lastavica CC, Wilson ML, Beradi VP, Spielman A, Deblinger RD (1989) Rapid emergence of a focal epidemic of Lyme disease in coastal Massachusetts. N Engl J Med 320:133–137

Latov N, Gross RB, Kastelman J et al. (1981) Complement fixing anti-peripheral nerve myelin antibodies in patients with inflammatory polyneuritis and with polyneuropathy and paraproteinemia. Neurology 31:1530–1534

Levy RL, Newkirk R, Ochoa J (1979) Treatment of chronic relapsing Guillain–Barré syndrome by plasma exchange. Lancet ii:741

Lewis RA, Sumner AJ (1982) The electrodiagnostic distinctions between chronic familial and acquired demyelinative neuropathies. Neurology 32:592–596

Lewis RA, Sumner AJ, Brown MJ, Asbury AK (1982) Multifocal demyelinating neuropathy with persistent conduction block. Neurology 32:958–964

Löffel NB, Rossi LN, Mumenthaler M (1977) The Landry–Guillain–Barré syndrome – complications, prognosis and natural history in 123 cases. J Neurol Sci 33:71–79

Maas AIR, Busch HFM, van der Heul C (1981) Plasma infusion and plasma exchange in chronic idiopathic polyneuropathy. N Engl J Med 305:344

Mancardi GL, Cadoni A, Zicca A et al. (1988) HLA-DR Schwann cell reactivity in peripheral neuropathies of different origins. Neurology 38:848–852

Matthews WB, Howell DA, Hughes RC (1970) Relapsing corticosteroid dependent polyneuritis. J Neurol Neurosurg Psychiatry 33:330–337

McCombe PA, McManis PG, Frith JA, Pollard JD, McLeod JG (1987a) Chronic inflammatory demyelinating polyradiculoneuropathy associated with pregnancy. Ann Neurol 21:102–104

McCombe PA, Pollard JD, McLeod JG (1987b) Chronic inflammatory demyelinating polyradiculo-
    neuropathy. Brain 110:1617–1630
McFarland HR, Heller GL, Arbor A (1966) Guillain–Barré disease complex. Arch Neurol 14:
    197–201
Mendell JR, Kolkin S, Kissel JT, Weiss KL, Hakeres CDW, Rammohan KW (1987) Evidence for
    central nervous system demyelination in chronic inflammatory demyelinating polyradiculoneur-
    opathy. Neurology 37:1291–1294
Mitchell GW, Bosch EP, Hart MN (1987) Response to immunosuppressive therapy in patients with
    HMSN and associated dysimmune neuromuscular disorders. Eur Neurol 27:188–196
Ninfo V, Rizzuto N, Terzian H (1967) Associazione anatomo-clinica di nevrite ipetrofica e sclerosi
    a placche. Acta Neurol (Napoli) 22:228–237
Nukada H, Pollock M, Haas LF (1989) Is ischemia implicated in chronic multifocal demyelinating
    neuropathy? Neurology 39:106–110
Nyland H, Aarli JA (1978) Guillain–Barré syndrome: demonstration of antibodies to peripheral
    nerve tissue. Acta Neurol Scand 58:35–43
Ormerod IEC, Waddy H, Kermode AG, Murray NMF, Thomas PK (1990) Involvement of the
    central nervous system in chronic inflammatory demyelinating polyneuropathy: a clinical, electro-
    physiological and magnetic resonance imaging study. J Neurol Neurosurg Psychiatry (in press)
Osler W (1892) Acute ascending (Landry's) paralysis. In: Principles and practice of medicine,
    Young J Pentland, Edinburgh and London
Osterman PO, Fagius J, Safwenberg J, Danersund A, Wallin B, Nordesjo LO (1982) Treatment of
    Guillain–Barré syndrome by plasmapheresis. Arch Neurol 39:148–154
Osuntokun BO, Prineas J, Field EJ (1966) Immunological study of chronic polyneuropathies of
    undetermined cause. J Neurol Neurosurg Psychiatry 29:456–458
Palmer KNV (1965) Polyradiculoneuropathy (Guillain–Barré syndrome) treated with 6-mercap-
    topurine. Lancet i:733–734
Palmer KNV (1966) Polyneuropathy treated with cytotoxic drugs. Lancet i:265
Parry GJ, Aminoff MJ (1987) Somatosensory evoked potentials in chronic acquired demyelinating
    peripheral neuropathy. Neurology 37:313–316
Pentland B (1980) Azathioprine in chronic relapsing idiopathic polyneuropathy. Postgrad Med J
    56:734–735
Pestronk A, Cornblath DR, Ilyas AA et al. (1988) A treatable multifocal motor neuropathy with
    antibodies to GM_1 ganglioside. Ann Neurol 24:73–78
Peterman AF, Daly DD, Dion FR, Keith HM (1959) Infectious neuronitis (Guillain–Barré syn-
    drome) in children. Neurology 9:533–539
Peters DK, Sevitt LH, Direkze M, Baykliss SG (1973) Landry–Guillain–Barré–Strohl polyneuritis
    and the nephrotic syndrome. Lancet i:1183–1184
Pleasure DE, Lovelace RE, Duvois RC (1969) The prognosis of acute polyradiculoneuritis. Neuro-
    logy (Minneapolis) 18:1143–1148
Poewe W, Sluga E, Aichner F (1981) Subacute-chronic polyneuritis. Acta Neuropathol 7:262–267
Pollard JD, Selby G (1978) Relapsing neuropathy due to tetanus toxoid. J Neurol Sci 37:113–125
Pollard JD, McLeod JG, Gatenby P, Kronenberg H (1983) Prediction of response to plasma ex-
    change in chronic relapsing polyneuropathy. A clinicopathological correlation. J Neurol Sci 58:
    269–287
Pollard JD, McCombe PA, Baverstock J, Gatenby PA, McCleod JG (1986) Class II antigen ex-
    pression and T lymphocyte subsets in chronic inflammatory demyelinating polyneuropathy. J
    Neuroimmunol 13:123–134
Prineas J (1970) Polyneuropathies of undetermined cause. Acta Neurol Scand 46 Suppl 44:1–97
Prineas JW, McLeod JG (1976) Chronic relapsing polyneuritis. J Neurol Sci 27:427–458
Prineas JW, McLeod JG, Wolfenden WH (1971) Endoneurial deposits of amyloid-like fibrils in a
    recurrent demyelinating neuropathy. J Neuropathol Exp Neurol 30:583–592
Ravn H (1967) The Landry–Guillain–Barré syndrome. A survey and a clinical report of 127 cases.
    Acta Neurol Scand 43: Suppl 30, 1–164
Rechthand E, Cornblath DR, Stern BJ, Meyerhoff JO (1984) Chronic demyelinating polyneur-
    opathy in systematic lupus erythematosus. Neurology 34:1375–1377
Rizzuto N, Moretto G, Monaco S, Martinelli P, Pazzaglia P (1982) Chronic relapsing polyneuritis.
    A light- and electron-microscopic study. Acta Neuropathol (Berl) 56:179–186
Ro YI, Alexander BE, Oh SJ (1983) Multiple sclerosis and hypertrophic demyelinating peripheral
    neuropathy. Muscle Nerve 6:312–316
Rosen AD, Vastola EF (1976) Clinical effects of cyclophosphamide in Guillain–Barré polyneuritis.
    J Neurol Sci 30:179–187

Rosenberg NL, Bourdette D (1983) Hypertrophic neuropathy and multiple sclerosis. Neurology 33:1361–1364

Rosenberg NL, Lacy JR, Kennaugh RC, Holers VM, Neville HE, Kotzin BL (1985) Treatment of refractory chronic demyelinating polyneuropathy with lymphoid irradiation. Muscle Nerve 8: 223–232

Rubin M, Karpati G, Carpenter S (1987) Combined central and peripheral myelinopathy. Neurology 37:1287–1290

Sahenk Z, Mendell JR, Rossio JL, Hurtubise P (1977) Polyradiculoneuropathy accompanying procainamide-induced lupus erythematosus: evidence for drug-induced enhanced sensitization to peripheral nerve myelin. Ann Neurol 1:378–384

Saida T, Saida K, Silberberg DH, Brown MK (1981) Experimental allergic neuritis induced by galactocerebroside. Ann Neurol 9 Suppl:87–101

Schapira AHV, Thomas PK (1986) A case of recurrent idiopathic ophthalmoplegic neuropathy (Miller Fisher syndrome). J Neurol Neurosurg Psychiatry 49:463–464

Schob F (1907) Ein Beitrag zur Pathologischen Anatomie der Multiplen Sklerose. Monatsschr Psychiatr Neurol 22:62–87

Schoene WC, Carpenter S, Behan PO, Geschwind N (1977) Onion bulb formations in the central peripheral nervous system in association with multiple sclerosis and hypertrophic polyneuropathy. Brain 100:755–773

Schott B, Michel D, Lejeune E et al. (1968) Polyradiculonévrites au cours de la maladie de Besnier–Boeck–Schaumann. J Méd Lyon 49:931

Seitz D (1966) Enquêtes catamnestiques sur les polynévrites inflammatoires. Rev Neurol (Paris) 115:845–857

Server AC, Lefkowith J, Braine H, McKhann GM (1979) Treatment of chronic relapsing inflammatory polyradiculoneuropathy by plasma exchange. Ann Neurol 6:258–261

Server AC, Stein SA, Braine H, Tandon DS, McKhann GM (1980) Experience with plasma exchange and cyclophosphamide in the treatment of chronic relapsing inflammatory polyradiculoneuropathy. Neurology 30:362

Seyal M, Ziegler DK, Couch JR (1978) Recurrent Guillain–Barré syndrome following influenza vaccine. Neurology 28:725–726

Sheremata WA, Rocklin RE, David J (1974) Cellular hypersensitivity in Guillain–Barré syndrome. Can Med Assoc J 110:1245–1247

Sibley WA (1972) Polyneuritis. Med Clin North Am 56:1299–1319

Stewart GJ, Pollard JD, McLeod JG, Wolnizer CM (1978) HLA antigens in the Landry–Guillain–Barré syndrome and chronic relapsing polyneuritis. Ann Neurol 4:285–289

Targowla J (1894) Polynévrite récidivante, envahissement des nerfs craniens et diplegie faciale. Rev Nuerol 2:465–472

Taylor WA, Hughes RAC (1988) Responsiveness to P_2 to blood and cauda equina derived lymphocytes in experimental allergic neuritis: Preliminary characterisation of a cauda equina derived P_2 specific T cell line. J Neuroimmunol 19:279–289

Taylor WA, Hughes RAC (1989) T lymphocyte activation antigens in Guillain–Barré syndrome and chronic idiopathic demyelinating polyradiculoneuropathy. J Neuroimmunol 24:33–39

Thomas PK, Lascelles RG (1969) Hypertrophic neuropathy. Q J Med 36:223–238

Thomas PK, Lascelles RG, Hallpike JF, Hewer RL (1969) Recurrent and chronic relapsing Guillain–Barré polyneuritis. Brain 92:589–606

Thomas PK, Walker RWH, Rudge P et al. (1987) Chronic demyelinating peripheral neuropathy associated with multifocal CNS demyelination. Brain 110:53–76

Thompson AJ, Kennard C, Swash M et al. (1989) Relative efficacy of intravenous methylprednisolone and ACTH in the treatment of acute relapse in MS. Neurology 39:969–971

Tindall RSA, Rollins J, Hall K (1989) Pilot study to assess the safety and effectiveness of cyclosporine A in the treatment of chronic inflammatory demyelinating polyneuropathy. Ann Neurol 22:168

Torvik A, Lundar T (1977) A case of chronic demyelinating polyneuropathy resembling the Guillain–Barré syndrome. J Neurol Sci 32:45–52

Toyka KV, Augspach R, Wietholter H et al. (1982) Plasma exchange in chronic inflammatory polyneuropathy: evidence suggestive of a pathogenic humoral factor. Muscle Nerve 5:479–484

Van der Meché FGA, Vermeulen M, Busch HFM (1989) Chronic inflammatory demyelinating neuropathy. Conduction failure before and during immunoglobulin or plasma therapy. Brain 112:1563–1571

van den Bergh P, Logigian EL, Kelly JJ (1989) Motor neuropathy with multifocal conduction blocks. Muscle Nerve 11:26–31

Vaughan R, Adam AM, Gray IA et al. (1990) Major histocompatability complex class I and class II polymorphism in chronic idiopathic demyelinating polyradiculoneuropathy. J Neuroimmunol 27:149–153

Vermeulen M, Van der Meche FGA, Speelman JD, Weber A, Busch HFM (1985) Plasma and gammaglobulin infusion in chronic inflammatory polyneuropathy. J Neurol Sci 70:317–326

Waddy HM, Misra VP, King RHM, Thomas PK, Middleton L, Ormerod IEC (1989) Focal cranial nerve involvement in chronic inflammatory demyelinating polyneuropathy: clinical and MRI evidence of peripheral and central lesions. J Neurol 236:400–405

Walker GL (1979) Progressive polyradiculoneuropathy: treatment with azathioprine. Aust NZ J Med 9:184–189

Whitaker JN, Sciabbarrasi J, King Engel W et al. (1973) Serum immunoglobulin and complement (C3) levels. A study in adults with idiopathic, chronic polyneuropathies and motor neurone diseases. Neurology 23:1164–1173

Wiederholt HM, Mulder DW, Lambert EH (1964) The Landry Guillain–Barré Strohl syndrome of polyradiculoneuropathy – historical review and report on 97 patients and present concepts. Mayo Clin Proc 49:427–451

Winer JB, Hughes RAC, Osmond C (1988) A prospective study of acute idiopathic neuropathy. I. Clinical features and their prognostic value. J Neurol Neurosurg Psychiatry 51:605–612

Witte AS, Burke JF (1987) Membranous glomerulonephritis associated with progressive demyelinating neuropathy. Neurology 37:342–345

Yuill GM, Swinburn WR, Liversedge LA (1970) Treatment of polyneuropathy with azathioprine. Lancet ii:854–856

Zweiman B, Rostami A, Lisak RP, Moskovitz AR, Pleasure DE (1983) Immune reactions to P_2 protein in human inflammatory demyelinative neuropathies. Neurology 33:234–237

Chapter 11

Animal Models of Guillain–Barré Syndrome

Marek's Disease

Marek's disease, the most interesting animal model of GBS, is important in the poultry industry. The disease was described by Marek in 1907 and shown to be due to a herpes virus just over 20 years ago (Churchill and Biggs 1967; Nazerian et al. 1968). There are different strains of the causative virus and different inherited susceptibilities of chicken breeds so that infection causes several clinical pictures. These include acute or chronic systemic lymphoma, transient encephalitis, and peripheral nerve disease (Payne et al. 1976; Stevens et al. 1981). With the discovery of the causative agent came a search for less virulent forms to produce a vaccine. It was found that turkey herpes virus was antigenically similar. A turkey herpes virus vaccine, routinely given to one-day-old chicks, has reduced the incidence of Marek's disease. Recent reviews of the pathogenesis of this fascinating condition are available (Calnek 1985; Payne 1988). This chapter will be confined to its neurological aspects.

In the classical form of Marek's disease, also called fowl paralysis, birds develop weakness of their wings and legs due to involvement of the brachial and sciatic nerves (Fig. 11.1). The disease usually affects young birds, between 12 and 24 weeks, but sometimes starts earlier or later. Other manifestations include torticollis and respiratory involvement which may cause death in about 10% of birds. The disease is spread by the virus being shed from the epithelium of the feather follicles as dander and probably inhaled (Calnek et al. 1970). The infection of the feather follicles is fully productive and leads to release of cell free virus. A restrictively productive infection occurs in the follicle epithelium and other cells, especially lymphocytes, in which viral polypeptides but not free virus are produced. The B cells are infected first and may undergo a cytolytic infection causing activation of T cells which become susceptible to infection and neoplastic transformation (Calneck 1985; Payne 1988). The incubation period varies from three weeks up to several months. Transient paralysis is quite different from fowl paralysis and is due to an encephalopathy. It consists of an illness ranging from mild ataxia to coma starting about ten days after infection. It is associated with an abnormal EEG, mild perivascular lymphocytic infiltration, extreme vacuolation of myelin sheaths, and extracellular oedema in the CNS (Lawn and Watson 1982; Kornegay et al. 1983; Kornegay and Gorgacz 1988, Swayne et al. 1988, 1989). The mechanism of the encephalitis is unclear but it is inherited as an autosomal recessive trait linked to the major histocompatability complex. Bursectomy or cyclophosphamide treatment both prevent its development but neither prevents Marek's disease virus replicating in the spleen.

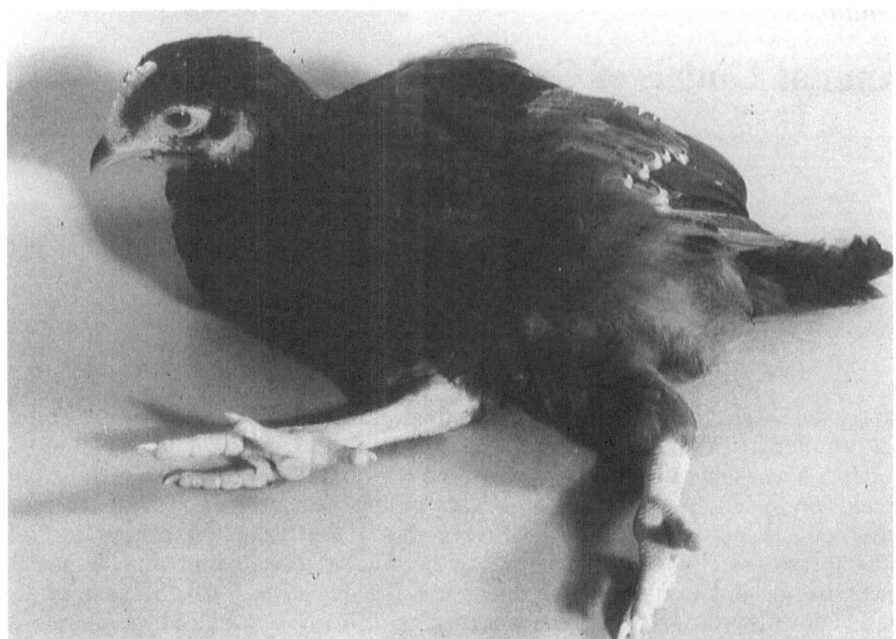

Fig. 11.1. Chicken with Marek's disease. (By courtesy of Dr A.M. Lawn.)

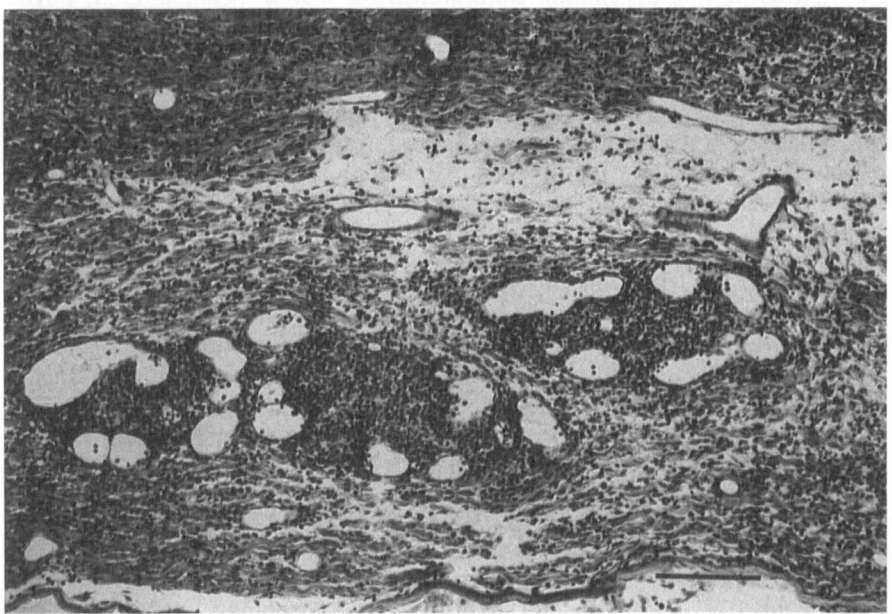

Fig. 11.2. Sciatic nerve of a chicken with Marek's disease showing focal lymphoid cell infiltration with surrounding oedema. 30 days after experimental virus infection. Bar = 100 μm. (Microphotograph by courtesy of Dr A.M. Lawn.)

This suggests that transient paralysis is caused by a B cell, possibly antibody-mediated, mechanism but it does not rule out a lymphokine-mediated pathogenesis (Parker and Schierman 1983).

The pathology of "fowl paralysis", the classical form of Marek's disease, is focused on the peripheral nerves which become markedly thickened and greyish. The swelling is due to infiltration by lymphoid cells consisting of small and large lymphocytes, principally T cells, with some plasma cells (Fig. 11.2). Lymphoid tumours also occur in other organs. In the so-called acute form of Marek's disease lymphomatous masses develop in the gonads, liver, lungs and other organs. There is overlap between this acute lymphomatous disease and classical fowl paralysis (Purchase 1972).

A distinction used to be drawn between two types of lesion in the nerve, type A, in which there are foci of lymphoma cell infiltration associated with distal axonal degeneration and type B, in which there is oedema, inflammation and demyelination (Purchase 1972). In the type B lesions multiple perivenous foci of

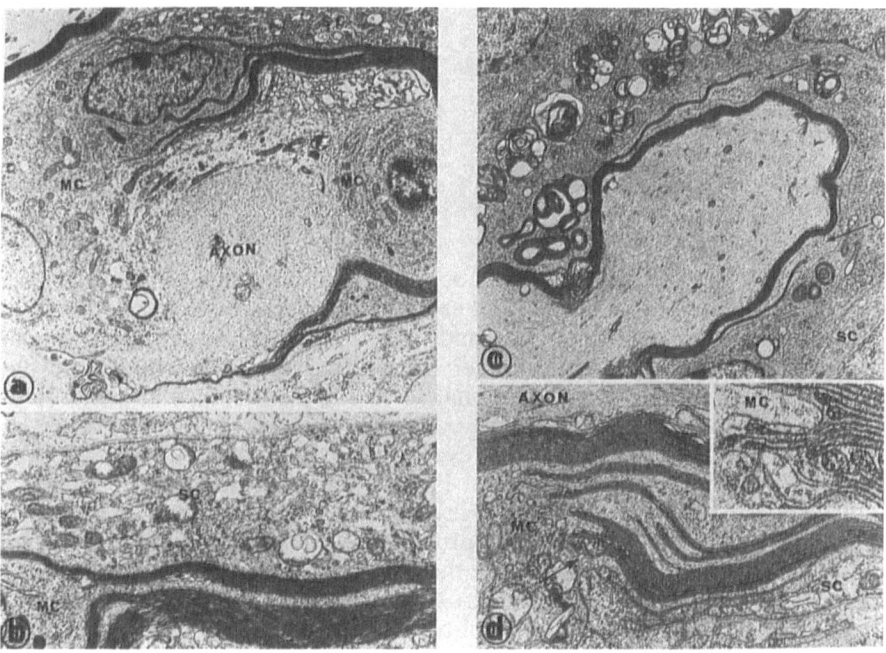

Fig. 11.3.   a EAN 21 days after sensitisation. Stripping of a damaged myelin sheath by mononuclear cells (MC) with abundant cytoplasm containing phagosomes suggestive of macrophages. The Schwann cell (SC) remained intact but shows reactive changes that are enlarged in **b** (×6750). **b** Higher magnification of the invading tongue of a mononuclear cell (MC) that is penetrating into a damaged myelin sheath. The covering Schwann cell (SC) shows reactive changes consisting of an increase in ribosomes, endoplasmic reticulum and mitochondria that show artefactual vacuolation (×20250). **c** Marek's disease 35 days after infection. Stripping of a myelin sheath by macrophages that penetrate the sheath (*arrows*) after lysis of the outermost lamellae. The Schwann cell (SC) is displaced by the invading cells. **d** Stripping of uniformly separated myelin lamellae at a node of Ranvier by cytoplasmic processes of an invading mononuclear cell (MC). The Schwann cell (SC) remained intact (×13613). The *arrow* points to an area that is enlarged in the inset showing intact cytoplasmic loops of terminating myelin lamellae (×40840). (From Lampert et al. (1977), with permission.)

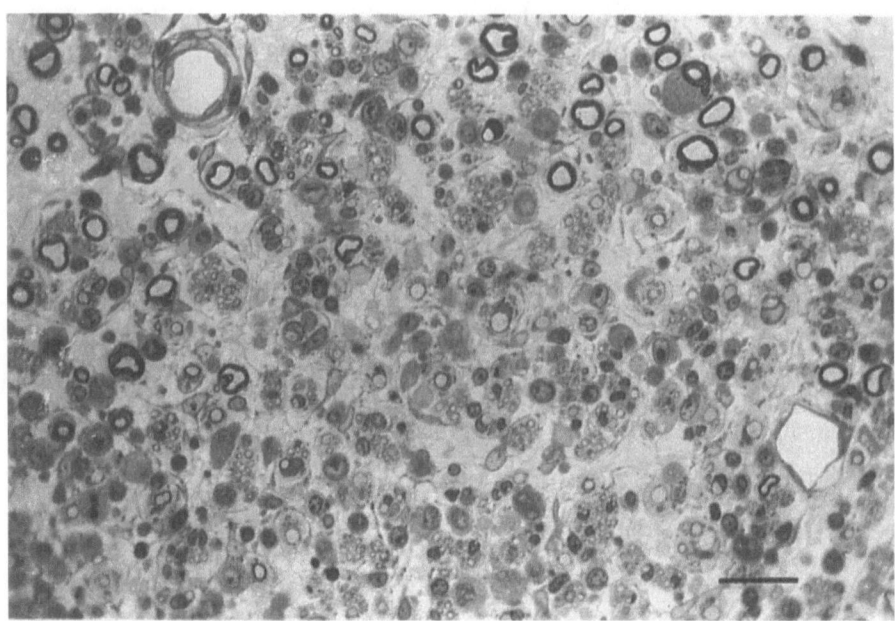

Fig. 11.4. Extensive demyelination in the sciatic nerve 51 days after experimental Marek's disease virus infection. Bar = 20 µm. (Microphotograph by courtesy of Dr A.M. Lawn.)

demyelination associated with lymphocytic and macrophage infiltration have been consistently observed. The light and electron microscopic appearances of these lesions resemble those of EAN and GBS. There is macrophage invasion of the Schwann cell basal lamina. Fingers of macrophage cytoplasm invade the myelin sheath and the macrophages phagocytose and digest myelin debris (Fig. 11.3). Large numbers of fibres become demyelinated (Fig. 11.4). The Schwann cells are pushed aside but survive and eventually proliferate and remyelinate the axons (Wight 1969; Prineas and Wright 1972). Lampert et al. (1977) undertook a study comparing EAN with Marek's disease in the chicken and confirmed that the morphology of the demyelinating process was similar in both. None of these studies demonstrated viral particles in the peripheral nerves. It has been argued that infection with Marek's disease virus induced autoimmunity to myelin antigens which had caused sensitised lymphocytes to invade the peripheral nerves as happens in EAN.

Lawn and Payne (1979) published a careful longitudinal study of the ultrastructure of the nerve lesions which challenges this earlier view. Their interpretation has been confirmed by Bourhy et al. (1988). In the first place the presence of lymphoid cells in chicken nerves has to be interpreted with the knowledge that foci of haemopoietic cells are often present in the nerves of normal animals (Fig. 11.5). The first abnormal accumulations of cells consisting of lymphocytes and macrophages were found about 7 days after experimental infection. During the second and third weeks the numbers of lymphoid cells infiltrating the nerves increased markedly, usually forming foci, which were often perivascular. Only sparse isolated demyelinated axons were present. These lesions corresponded to the cellular lymphoid cell accumulations pre-

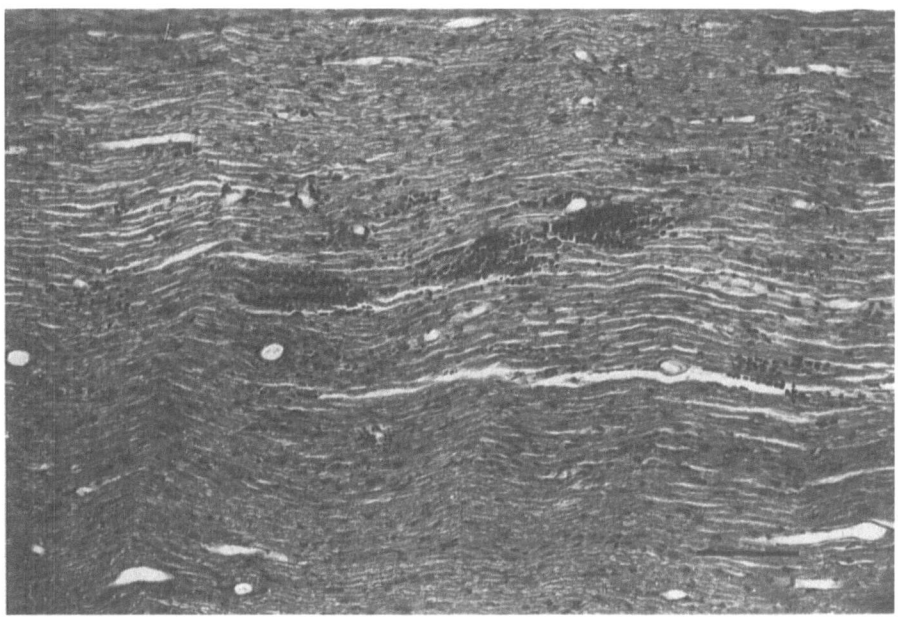

Fig. 11.5. Haemopoietic focus in the sciatic nerve of a normal chicken. Bar = 100 μm. (From Lawn (1979), with permission.)

viously called type A. During the fourth week the cellular infiltration reached a plateau, plasma cells were abundant and numerous demyelinated axons were present. During the next few weeks the lesions in the nerves became much more variable. Some lesions were densely cellular but others, corresponding to the type B lesion, showed a reduction in cellularity with many demyelinated or remyelinating axons (Figs. 11.2 and 11.4). This study thus demonstrated that cellular infiltration of the nerves occurred about two weeks before the onset of demyelination and was not consistent with the idea that the cells enter the nerve by reason of being sensitised to myelin antigens. In rodent EAN, by contrast, demyelination is seen immediately after or with the first invasion of cells into the nerve roots.

Nevertheless there is other evidence that immune responses, and possibly autoimmune responses, are important in producing the nerve lesions in Marek's disease. Chickens with Marek's disease have delayed hypersensitivity, detected by a skin test, to myelin (Schmahl et al. 1975; Pepose et al. 1981) and IgG antibodies to myelin detected with an indirect immunofluorescent test (Pepose et al. 1981). Neither test showed a close correlation with presence or severity of paralysis.

Although the morphology of the lymphomas in Marek's disease shows a mixture of cells including macrophages, the majority are T cells (Payne 1977). The target cell of the virus is the T cell, and the other cells are presumably attracted by the activated T cells and release of lymphokines. According to some, but not all, reports, neonatal thymectomy reduces the incidence of Marek's disease but removal of the bursa of Fabricius, from which B cells originate, has no effect (Goto et al. 1979). It is of interest that the genetic resistance of different strains of chicken is related to genes which are present on the T cell surface. These include one gene which forms part of the chicken major histocompatibility B locus and another which codes for a T-cell marker outside the major histocompatibility antigen complex. T cells are much more susceptible to lytic infection by Marek's disease virus than other cells such as fibroblasts. These observations can all be explained if the Marek's disease virus attaches to a receptor on the surface of the T cell. This may cause activation of the T cells. According to the proposal that activated T cells cross the blood–brain barrier (Wekerle et al. 1986) and blood–nerve barrier more readily than non-activated T cells, the activated T cells may localise in the neural parenchyma and secrete lymphokines which recruit B cells and macrophages to accumulate around them.

The explanation for the ensuing demyelination might either be a non-specific consequence of the accumulation of highly activated cells or because of superimposed autoimmunity. In other human and veterinary conditions and in experimental models non-specific accumulation of activated T cells does not cause demyelination (Chapters 3 and 12). The other possibility that the demyelination is caused by an immune-mediated attack on Schwann cells expressing viral antigens seems less likely. Viral particles have not been identified in infected nerves by electron microscopy. This does not exclude the possibility that viral genome is present. When viral DNA was sought by in situ hybridisation on sections of infected nerve, virus was found in infiltrating lymphoid cells but not in Schwann cells (Ross et al. 1981). Virus could be cultured from spinal ganglia of infected chickens and viral particles were demonstrated in non-myelinating Schwann cells which had been maintained in culture for two or more days (Pepose et al. 1981). If the demyelination were caused by an immune attack on viral antigens on or in the Schwann cell, one would expect necrotic Schwann cells to be prominent, whereas they are conspicuous by their absence according to Prineas and Wright (1972). Finally it is worth pointing out that the demyelination in Marek's disease is specific for the PNS whereas cellular infiltration also occurs in the CNS (Lawn and Watson personal communication). If the demyelination were a non-specific consequence of the presence of activated T cells and the B cells and macrophages which they have attracted, demyelination would be expected in the CNS as well as the peripheral nerves. Accordingly it seems likely that the demyelination in Marek's disease T cells which have been activated by Marek's disease virus cross the blood–nerve barrier, proliferate and attract B lymphocytes and macrophages but some additional autoimmune mechanism, T cell mediated, antibody mediated, or both, causes the demyelination. In view of the association between GBS and herpes viruses, the link between Marek's disease virus and demyelination would repay further study.

Coonhound Paralysis

Another relevant, but extraordinary and incompletely understood illness is coonhound paralysis. This is an acute paralysing illness, originally described by Kingma and Catcott (1954) which affects coonhounds and other dogs soon after exposure to a raccoon. The interval between exposure, which usually involves

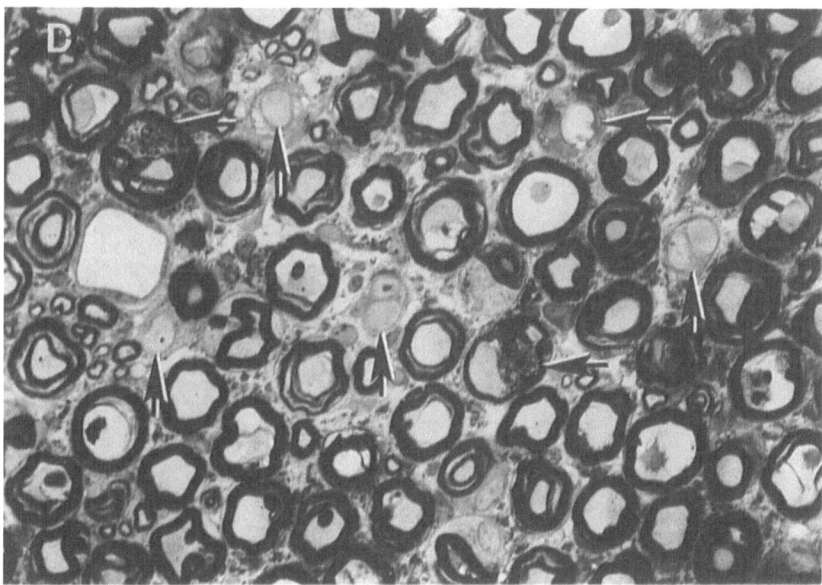

Fig. 11.6. Coonhound paralysis. Lumbar ventral root. Some myelin sheaths have been invaded by macrophages (*horizontal arrows*). Note demyelinated axons (*vertical arrows*). (From Cummings et al. (1982), with permission.)

being bitten, and rapid development of a flaccid areflexic tetraparesis is seven to ten days. The spinal nerve roots have been shown to bear the brunt of the pathology (Cummings and Haas 1967; Cummings and Haas 1972). The ventral roots were more affected than the dorsal roots and showed varying degrees of leucocytic infiltration and demyelination with axonal preservation comparable with GBS (Fig. 11.6). Subsequent more detailed studies have borne out this comparison, although it has become clear that there is a considerable amount of axonal damage in coonhound paralysis (Cummings et al. 1982). Teased fibre studies showed segmental demyelination in some fibres and axonal degeneration in others. The inflammatory infiltrate varied in intensity but usually consisted predominantly of cells of the "monocyte-macrophage" series. Both lymphocytes and, unlike GBS, plasma cells were also present although in small numbers (Fig. 11.7). On electron microscopy macrophages were frequently encountered lying within myelin sheaths but macrophages stripping myelin lamellae were rare. Although disintegrating or split myelin sheaths were encountered remote from infiltrating cells, the possibility that these were due to age-related non-specific changes or fixation could not be excluded. The histological picture does indeed bear comparison with GBS and also EAN in the dog. This conclusion was confirmed by electrophysiological studies showing a mixture of demyelination and axonal degeneration with denervation changes. The cerebrospinal fluid showed an increased CSF protein with normal cell count in two of three dogs with coonhound paralysis (Cummings et al. 1982).

Susceptibility to coonhound paralysis appears to be under genetic control. The disease is relatively rare so that if a pack of hounds are bitten by the same

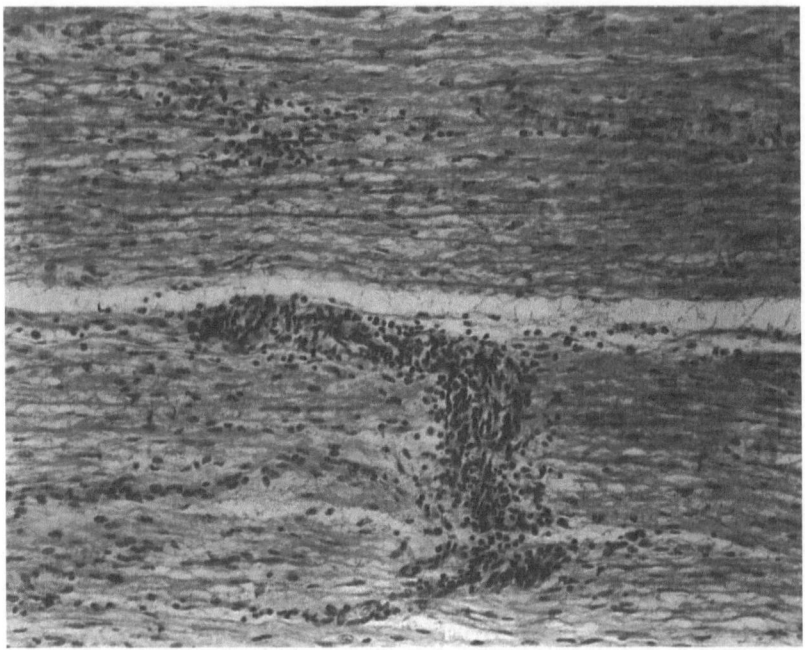

Fig. 11.7. Coonhound paralysis. Cervical ventral root from a coonhound dying 90 hours after the onset of signs. (From Cummings and Haas (1967), with permission.)

raccoon, only one may be affected. Several examples have been reported of coonhound paralysis occurring in related dogs. Also dogs do not become immune following coonhound paralysis but may develop further attacks if they encounter a raccoon again (Cummings and Holmes 1979).

One transmission experiment suggested that the agent responsible for coonhound paralysis is present in raccoon saliva (Holmes et al. 1979). Two dogs which had had attacks of coonhound paralysis were kept in quarantine for a year and then received 1.0 ml of raccoon saliva by subcutaneous injection. After nine days one of the two dogs developed typical coonhound paralysis requiring artificial ventilation within three days followed by gradual recovery so that ventilation was stopped after a week, the dog could walk after eight weeks and eventually recovered completely. The nature of the transmitted agent has not been discovered. It seems most likely that the agent is a virus, although none has been identified. Most other infective organisms, such as toxoplasma, would be readily identified on histological sections (Cummings et al. 1988).

Animal Idiopathic Polyradiculoneuropathy

There have been several reports of an illness occurring spontaneously in dogs which resembles GBS and which has not been associated with raccoon bites (reviewed by Northington and Brown 1982). One study from Switzerland reported four dogs with such an illness in which there was prominent primary demyelination affecting the nerve roots more than the distal nerves. There was extensive perivascular inflammation in every case (Vandervelde et al. 1981). Northington and Brown (1982) conducted a retrospective study of 14 dogs which had been admitted to their veterinary service with acute paralysis. In two dogs the illness pursued a relapsing course resembling CIDP but in the rest the illness appeared monophasic progressing over between one and 21 days and usually recovering over about five weeks. Three dogs died. Motor nerve conduction was often severely slowed, sometimes absent, while sensory conduction was less abnormal. The ten biopsies usually showed Wallerian degeneration in the distal nerves sampled. The autopsy material only showed myelin loss in nerves and roots. Perivascular cellular infiltration was only found in one of the three cases. In two cases in which the clinical picture was of a subacute cauda equina syndrome there was extensive perivascular and diffuse infiltration of the cauda equina roots. Other roots and some cranial nerves were also involved. Both degenerated and demyelinated or remyelinated fibres were present. Antibodies to P_2 protein were absent (Griffiths et al. 1983).

It seems likely that inflammatory polyradiculoneuropathy occurs in many different species. The diagnosis was made in a goat based on a post-mortem examination (MacLachlan et al. 1982). In chickens sporadic cases of inflammatory polyradiculoneuropathy occur which bear no relationship to Marek's disease. Biggs et al. (1982) reported such cases in a flock of Rhode Island Red chickens which had been housed in isolation for eight years and in which there was no evidence of Marek's disease virus infection. The histological appearances of multifocal lymphocyte, inflammatory cell infiltration and macrophage mediated demyelination resembled those reported by the same authors in chicken EAN and in the later stages of Marek's disease.

Cauda Equina Neuritis of the Horse

Horses have a strange condition causing a progressive cauda equina syndrome which is called cauda equina neuritis. This causes progressive wasting and weakness of the muscles innervated by the sacral and coccygeal roots. The brunt of the pathology falls on the cauda equina in which the roots become thickened and matted by a fibrous chronic inflammatory infiltrate with giant cells (Figs. 11.8 and 11.9). Although fungal and tuberculous causes have been sought none have been identified. Inflammation may extend into nerves in other parts of the body, including the cranial nerves (Greenwood et al. 1973; Manning and Gosser 1973; Cummings et al. 1979). Antibody to P_2 was demonstrated in the sera of the four horses with this condition which were tested by Kadlubowski and Ingram (1981). In a study of 27 horses with various forms of neuropathy 14 had antibodies to P_2 and 12 of these had cauda equina neuritis (Fordyce et al. 1987). Equine adenovirus 1 was rescued by co-culture of sensory ganglia or lumbar spinal cord with equine embryonic kidney cells in two of three cases but none of six abattoir horses (Edington et al. 1984). This observation needs to be confirmed because the pathology of cauda equina neuritis is quite unlike any

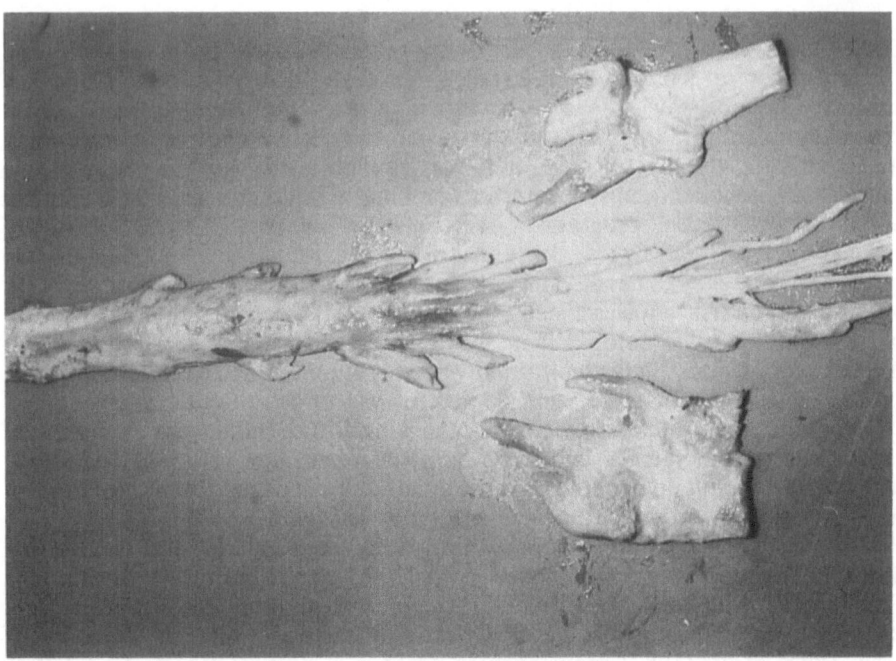

Fig. 11.8. Cauda equina neuritis of the horse. The conus medullaris and cauda equina are markedly thickened and matted by a chronic granuloma (see Fig. 11.9). (By courtesy of the late Dr P.L. Ingram.)

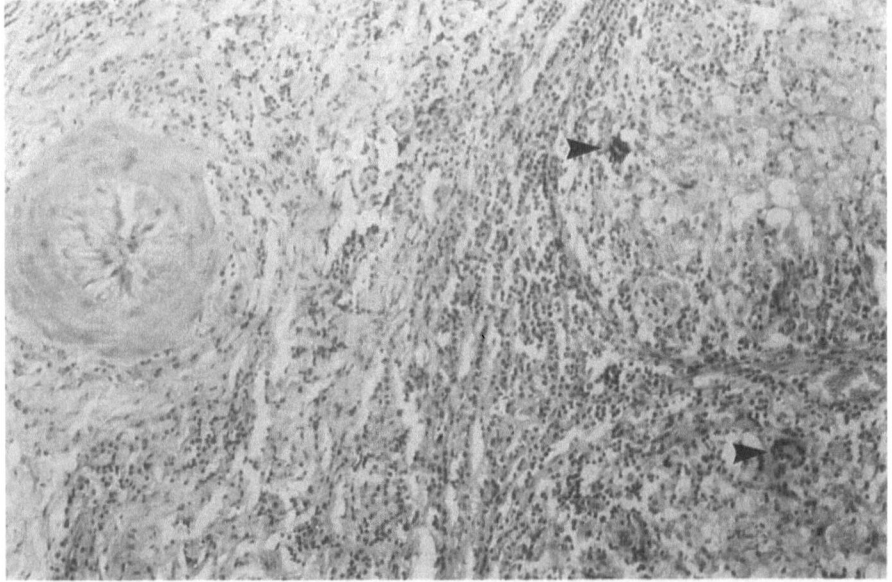

Fig. 11.9. Cauda equina neuritis of the horse. Light micrograph of chronic inflammatory infiltrate with giant cell (*arrowheads*) formation.

other viral infection. Recently Fordyce (personal communication) has reproduced the pathological changes of cauda equina neuritis by immunising with P_2
two ponies which had previously had intrathecal transfer of equine adenovirus.

The granuloma in cauda equina neuritis might be regarded as a Freund's
adjuvant type of granuloma in which macrophages digest and process myelin
antigens which could then be presented to T helper cells. An autoimmune
response to P_2 might be an epiphenomenon but might also contribute to the
pathogenesis of more distant lesions such as occur in the cranial nerves. Cauda
equina neuritis remains the only naturally occurring condition in which antibodies to P_2 are regularly found in high titre.

Summary

Infection with Marek's disease herpes virus in chickens rapidly induces infiltration of many tissues but particularly the nerves with T lymphoma cells. Two or
three weeks later macrophages invade the myelin sheaths and induce demyelination which resembles that in EAN and GBS. At the same time both antibodies
and delayed hypersensitivity to myelin can be demonstrated. Viral particles
cannot be found in the nerves but virus can be rescued from dorsal root ganglia
by co-culture. The immune response to myelin antigens may be important in the
pathogenesis of the demyelination. Spontaneously occurring inflammatory polyradiculoneuropathy has been reported in other species, especially dogs. Some
dogs have an inherited susceptibility to develop this condition after being bitten
by a raccoon. This is due to some factor in raccoon saliva; its nature remains
unknown but a viral cause is possible. The only condition in which antibodies to
a defined myelin protein, P_2, have been identified is cauda equina neuritis of the
horse. This is a granulomatous condition of the cauda equina and is associated
with more distant inflammatory lesions affecting peripheral nerves. These
animal conditions resembling GBS, especially Marek's disease, would repay
further study.

References

Biggs PM, Shillito RFW, Lawn AM, Cooper DM (1982) Idiopathic polyneuritis in SPF chickens.
    Avian Pathol 11:163–178
Bourhy H, Wyers M, Guittet M, Bennejean G, Le Coq H (1988) Etude statistique de l'évolution des
    lesions histologiques au cours de l'infection experimentale du poulet par le virus de la maladie de
    Marek. Avian Pathol 17:689–701
Calnek BW (1985) Review. Pathogenesis of Marek's disease. In: Calnek BW, Spencer JL (eds)
    International symposium on Marek's disease. American Association of Animal Pathologists,
    Kennett Square PA, pp 374–390
Calnek BW, Aldinger HK, Kahn DE (1970) Feather follicle epithelium: a source of enveloped and
    infectious cell-free herpes virus from Marek's disease. Avian Dis 14:219–233
Churchill AE, Biggs PM (1967) Agent of Marek's disease in tissue culture. Nature 215:528–530
Cummings JF, Haas DC (1967) Coonhound paralysis, an acute idiopathic polyradiculoneuritis in
    dogs resembling the Landry–Guillain–Barré sydrome. J Neurol Sci 4:51–81

Cummings JF, Haas DC (1972) Animal model for human disease: idiopathic neuritis, Guillain–
  Barré syndrome. Animal model: coonhound paralysis, idiopathic polyradiculoneuritis of coon-
  hounds. Am J Pathol 66:189–192
Cummings JF, Holmes DF (1979) Coonhound paralysis–Guillain–Barré syndrome. In Andrews EJ,
  Ward BC, Altman NH (eds) Spontaneous animal models of human disease. Academic Press, New
  York, pp 174–175
Cummings JF, de Lahunta A, Timoney JF (1979) Neuritis of the cauda equina, a chronic idiopathic
  polyradiculoneuritis in the horse. Acta Neuropathol 46:17–24
Cummings JF, de Lahunta A, Holmes DF, Schultz RD (1982) Coonhound paralysis – further clinical
  studies and electron microscopic observations. Acta Neuropathol 56:167–181
Cummings JF, de Lahunta A, Suter MM, Jacobson RH (1988) Canine protozoan polyradiculoneuri-
  tis. Acta Neuropathol 76:46–54
Edington N, Wright JA, Patel JR, Edwards GB, Griffiths L (1984) Equine adenovirus 1 isolated
  from cauda equina neuritis. Res Vet Sci 37:252–254
Fordyce PS, Edington N, Bridges GC, Wright JA, Edwards GB (1987) Use of an ELISA in the
  differential diagnosis of cauda equina. Equine Vet J 190:55–59
Goto N, Fujimoto Y, Okada N, Mikami T, Kodama H, Ichijo K (1979) Effect of thymectomy on the
  initial cytolytic lesions and nerve demyelination of Marek's disease. Zentralbl Veterinarmed [B]
  26:61–72
Greenwood AG, Barker J, McLeish I (1973) Neuritis of the cauda equina in a horse. Equine Vet J
  5:111–115
Griffiths IR, Carmichael S, Mayer SJ, Sharp NJ (1983) Polyradiculoneuritis in two dogs presenting
  as neuritis of the cauda equina. Vet Rec 1120:360–361
Holmes DF, Schultz RD, Cummings JF, de Lahunta A (1979) Experimental coonhound paralysis:
  animal models of Guillain–Barré syndrome. Neurology 29:1186–1187
Kadlubowski M, Ingram PL (1981) Circulating antibodies to the neurotigenic myelin protein, P_2, in
  neuritis of the cauda equina of the horse. Nature 293:299–300
Kingma FJ, Catcott EJ (1954) A paralytic syndrome in coonhounds. N Am Vet 35:115–117
Kornegay JN, Gorgacz EJ (1988) Marek's disease virus-induced transient paralysis in chickens. Acta
  Neuropathol 75:604–611
Kornegay JN, Gorgacz EJ, Parker MA, Brown J, Schierman LW (1983) Marek's disease virus-
  induced transient paralysis Am J Vet Res 44:1541–1544
Lampert PW, Garrett R, Powell HC (1977) Demyelination in allergic and Marek's disease virus
  induced neuritis comparative microscopic studies. Acta Neuropathol 40:103–110
Lawn AM (1979) Haemopoietic cells in peripheral nerves of SPF chickens. Avian Pathol 8:477–481
Lawn AM, Payne AL (1979) Chronological study of ultrastructural changes in the peripheral nerves
  in Marek's Disease. Neuropathol Appl Neurobiol 5:485–497
Lawn AM, Watson JS (1982) Ultrastructure of the central nervous system in Marek's disease and the
  effect of route of infection on lesion incidence in the central nervous system. Avian Pathol
  11:213–215
MacLachlan NMJ, Gribble DH, East NE (1982) Polyradiculoneuritis in a goat. J Am Vet Med
  Assoc 180:2–7
Manning JP, Gosser HS (1973) Neuritis of the cauda equina in horses. Vet Med/Small Anim Clin
  1162–1165
Marek J (1907) Multiple Nerventzündung (Polyneuritis) bei Hühnen. DTW 15:417–421
Nazerian K, Solomon JJ, Witter RL, Burmester BR (1968) Studies on the etiology of Marek's
  disease. II. Finding of a herpes virus in cell culture. Proc Soc Exp Biol Med 127:177–182
Northington JW, Brown MJ (1982) Acute canine idiopathic polyneuropathy. A Guillain–Barré-like
  disorder. J Neurol Sci 56:259–273
Parker MA, Schierman LW (1983) Suppression of humoral immunity in chickens prevents transient
  paralysis causes by a herpes virus. J Immunol 130:2000–2001
Payne LN (1977) Viral lymphomagenesis in the domestic fowl: a review. Proc Soc Med 70:559–562
Payne LN (1988) Pathogenesis of Marek's disease. Third international symposium on Marek's
  disease. Research Institute for Microbial diseases, Osaka University, Japan, pp 307–316
Payne LN, Frazier JA, Powell PC (1976) Pathogenesis of Marek's disease. Int Rev Exp Pathol
  16:59–154
Pepose JS, Stevens JG, Cook ML, Lampert PW (1981) Marek's disease as a model for the Landry–
  Guillain–Barré syndrome: latent viral infection in non-neuronal cells accompanied by specific
  immune responses to peripheral nerve and myelin. Am J Pathol 103:309–320
Prineas JW, Wright RG (1972) The fine structure of peripheral nerve lesions in a virus induced
  demyelinating disease in fowl (Marek's disease). Lab Invest 26:548–557

Purchase HG (1972) Recent advances in the knowledge of Marek's disease. Adv Vet Sci Comp Med 16:223–257

Ross NL, Delorbe W, Varmus HE (1981) Persistence and expression of Marek's disease virus DNA in tumour. J Gen Virol 57:285–96

Schmahl W, Hoffman-Fezer G, Hoffman R (1975) Zur Pathogenese der Nervenlesionen bei Marekscher Krankheit des Hühnes 1. Allergische Hautreaktion gegen Myelin peripherer Nerven. Z Immunitätsforsch Exp Klin Immunol 150:175–183

Stevens JG, Pepose JS, Cook ML (1981) Marek's disease: a natural model for the Landry–Guillain–Barré syndrome. Ann Neurol 9:102–106

Swayne DE, Fletcher OJ, Schierman LW (1988) Marek's disease virus-induced transient paralysis in chickens. Acta Neuropathol 76:287–291

Swayne DE, Fletcher OJ, Schierman LW (1989) Marek's disease virus-induced transient paralysis in chickens. 2. Ultrastructure of central nervous system. Avian Pathol 18:397–412

Vandervelde VM, Oettli P, Fatzer F, Rohr M (1981) Polyradikuloneuritis beim Hund – Klinische, histologische und ultrastrukturelle Beobachtungen. Schweiz Arch Tierheilkd 123:207–217

Wekerle H, Linington C, Lassmann H, Meyermann R (1986) Cellular immune reactivity within the CNS. Trends Neurol Sci 9:271–277

Wight PAL (1969) The ultrastructure of sciatic nerves affected by fowl paralysis (Marek's disease). J Comp Pathol 79:563–570

Chapter 12

Other Inflammatory or Demyelinating Neuropathies

Vasculitic Neuropathy

Classification of Vasculitis

Kussmaul and Maier (1866) described a patient with multiple mononeuropathy and nodules along medium-sized arteries due to necrotising vasculitis which they called "periarteritis nodosa". Wegener (1936) described the association of vasculitis of small vessels affecting the kidneys and extensive granulomas affecting the sinuses and lungs. By 1947 Klemperer was describing vasculitides, systemic lupus erythematosus and some other disorders which affected the connective tissues as "diseases of the collagen system". Zeek (1953) recognised five types of vasculitis: hypersensitivity angiitis with necrosis and inflammation of small vessels, periarteritis nodosa affecting medium-sized arteries, rheumatic arteritis affecting small arteries associated with rheumatic carditis, allergic granulomatous angiitis associated with asthma and other allergic conditions, and temporal arteritis. Fauci et al. (1978) introduced a classification which has become more popular including polyarteritis nodosa, allergic granulomatous angiitis, hypersensitivity vasculitis, Wegener's granulomatosis and giant cell arteritis. Satisfactory histological or laboratory markers have still not been developed to help resolve what appears to be a spectrum of clinical and histological patterns amongst which the classical syndromes can be discerned with varying degrees of difficulty (Case records of the Massachussetts General Hospital 1987). The pathological variability can helpfully be considered as a spectrum of changes varying along two separate axes. Along one axis is the size of vessel, from small capillaries and vessels in the case of hypersensitivity vasculitis, through medium-sized arteries in polyarteritis nodosa, to large arteries in Takayasu's arteritis. Along the second axis is granuloma formation which may be minimal in hypersensitivity vasculitis, or prominent in Wegener's granulomatosis (Leibowitz and Hughes 1983). The vasculitides may affect any organ in the body. Since small lesions in the nervous system may produce prominent symptoms, neurological presentations of these disorders are quite common.

Presentation of Vasculitic Neuropathy

When vasculitis affects the peripheral nervous system the usual presentation is with the acute or subacute onset of tingling paraesthesiae, impairment of sensation and motor deficit in the distribution of one or several peripheral or cranial nerves. The symptoms of individual nerve lesions usually evolve rapidly over several hours or a few days, but, as more nerves or parts of the same nerve become involved, the story may be one of stepwise progression over a few weeks. Pain may be a prominent and distressing symptom, usually most severe at the onset and gradually improving to leave areas of diminished or absent sensation but sometimes proving distressingly persistent. The distribution of nerve lesions is extremely variable. The pattern of deficit may implicate individual distal nerves, such as ulnar, median or radial in the upper limb and common peroneal, sural or plantar nerves in the lower limb. This apparently distal distribution might be the cumulative effect of more proximal lesions in the nerve trunks. Although multiple mononeuropathy is the characteristic clinical picture, vasculitis also enters the differential diagnosis of symmetrical polyneuropathy. In a recent series of 100 patients with vasculitic neuropathy 19 presented with symmetrical sensory or sensory and motor polyneuropathy (Said et al. 1988), and similar proportions have been discovered by others in smaller series.

Histology of Vasculitic Neuropathy

The histological hallmarks of vasculitis are transmural cellular infiltration of the vessel wall, fibrin deposition and occlusion of the lumen, and fibrinoid necrosis (Fig. 12.1). The necrosis destroys the internal elastic lamina which is not repaired and provides persistent evidence of vasculitic scarring even when repair and recanalisation have occurred (Fig. 12.2). The necrotic changes in the vessel wall permit the escape of red cells and cause the deposition of haemosiderin. The detection of iron with Perl's stain is useful in demonstrating the previous leakage of blood into the surrounding tissues, which provides supportive evidence of vasculitis (Adams et al. 1989). In clinical practice histological changes can be sought by biopsy of accessible nerves, such as the sural nerve, superficial radial nerve or superficial branch of the peroneal nerve. It is best to choose the nerve which has been most recently affected. A full thickness biopsy of the sural nerve is usually recommended since that will increase the size of the sample of blood vessels available for inspection. This goal has to be weighed against the severity of pre-existing deficit and possibility of recovery, which will presumably be less complete if the whole nerve has been removed rather than a fascicle. There is, however, a surprising observation that the deficit was no greater after full thickness than fascicular sural nerve biopsies (Pollock et al. 1983). It is difficult to believe that the extent and duration of the deficit does not depend on the number and size of fascicles removed. In biopsies of affected patients with suggestive clinical pictures, the yield of biopsies with histological findings clinching a diagnosis of vasculitis is high, but biopsies showing merely axonal loss and axonal degeneration may be encountered. These usually represent the problems of sampling inherent within the clinical constraints of taking only a 2–3 cm length from nerve fibres which are a metre long. In a series of 100 biopsies from patients with vasculitic neuropathy, muscle biopsy, usually of the peroneus brevis muscle, provided evidence of vasculitis more frequently (80%) than biopsy of the superficial peroneal nerve (55%) (Said et al. 1988).

Other changes in the nerve which support but do not clinch a diagnosis of vasculitis are perivascular inflammation and a non-uniform pattern of nerve fibre

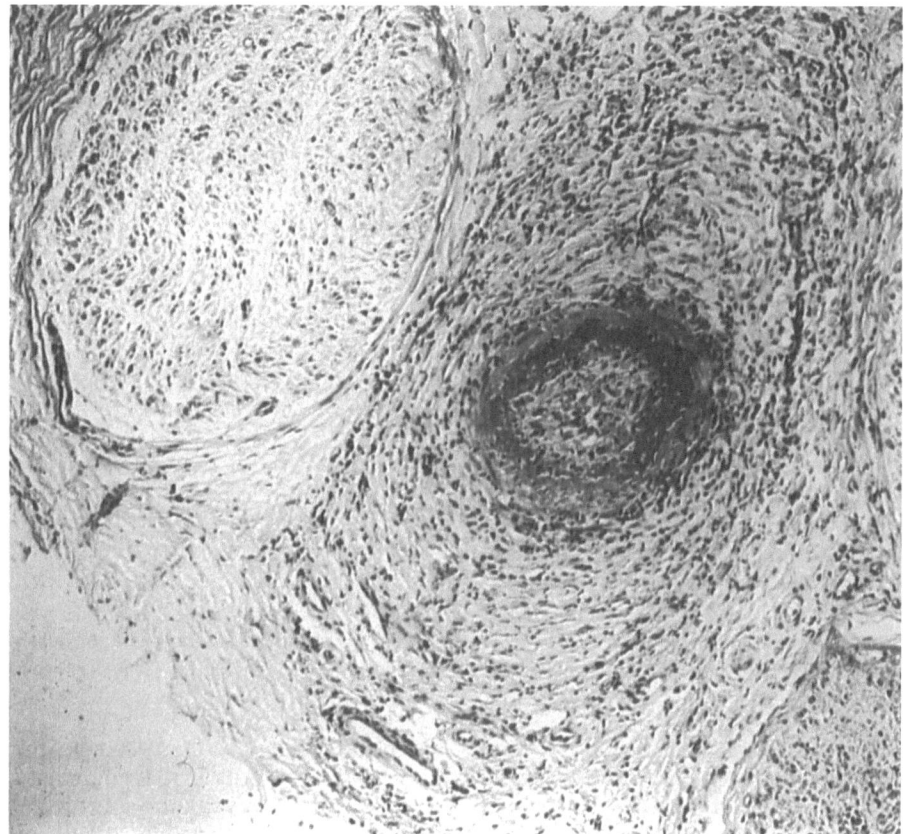

Fig. 12.1. Vasculitic neuropathy associated with SLE. Fibrinoid necrosis of an epineurial arteriole.

loss. Perivascular accumulations of lymphocytes around epineurial vessels are not uncommon in nerve biopsies and may be difficult to interpret. They may represent the edge of a vasculitic focus out of the plane of the section. In the absence of inflammation of the vessel wall itself such perivascular infiltration cannot be taken as evidence of vasculitis. Following proximal root compression, neuronopathy or axonal neuropathy there is usually a uniform depletion of nerve fibres. Vasculitic lesions cause ischaemia of the endoneurium which may affect some fascicles but not others and produce a centrifascicular distribution of fibre loss, probably because the nerve fibres at the periphery of the fascicles receive a blood supply from the perineurial vessels. In a patient with neuropathy and rheumatoid arthritis post-mortem showed that the most severe fibre degeneration was in the portions of the ulnar and median nerves in the middle of the upper arm and in the sciatic nerves in the middle of the thigh: the authors proposed that these represent watershed zones where the perfusion of the nerves is poorest (Dyck et al. 1972). Experiments to induce ischaemic lesions in animals have produced similar patterns of fascicular or centrifascicular fibre degeneration. Necrosis and infarction do not occur, probably because the rich anasto-

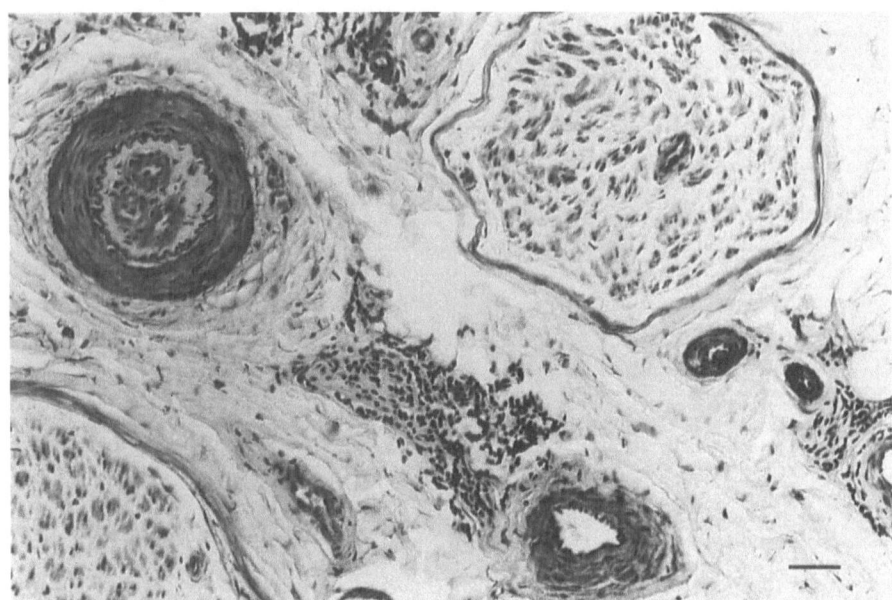

Fig. 12.2. Wegener's granulomatosis with vasculitic neuropathy. Recanalisation of a previously thrombosed epineurial arteriole and perivascular inflammation. Bar = 40 μm.

motic blood supply protects the nerve (Conn and Dyck 1984). Although the major change in the nerve fibres associated with vasculitis is axonal degeneration and fibre loss, a few fibres may exhibit paranodal demyelination probably secondary to axonal atrophy (Conn and Dyck 1984; Osler and Sidell 1960).

The underlying pathogenetic mechanism for vasculitis is not established but there is evidence that immune complex deposition may be responsible. Similar clinical and histological pictures occur following injection of foreign protein into immunised people causing serum sickness, in which circulating immune complexes incorporating the foreign protein are present in the blood (Vaughan et al. 1967). If neurological damage occurs in serum sickness it usually takes the form of a painful brachial plexopathy (neuralgic amyotrophy) not a polyneuropathy. Vasculitis is characteristic of essential mixed cryoglobulinaemia in which the cryoglobulin represents circulating immune complexes made up of immunoglobulin and rheumatoid factor. The vasculitis usually affects the skin and is associated with Raynaud's phenomenon, but 10%–20% of patients develop multiple mononeuropathy which has been shown to be based on necrotising vasculitis (Cream et al. 1974). Attempts to identify immune complexes in the blood of patients with active vasculitis are not uniformly successful. Although complement and immunoglobulins may be demonstrated immunohistochemically in the walls of affected vessels their presence does not mean that they triggered the lesion (Kauffmann et al. 1980; Neild and Williams 1987). In rodents and rabbits, hyperimmunisation with foreign protein does produce a form of experimental serum sickness with deposition of immune complexes in the renal glomerulus and choroid plexus but not the multifocal necrotising vasculitis occurring in human vasculitis (Leibowitz and Hughes 1983). This experimental illness is considered to be due to medium-sized immune complexes of antibody and antigen in slight antigen excess in the blood being deposited in the vessel wall. These immune complexes fix and activate complement-generating components which increase vascular permeability and attract and activate neutrophils and mast cells. The subcutaneous injection of a foreign protein into a hyperimmunised animal rapidly produces a florid local inflammatory reaction, the Arthus phenomenon, which also causes a necrotising vasculitis. The relevance of these models to human disease remains unclear.

Many patients with Wegener's granulomatosis and hypersensitivity vasculitis have antibodies which react with cytoplasmic antigens in neutrophils. These have only been demonstrated recently and their importance is not yet clear. They can be detected by adding the test serum to formalin–acetone fixed neutrophils and then detecting bound immunoglobulin with fluorescein-labelled anti-immunoglobulin antibody. An enzyme-linked immunosorbent assay has been developed with neutrophil granule preparations as antigen. The test is relatively specific for the group of patients with "vasculitis" as a whole but does not sharply delimit any particular subgroup (Falk and Jennette 1988). An antibody which reacts with neutrophil contents might provide a common pathogenetic mechanism by initiating or contributing to the release of vasoactive peptides and inflammatory mediators.

Polyarteritis Nodosa and Related Disorders

Classical polyarteritis nodosa usually occurs in young adults, more commonly men, and presents with fever, muscle and joint pains, and malaise. Almost any organ system may be involved, including the peripheral nerves in 20%–50%, the muscles in 40%–80%, but the CNS in only 10%–20% (Goetz 1980). In one series of 16 patients, six had circulating complexes containing hepatitis B surface antigen in the blood. In two cases the antigen was demonstrated immunohistochemically in the walls of blood vessels which provided some evidence that it might be pathogenetically important (Gocke et al. 1971). Subsequent series of patients with polyarteritis nodosa have consistently shown a small proportion with positive tests of hepatitis B surface antigen in their blood. Other possible precipitants of classical or microscopic polyarteritis nodosa (hypersensitivity vasculitis) have been drugs, including sulphonamides and amphetamines.

The characteristic feature of Wegener's granulomatosis is granulomatous involvement of the nasal sinuses and lungs in association with microscopic polyarteritis nodosa. Peripheral nervous system involvement in Wegener's granulomatosis is similar to that in polyarteritis nodosa with necrotising vasculitis in peripheral nerves (MacFadyen 1960). The only special features relate to the nasal granulomas which may invade the base of the skull and produce cranial nerve palsies (Drachman 1963; Stern 1980). Relapses of Wegener's granulomatosis may be precipitated by bacterial or viral infections (Gocke et al. 1971). A granuloma of the orbit, orbital pseudotumour, and a syndrome of painful ophthalmoplegia with arteritis affecting the internal carotid artery (Tolosa-Hunt syndrome) may be related to Wegener's granulomatosis (Leibowitz and Hughes 1983).

In Churg–Strauss syndrome, allergic granulomatosis, a granulomatous vasculitic picture with a market blood eosinophilia develops on a background of allergic rhinitis and asthma. Churg and Strauss (1951) described 13 patients including nine who had a peripheral neuropathy. They demonstrated necrotising vasculitis in a nerve of one of their patients. The occurrence of multiple mononeuropathy in 10%–20% of patients with essential mixed cryoglobulinaemia has been mentioned above. Temporal or giant cell arteritis is well known as a serious cause of headache and blindness in old people and as an association with polymyalgia rheumatica. Mononeuropathy, multiple mononeuropathy and symmetrical polyneuropathy are not uncommon complications. In a recent large series 14% had peripheral nerve involvement coincident with active temporal arteritis, including 11 with symmetrical peripheral neuropathy, nine with multiple mononeuropathy and three with mononeuropathy. In two cases angiograms showed evidence of large vessel vasculitis in the lower limbs which was confirmed in an amputation specimen in one (Caselli et al. 1988).

Hypereosinophilic Syndrome

The hypereosinophilic syndrome is a rare multisystem disorder with a marked blood eosinophilia, but without any of the recognised causes for a raised eosinophil count, such as parasitic infection, vasculitis or leukaemia. It is commonly associated with endocarditis and bilateral cerebral infarction (Gardner-Thorpe et al. 1971). In an early review of the syndrome four of 57 patients had peripheral nerve involvement (Chusid et al. 1975). Three patients with this syndrome had a very rapid onset of severe generalised symmetrical or asymmetrical neuropathy followed by slow improvement. Nerve biopsies showed axonal degeneration and not vasculitis (Dorfman et al. 1983). In another case of hypereosinophilia with subacute neuropathy, vasculitis was identified in a sural nerve biopsy: at post-mortem examination vasculitis was confined to epineurial vessels and spared other organs (Grisold and Jellinger 1985).

Systemic Lupus Erythematosus

Systemic lupus erythematosus (SLE) affects the peripheral nervous system relatively infrequently compared with polyarteritis nodosa, but more commonly involves the CNS. The frequency of peripheral nerve involvement has ranged from 12% (Dubois 1974) to 29% (Feinglass et al. 1976; Gibson and Myers 1976). Clinical pictures have included acute sensory neuronopathy (Sadeh et al. 1980), acute, subacute or chronic predominantly motor polyradiculoneuropathy, carpal tunnel syndrome (Sidiq et al. 1972) and multiple mononeuropathy (Johnson and Richardson 1968; Richardson 1980). One patient with a subacute sensory and motor neuropathy associated with a facial rash later developed multiple organ involvement and died: post-mortem examination revealed occluded epineurial vessels and surrounding chronic inflammatory cells indicating chronic arteritis (Heptinstall and Sowry 1952). Similar arteritis has been documented in some other cases of symmetrical neuropathy (Bailey et al. 1956) and multiple mononeuropathy (Hughes et al. 1982). The mononeuropathy may affect any peripheral or cranial nerve and may also affect the brachial plexus (Bloch et al. 1979). When isolated nerve lesions occur in the context of other clinical features or serological tests suggesting SLE, it is likely that vasculitis is the underlying mechanism. In cases of symmetrical polyneuropathy the underlying pathogenesis is less clear. Rechthand et al. (1984) presented one case of chronic sensory and motor polyneuropathy with an increased CSF protein and markedly slowed nerve conduction strongly suggestive of a demyelinating process. A sural nerve biopsy from this case showed that myelinated nerve fibres were reduced in number and some thinly remyelinated fibres were present. Sahenk et al. (1977) reported a man with chronic predominantly motor polyradiculoneuropathy with only slight slowing of nerve conduction, in whom a nerve biopsy also showed a reduction of myelinated nerve fibres and some thinly remyelinated fibres. On the basis of these and a few less well documented cases it has been proposed that SLE is associated with CIDP (Rechthand et al. 1984). In a few reports acute monophasic neuropathy resembling GBS has occurred in SLE (Richardson 1980). In several other reports SLE has been associated with subacute neuropathy, resembling our definition of subacute CIDP (Heptinstall and Sowry 1952; case 2 of Rechthand et al. 1984). In these cases an acute possibly autoimmune demyelinating process has been invoked but not yet proved histologically. When cases have been studied in the acute stage, active vasculitis has been found (Hughes et al. 1982). Even in cases where axonal degeneration has been the only finding, underlying vasculitis might have been missed in the sample taken. The idea that SLE is associated with acute or chronic macrophage-mediated demyelinating neuropathy awaits pathological proof.

Rheumatoid Arthritis

Rheumatoid arthritis is the commonest of the connective tissue disorders affecting 2% of the male and 5% of the female adult population in the UK (Copeman 1964) and is complicated by neuropathy in about 1% of cases (Conn and Dyck 1984). Patterns of neuropathy include the familiar compression palsies related to swollen and disrupted joints, e.g. carpal tunnel syndrome and ulnar lesions at the elbow, mild distal sensory neuropathy, rapidly progressive sensory and motor neuropathy, and multiple mononeuropathy (Pallis and Scott 1965). The mild distal sensory neuropathy is usually associated with electrophysiological and, where sought, biopsy evidence of axonal degeneration, possibly with some associated segmental demyelination and remyelination. Vascular changes are limited to non-specific thickening of vessel walls and active vasculitis is not seen (Weller et al. 1970). In recently involved nerves affected by acute neuropathy or multiple mononeuropathy florid vasculitis with fibrinoid necrosis is found (Conn and Dyck 1984). In a detailed autopsy study of such a patient multiple vasculitic lesions affecting the nerves in the mid portions of the upper arm and thigh were discovered. These lesions did not cause frank necrotic infarcts but did lead to a centrifascicular pattern of degeneration in more distal nerves (Dyck et al. 1972).

In the early days of steroid treatment cases of severe neuropathy seemed to occur in patients who were taking steroids so that blame was sometimes attached to the steroids for causing a vasculitic neuropathy. It is now clear that vasculitic neuropathy occurs in patients who have not been treated with steroids and that reduction of steroid dose may trigger a vasculitic neuropathy. The suggestion of a relationship between steroids and vasculitic neuropathy arose because more severe cases of rheumatoid arthritis were more likely to be treated with steroids. Treatment of vasculitic neuropathy associated with rheumatoid arthritis is usually with large doses of steroids. I also use azathioprine 2.5 mg/kg but others favour cyclophosphamide. There is no evidence other than clinical experience whether these or other cytotoxic regimes or plasma exchange are superior to steroids alone. Although the occurrence of vasculitic neuropathy indicates that the disease is very active and other vital organs are at risk of vasculitis, the neuropathic lesions are usually arrested by such treatment and slow improvement usually follows. The extent of improvement depends on the severity of the deficit which had developed before treatment was started.

Sjögren's Syndrome

Peripheral neuropathy occurs in about 10% of cases of primary Sjögren's syndrome (sicca syndrome) and secondary Sjögren's syndrome (sicca syndrome associated with a connective tissue disorder such as rheumatoid arthritis) (Kaltreider and Talal 1969). The neuropathy is usually entirely or predominantly sensory, subacute or chronic and fairly mild. Electrophysiological and biopsy evidence suggest that axonal degeneration secondary to vasculitis is the underlying cause. The diagnosis is made on clinical grounds, but confirmatory tests include the Rose Bengal test for a defective tear film, the Schirmer test for defective tear production, minor labial salivary gland biopsy, salivary flow measurement, antibodies to salivary gland epithelium detected by immunofluorescence, and antinuclear antibodies to Ro and La antigens (complexes of protein and RNA). Steroids have been helpful in treatment (Mellgren et al. 1989). In addition to peripheral neuropathy, trigeminal neuropathy is particularly associated with Sjögren's syndrome (Lecky et al. 1987). The symptoms of dry eyes and dry mouth may not be obtrusive and should be sought when considering the cause of an unexplained neuropathy.

Vasculitic Neuropathy Without Clinical Evidence of Other Organ Involvement

Of particular importance to the neurologist are those patients whose neuropathy is due to vasculitis in the absence of any evidence of a connective tissue disorder or vasculitis affecting other organs. Surprisingly in a study of 33 patients in whom sural nerve biopsy revealed vasculitis, the clinical presentation was more commonly a symmetrical sensory and motor polyneuropathy (26 cases) than multiple mononeuropathy (seven cases) (Harati and Niakan 1986). Only five of these 33 cases had an associated collagen vascular disease and three had a malignant tumour. In another series of 16 nerve biopsies showing vasculitis, six had a distal symmetrical polyneuropathy, eight multiple mononeuropathy with overlapping involvement, somewhat resembling symmetrical polyneuropathy, and only two obvious multiple mononeuropathy. Underlying collagen vascular disease was identified in only four cases (Kissel et al. 1985). Dyck et al. (1987) described 65 patients with neuropathy and vasculitis: 45 patients had an associated systemic vasculitic disorder, usually polyarteritis or hypersensitivity vasculitis, and 20 patients had vasculitis revealed only on nerve biopsy. The clinical pattern in these 20 patients was usually a multiple mononeuropathy but eight had a distal or asymmetrical polyneuropathy usually affecting motor and sensory fibres.

These series of patients raise the question whether vasculitis confined to nerves is a different disorder from conventional systemic vasculitis or whether it represents a mild form of polyarteritis nodosa which happens to have affected the peripheral nerves and spared other organs. The nerves are eloquent sites where tiny lesions cause paraesthesiae and deficit. Similar-sized lesions in muscles, liver, or kidneys might be clinically silent. Dyck et al. (1987) argue against this possibility because their patients had been followed on average for more than a decade, none had developed systemic involvement, and all had survived. However, involvement of other organs would have led to reclassification of such a case as systemic. Said et al. (1988) studying a larger series of 100 patients with vasculitis and neuropathy, identified 32 patients in whom there was no associated systemic disease but argued that this was not an organ-specific vasculitis confined to the PNS. These authors undertook biopsies of the superficial peroneal nerve, rather than the more conventional sural nerve, and sampled the short peroneal muscle at the same time. Most (86%) of their 32 patients with vasculitis apparently confined to nerves also had necrotising arteritis in the muscle biopsy. However, the clinical observation remains true that in some cases vasculitis is apparently confined to the peripheral nerves and has a relatively good prognosis compared with the typical case of systemic vasculitis with multiple organ involvement. On the basis of personal and reported clinical experience, steroid treatment is probably helpful (Dyck et al. 1987). The facts that nerve or muscle biopsy may provide the only evidence of vasculitis in an otherwise unexplained neuropathy, and that vasculitis can be treated provide the principal clinical justification for undertaking a nerve biopsy in a case of undiagnosed neuropathy.

Treatment of Vasculitic Neuropathy

Although many papers are devoted to the clinical, electrophysiological and neuropathological features of vasculitic neuropathy, little has been written about treatment. This is not unusual in neurology and neurologists have an unfortunate tendency for passing on neurological disorders for which treatment becomes available to colleagues in other specialties. One principle behind the management of vasculitis affecting the nervous system must be recognised: treatment is urgent. Each day which is lost in organising diagnostic tests is another day during which fresh ischaemic lesions may occur in the nerves (or CNS) which will delay recovery and increase the risk of permanent residual deficit. Steroids are the mainstay of treatment and are usually used in large single oral daily doses, such as prednisolone 60 mg/day for the first week at least. There is a wide variation in different clinicians' practices round the world and no controlled comparative trials on which to base judgements. Alternatively high dose pulse therapy with intravenous methylprednisolone 500–1000 mg daily for three to five days, a regime used in the management of transplant rejection, has been tried. There is no evidence that such a regime is any worse or better than oral steroids and it would probably be necessary to continue oral steroids after such a course. There is much evidence that immunosuppressive drugs have improved the prognosis of Wegener's granulomatosis (Editorial 1972) and it is common practice to use these in the event of a recurrence of vasculitis. I believe it is so important to achieve control as far as possible that I start combined treatment with azathioprine 2.5 mg/kg, and prednisolone 60 mg daily. The side-effects of azathioprine include gastric intolerance, bone marrow depression and small increased risk of malignancy, especially non-Hodgkin lymphoma (Chapter 10). Fauci et al. (1979) consider that cyclophosphamide is more effective than azathioprine, but the evidence is based on clinical anecdote not controlled trials. Cyclophosphamide has the disadvantage of causing alopecia, haemorrhagic cystitis, and an even greater increase in risk of malignancy, especially affecting the bladder compared with azathioprine.

Sarcoid Neuropathy

Unlike GBS, which has a clinical definition, sarcoidosis is defined pathologically from the presence of non-caseating granulomas forming tubercles in several affected organs throughout the body. The tubercles are composed of accumulations of large "epithelioid" cells with abundant cytoplasm. The centre of the tubercles may undergo fibrinoid necrosis. The margins have an admixture of variable numbers of lymphocytes and plasma cells. The epithelioid cells were named because of their rather vague resemblance to epithelial cells but represent mature stimulated macrophages having the properties of secretory cells with abundant cytoplasm, endoplasmic reticulum and lysosomes but no phagocytic function. The macrophages may also fuse to form multinucleate Langhans' giant cells with peripherally disposed nuclei, or foreign body giant cells in which the nuclei are dispersed throughout the cytoplasm. Boeck recognised that the disease spread beyond the skin and applied the term "sarcoid", derived from the Greek word for flesh, under the impression that the tubercles were derived from connective tissue. Sarcoidosis most commonly affects the lungs causing diffuse infiltration or hilar lymphadenopathy.

The prevalence of sarcoidosis is not clearly established and is probably variable. In the UK the prevalence of radiographic abnormalities suggesting sarcoidosis is in the range 4–40 per 100 000 (Scadding and Mitchell 1985). The nervous system is affected in about 5% of cases and the peripheral nervous system is involved in two-thirds of these (Matthews 1984, 1987). CNS involvement, though less common, is much more serious and can produce many different manifestations depending on the situation of the granulomas in the cerebral hemispheres, brainstem or spinal cord, or alternatively on the production of a basal meningoencephalitis (Scadding and Mitchell 1985). The most common pattern of peripheral nervous system involvement is a syndrome of multiple fluctuating cranial nerve palsies. This may be associated with evidence of multiple mononeuropathy elsewhere in the body. The facial nerve is the nerve most commonly affected, and both sides of the face may be involved, not necessarily synchronously. According to Matthews (1984) the severity of the facial palsy is greater than is usual with Bell's palsy, and deficit and abnormal movements are more likely to persist. Of the other cranial nerves, the fifth, eighth, tenth, eleventh and twelfth are commonly involved, but it is unusual for ocular palsies to occur. Nevertheless the pupils may be involved and the sixth nerve may be affected, perhaps due to raised intracranial pressure associated with meningeal or CNS sarcoid. Although well recognised, conventional peripheral neuropathy is relatively uncommon compared with cranial nerve involvement.

When the peripheral nerves are involved it is usually in the form of a multiple mononeuropathy. This may occur in isolation or with CNS or cranial nerve involvement. Any nerve, motor or sensory, may be affected but a rather characteristic feature is the loss of patches of sensation on the trunk which was first noted by Heerfordt (1909). These may be accompanied by dysaesthesiae and pain, and, when the abdominal wall is involved, this can lead to diagnostic confusion with surgical conditions. Matthews (1984) reports that areas involved have included the entire posterior chest wall, or the skin of the whole abdomen. The nerve lesions may improve spontaneously although their healing is probably hastened by steroid treatment which may also prevent recurrence. The prognosis of sarcoidosis of the peripheral nervous system is much more favourable than that of CNS sarcoid. Probably for that reason there is little information about the pathology and pathogenesis of the peripheral nerve lesions.

Symmetrical polyneuropathy also occurs but less commonly than multiple mononeuropathy. It is usually subacute or chronic and may be relapsing or recurrent. A rare case presenting as an acute polyneuropathy of GBS type was described in a boy aged 11 (Leluyer et al. 1983). Another case of typical GBS with albuminocytological dissociation of the CSF was described in a man aged 25: his disease started during treatment of pulmonary sarcoid with steroids. Steroids were continued and he gradually recovered completely (Schott et al. 1968) (case 2). Another case in a young woman with pulmonary sarcoidosis was called GBS by the authors (Strickland and Moser 1967) but the progressive phase was more prolonged than our working definition of GBS reaching its nadir within 4 weeks allows: progressive painful sensory and motor polyneuropathy with areflexia progressed over at least 10 weeks to the point that the patient

Table 12.1. Sarcoid neuropathy with histological examination

	Age (years)	Sex	Duration (months)	Type	CSF	MCV	Biopsy	Granuloma	Steroid response
Personal case	74	F	2	SM	N	ax	ax	epi, peri endo	+
Nemni (1981)	20	F	2	SM	N	de	mixed	endo	+
Galassi et al. (1984)	70	M	3	SM	N	ax	ax	epi, peri	+
Vital et al. (1982)	44	M	4	S	–	–	N	epi	Not given
	40	F	5	S	–	–	ax	epi	+
Oh (1979)	58	F	6	SM	N	ax	ax	epi, peri	++
Galassi et al. (1984)	54	M	18	SM	N	ax	ax	epi, peri	+

This table lists all the cases of sarcoid neuropathy in which nerve biopsy appearances have been published. Abbreviations: N, normal; MCV, motor conduction velocity; ax, axonal neuropathy; de, demyelinating neuropathy; mixed, mixed axonal/demyelinating neuropathy; granuloma: epi, epineurial; peri, perineurial; endo, endoneurial; steroid response: +, improvement; ++, marked improvement, cure.

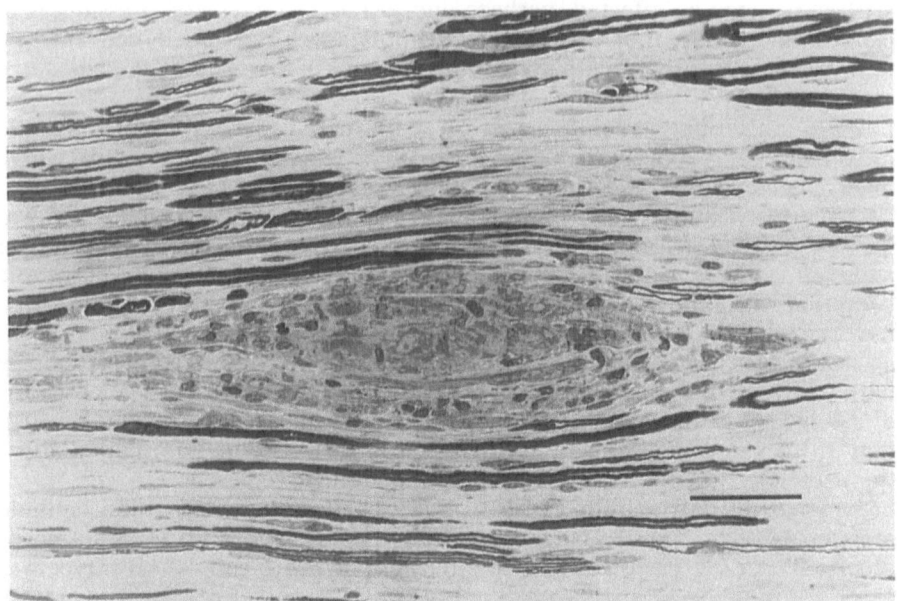

a

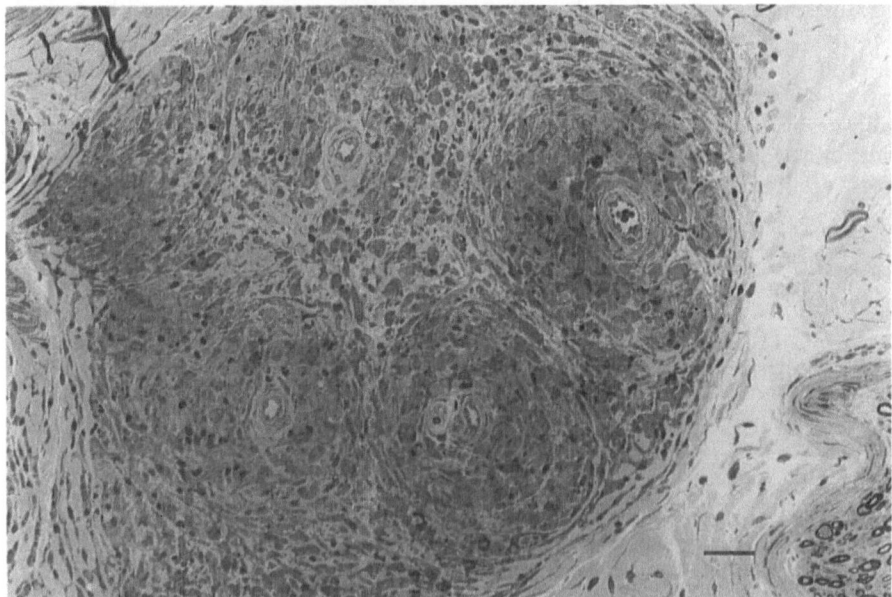

b

Fig. 12.3. Sarcoid neuropathy. Resin sections (1 μm) stained with thionin and acridine orange. **a** Endoneurial granuloma. Bar = 40 μm. **b** Epineurial granuloma. Bar = 40 μm. (**a** is reproduced from Gainsborough et al. (1990), with permission.)

could not walk. The CSF contained no cells but had an increased protein content. Both her neuropathy and pulmonary sarcoid improved rapidly with oral prednisolone. Many similar cases have been reported but the pathology and pathogenesis have only recently been investigated. Oh (1979) reported an elderly woman with a progressive sensory and motor neuropathy whose sural nerve biopsy demonstrated axonal loss and ongoing axonal degeneration and epineurial and perineurial granulomas. Similar findings have been reported in four other cases (Table 12.1). The perivascular infiltration round epineurial and sometimes endoneurial vessels and occasional invasion of the vessel wall by the granulomas have led to the suggestion that the nerve damage has an ischaemic basis. This might be true, but the distribution of nerve fibre loss has never been centrifascicular or non-uniform between fascicles as is characteristic of neuropathy associated with vasculitis. in one published case nerve conduction was markedly slowed in the lower limbs and the sural nerve contained a mixture of fibres undergoing axonal degeneration and rather more undergoing segmental demyelination (Nemni et al. 1981). In our own case of a subacute predominantly sensory polyneuropathy in an elderly woman there were florid granulomas in the epineurium and endoneurium (Fig. 12.3). The granulomas consisted of a mixture of epithelioid cells with characteristically abundant cytoplasm packed with mitochondria and endoplasmic reticulum (Fig. 12.4) which did not express MHC class II antigen, macrophages which did express this antigen plasma cells and lymphocytes. Fibres in the immediate vicinity of the endoneurial granulomas swerved round and were distorted by the endoneurial granulomas but the

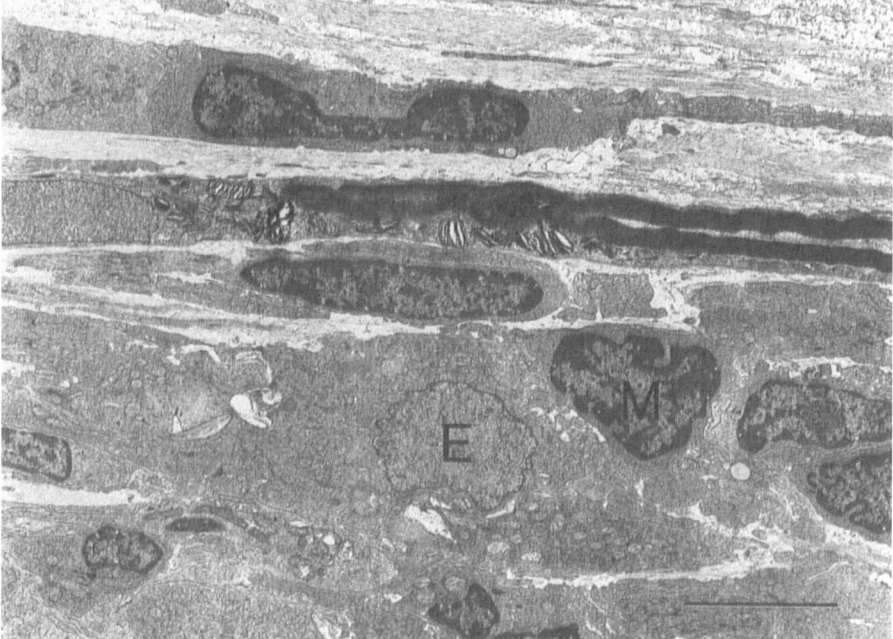

Fig. 12.4. Sarcoid neuropathy. Electron micrograph of the edge of an endoneurial granuloma showing mononuclear cells (M) associated with epithelioid cells (E). Bar = 10 μm. (From Gainsborough et al. (1990), with permission.)

majority remained normal. Many adjacent fibres were undergoing axonal degeneration and a few showed extensive paranodal demyelination consistent with the effect of compression. Invasion of the Schwann cell basal lamina by macrophages such as occurs in GBS was not seen (Gainsborough et al. 1990). This point is important because if non-specifically activated lymphocytes and macrophages were an important cause of primary demyelination such a granuloma might have been expected to cause extensive primary demyelination.

The prognosis of sarcoid polyneuropathy is variable. Most cases seem to recover, with (Schott et al. 1968) or without the use of steroids (Wiederholt and Siekert 1965; Rinne 1967) so that assessing their role in difficult. However, chronic, progressive and ultimately fatal courses have been reported (Garland and Thompson 1953).

Plasma Cell Dyscrasias and Neuropathy

Of 279 patients with peripheral neuropathy of unknown cause 10% had a paraproteinaemia (Kelly et al. 1981b) whereas the percentage in a normal population is only about 1% over age 50 and 3% over age 70 (Meier 1985). Paraproteins are abnormal proteins detected in body fluids and are usually monoclonal (M) proteins consisting of homogeneous immunoglobulin molecules produced by one clone of plasma cells. The monoclonal protein is detected by serum protein electrophoresis on which a single M band is identified usually in the gammaglobulin region. The nature of the M band can be identified by immunoelectrophoresis or more sensitively by immunofixation (Kelly et al. 1987). Sometimes, especially with IgD myeloma, the paraprotein consists only of light chains which are not identified by conventional serum protein electrophoresis but can be detected by examination of concentrated urine for Bence-Jones protein. Benign plasma cell clones may persist for many years without damaging health. Such patients have benign monoclonal gammopathy which is better called monoclonal gammopathy of undetermined significance. They have less than 30 g/l paraprotein

Table 12.2. Approximate percentage of cases with clinical neuropathy

	Percentage
Monoclonal gammopathy of undetermined significance	
IgM	50
IgG	rare
IgA	rare
Malignant plasma cell dyscrasia	
Multiple myeloma: osteolytic	3–11
Multiple myeloma: osteosclerotic	50
Solitary plasmacytoma	occurs
Waldenstrom's macroglobulinaemia	50
Amyloidosis	
1. Primary	20
2. With multiple myeloma	20

Percentages are based on the review by Meier (1985) and the series of Kelly et al. (1987).

in the serum, no Bence-Jones protein, less than 5% plasma cells in the bone marrow and no evidence of systemic disease. Over a ten-year period at least 20% of patients with an M protein develop more serious manifestations of B cell malignancy, such as multiple myeloma, Waldenstrom's macroglobulinaemia or amyloidosis. Neuropathies occur in association with all these conditions at all stages and for several different reasons. Helpful recent reviews are available (Meier 1985; Kelly et al. 1987). The frequency of neuropathy in these conditions is estimated in Table 12.2. Only the most important features of these neuropathies will be reviewed with emphasis on those in which there is evidence for the presence of antibodies directed against nerve antigens which may be relevant to the pathogenesis of nerve damage.

Monoclonal Gammopathy of Undetermined Significance

Monoclonal gammopathy of undetermined significance (MGUS) is often called benign monoclonal gammopathy but this adjective benign belies the known evolution of the clone into a malignant plasma cell dyscrasia. About half patients

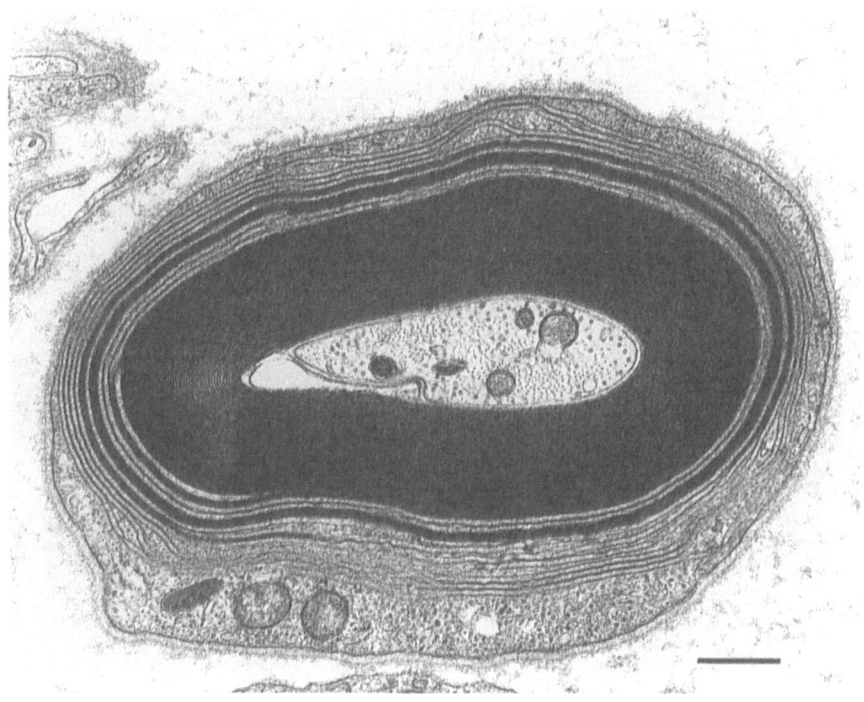

Fig. 12.5. Demyelinating neuropathy associated with IgM$_k$ paraprotein and antibodies myelin associated glycoprotein. Electron micrographs to show the abnormal wide spacing of myelin lamellae. **a** Bar = 0.5 μm. **b** Bar = 0.2 μm. **c** Bar = 1 μm. (The photograph in **c** is by courtesy of Dr S.M. Hall.)

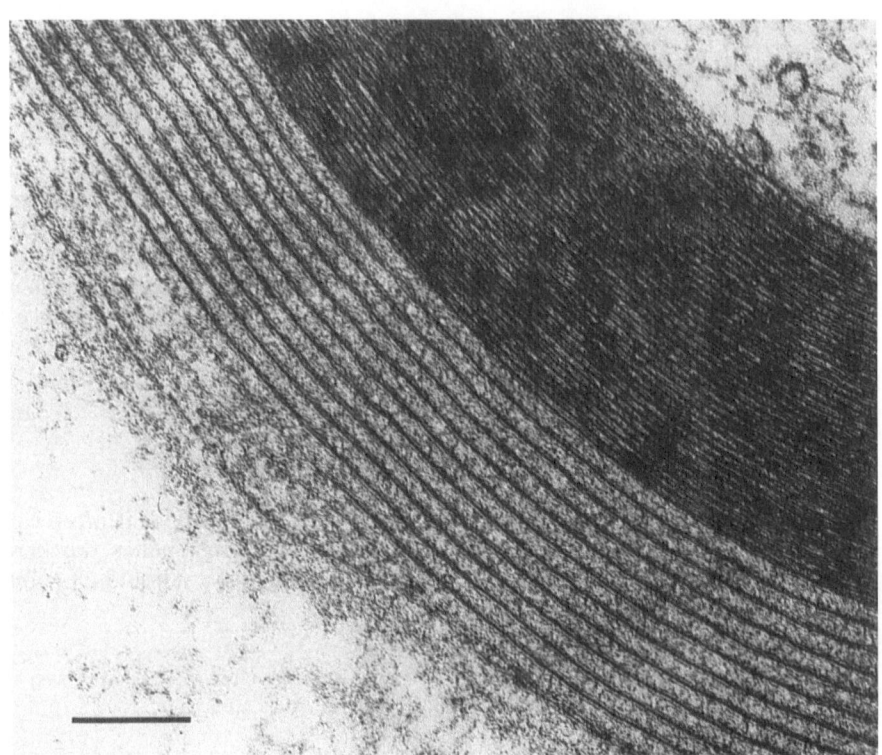

b

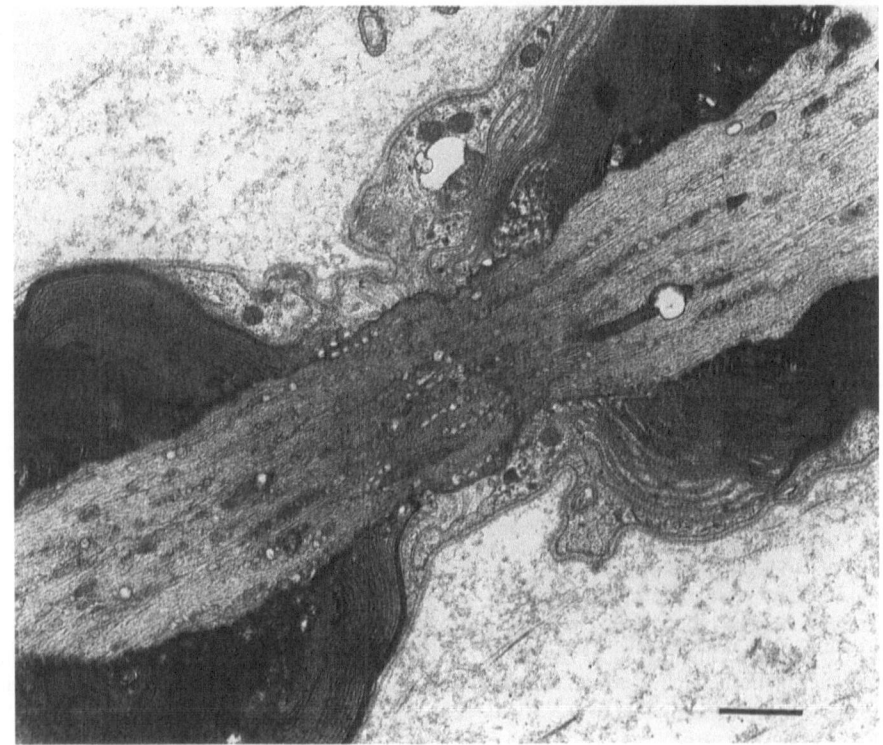

c

with IgM MGUS have antibodies against myelin-associated glycoprotein which is present in small quantities in CNS and PNS myelin (Chapter 2). The clinical picture is fairly homogeneous (Kahn et al. 1980; Latov et al. 1980; Smith et al. 1983). Most patients are male and develop a progressive, predominantly sensory, polyneuropathy in middle or old age. Distal paraesthesiae may be prominent and out of proportion to sensory loss. The cranial nerves are usually spared. The course is usually extremely insidious, evolving over years, but sometimes eventually causing considerable disability. The electrophysiological findings indicate a demyelinating neuropathy. Nerve biopsies have revealed marked depletion of myelinated (but not unmyelinated) nerve fibres and a mixture of demyelination, remyelination (sometimes with onion bulb formation), and axonal degeneration (Smith et al. 1983). A characteristic feature is widening of the minor dense line of the myelin lamellae (Fig. 12.5). Although characteristic of IgM_k MGUS neuropathy with anti-MAG antibodies this abnormal spacing also occurs in neuropathy associated with Waldenstrom's macroglobulinaemia, CIDP, and, rarely with benign IgG paraproteinaemia and in one case non-Hodgkin lymphoma without paraproteinaemia (King and Thomas 1984; Meier 1985; Vital et al. 1983). A similar separation of the minor dense line has been produced experimentally by applying anti-myelin serum without complement to myelinated tissue cultures (Chapter 2).

The unifying feature of these patients is the property of the IgM M protein to react with antigens in peripheral nerve myelin. These can be demonstrated quite simply with a complement fixation assay with human nerve or myelin as antigen: the sera uniformly give positive results at high dilutions. The antibody also reacts with peripheral nerve, but not CNS, myelin and with the pi granules of Schwann cells (Leibowitz et al. 1983). The chemical nature of the antigen was at first identified by immunoblot as MAG (Latov 1981). MAG is present in CNS as well as PNS myelin. However, only peripheral nerve and not CNS demyelination occurs in this condition. This anomaly was eventually explained by the fact that the antibody cross-reacts with novel glycolipids which are specific to peripheral nerve myelin. The glycolipids have been identified as sulphated glucuronic acid containing paragloboside and sulphated glucuronic acid containing lactosaminosyl paragloboside (Kelly et al. 1987; McGinnis et al. 1988). Although there is a very tight association between high titres of anti-MAG antibodies and presence of demyelinating neuropathy, formal proof that the antibodies cause the demyelination has been difficult to obtain. IgM is present on the myelin sheaths of affected individuals and immunohistochemical studies of plastic embedded sections show staining especially at the nodes of Ranvier and Schmidt–Lantermann incisures, sites where MAG is abundant (Meier et al. 1983; Mendell et al. 1985). Injection of human anti-MAG serum into the sciatic nerve of the cat, a species which shares the same antigenic epitope as human MAG (unlike rodent MAG which does not), is followed by acute demyelination (Hays et al. 1987). Removal of the paraprotein with plasma exchange and cytotoxic treatment has been tried but only a few reports are available (Sherman et al. 1984; Haas and Tatum 1988; Kelly et al. 1987) and the insidious nature of the disease makes assessment of any therapeutic benefit difficult. The author has had success treating one patient with plasma exchange, melphalan and steroids, and in another with chlorambucil alone. In most of my other cases the disease has been too insidious to require treatment or too advanced to make interference worthwhile.

Half the patients with IgM MGUS and neuropathy do not have antibodies to MAG. In most cases the nature of the antigen to which the M protein reacts has not been identified. In two patients with axonal neuropathy, the IgM paraprotein reacted with chondroitin sulphate which is a component of endoneurial matrix. This reaction might indicate a pathogenetically important antibody (Sherman et al. 1983).

Several case reports of chronic progressive or relapsing peripheral neuropathy associated with monoclonal IgG paraproteins have been published (Contamin et al. 1976; Read et al. 1978; Noring et al. 1980; Dalakas and Engel 1981; Ohnishi and Hikano 1981; Sewell et al. 1981; Bosch 1982). The clinical picture is heterogeneous. Most biopsies are reported as showing a mixture of axonal degeneration and demyelination. In some cases IgG of a single light chain type has been identified on the myelin sheaths in biopsies. However, specific myelin antigens against which these paraproteins might be directed have not been discovered.

Of the few reported cases of neuropathy associated with IgA, two have had a purely motor neuropathy mimicking motor neuron disease (Kahn et al. 1980; Chazot et al. 1976).

Lower Motor Neuron Syndrome and IgM Monoclonal Gammopathy of Undetermined Significance

Since the first report (Rowland et al. 1982), several cases of a lower motor neuron syndrome with purely lower motor neuron features and serum IgM M proteins have been published. The serum of these patients has been shown to react with gangliosides GM_1 and GD_{1b} (Rudnicki et al. 1987; Nardelli et al. 1988). The antibodies are specific for the Gal(β1–3)GalNAc or Gal(β1–3) GlucNAc residues (Latov et al. 1988; Ito and Latov 1988). The serum reacted with both grey matter and myelin of human spinal cord on an indirect immunofluorescence test (Latov et al. 1988). Treatment of the paraprotein with plasma exchange and cytotoxic drugs has appeared beneficial in some (Latov et al. 1988) but not all cases (Rudnicki et al. 1987). The anti-ganglioside antibodies might bind to epitopes on the external surface of synaptic terminals or could be taken up at motor end plates and transported centripetally to the cell body where they might interfere with synaptic function or metabolic processes. This remains an area of active research.

Polyclonal antibodies to ganglioside GM_1 are found in low titre in normal people, in higher titre in patients with a variety of autoimmune diseases and in ordinary motor neuron disease, and in even higher titre in patients with motor neuron disease with predominantly lower motor neuron features (Pestronk et al. 1988a). Rare patients with high titres of polyclonal antibodies to GM_1 and a lower motor neuron syndrome with conduction block, indicating a multifocal demyelinating motor neuropathy, have shown remarkable improvement following a course of high-dose intravenous cyclophosphamide (Pestronk et al. 1988b; Shy et al. 1988).

Waldenstrom's Macroglobulinaemia

Waldenstrom's macroglobulinaemia is a malignant plasma cell dyscrasia of IgM producing cells which has to be distinguished from IgM MGUS by the higher concentrations of paraprotein (>30 g/l) and malignant plasma cell infiltration of the bone marrow. It affects older men and usually causes malaise, weakness, bleeding, anaemia, hepatosplenomegaly and lymphadenopathy. Hyperviscosity caused by the paraprotein produces retinal lesions, dizziness, deafness, and cerebrovascular disease. The bleeding diathesis may cause subarachnoid haemorrhage. About a quarter of cases have such neurological complications. In addition to the CNS features, several forms of neuropathy, including a clinical picture resembling GBS, chronic progressive sensory and motor neuropathy and multiple mononeuropathy have been reported in 76% of cases (Logothetis et al. 1960). The cause of these neuropathies is heterogeneous. The cases resembling GBS were only briefly described and their pathological basis was not mentioned. Most patients with chronic progressive sensory and motor neuropathy probably have a demyelinating neuropathy and abnormal myelin lamellar spacing of surviving myelinated fibres (e.g. case 11 of Powell et al. (1984)). These patients are usually found to have IgM with either kappa or lambda light chains on their myelin sheaths and antibodies to myelin similar to those which are found in IgM MGUS associated with neuropathy. In some reported series of patients with Waldenstrom's macroglobulinaemia and neuropathy the concentration of IgM paraprotein has been too low to fit the conventional definition of Waldenstrom's macroglobulinaemia (Dellagi et al. 1983). Since MGUS is known to evolve into Waldenstrom's macroglobulinaemia in at least 20% of cases, it would be expected that some of those with MGUS and neuropathy would develop malignant plasma cell dyscrasia. An additional or contributory cause of progressive neuropathy in Waldenstrom's macroglobulinaemia is amyloid deposition.

The walls of endoneurial blood vessels are abnormally thickened in neuropathy associated with several types of paraproteinaemia, and abnormalities of endoneurial blood vessel perfusion might contribute to damage to nerve fibres. The thickening is due to a combination of remyelination of the basal lamina and hypertrophic changes of the endothelial cells which contain abnormal accumulations of microfilaments (Powell et al. 1984). Thickening of endoneurial blood vessel walls may be non-specific, occurring in other forms of chronic neuropathy especially in the elderly. More unusual abnormalities consisting of endoneurial perivascular deposits of felt-like or granular material have been described in three cases (Meier 1985; Lamaria et al. 1987). This material stained positively for IgM and probably represents precipitated paraprotein.

Multiple Myeloma

Multiple myeloma was first clearly described by McIntyre in London in 1846. McIntyre recognised the unique properties of a protein in the urine which would precipitate on warming and dissolve on boiling. He sent the urine to Henry Bence-Jones who ran a presumably lucrative clinical pathology service in Harley Street. Bence-Jones described the properties of the protein in the *Lancet* in 1847 and achieved eponymous fame. It is amusing to note that even 150 years ago doctors enjoyed publishing as many papers as possible. Four publications were obtained from this one case and the pathology was written up even before the clinical and chemical reports had left the press. A fifth publication was obtained from the same case by a diligent medical historian who discovered from the archives of the Marylebone Registrar of Births and Deaths for the relevant period that the patient's name was McBain (Clamp 1967). Multiple myeloma should really be called McBain's disease and Bence-Jones protein McIntyre's protein!

Neuropathy is a rare, although well-recognised, complication of multiple myeloma. More common neurological manifestations are spinal cord compression and painful radiculopathy associated with osteolytic vertebral lesions. In a series of 277 patients bone pain was a feature in 60% (Silverstein and Doniger 1963), cord/cauda compression in 10%, root compression in 20% and peripheral neuropathy in only 3%. According to one review neuropathy associated with the usual osteolytic myeloma occurs in 3%–5% of cases (Kelly et al. 1981a). The pattern is heterogeneous, mild sensory and motor, pure sensory or severe motor. Some cases are associated with amyloid deposition in the nerves causing a painful neuropathy with autonomic involvement. Carpal tunnel syndrome due to amyloid deposition is rather characteristic.

The rare osteosclerotic form of myeloma is complicated by neuropathy in 50% of cases more often than in the usual osteolytic myeloma (Kelly et al. 1983). Clinically the picture is homogeneous; a severe, progressive predominantly motor neuropathy. Nerve conduction is markedly slowed and the CSF protein is raised moderately or markedly. Sural nerve biopsies show a mixture of axonal degeneration and segmental demyelination with some mononuclear infiltration of the epineurium. These are important patients to diagnose since immunosuppressive treatment, and especially irradiation of osteosclerotic lesions, particularly if they are solitary, has usually been beneficial (Kelly et al. 1983; Milanese et al. 1982).

Some patients with osteosclerotic myeloma have a combination of systemic features involving polyneuropathy, organomegaly affecting the liver and spleen, endocrinopathy (hypogonadism and hypothyroidism), myeloma and skin changes (hyperpigmentation, hypertrichosis and hyperhidrosis) giving the acronym POEMS syndrome (Solomons 1982). This syndrome has also been called the Crow–Fukase syndrome (Fukase et al. 1969; Nakanski et al. 1984). In a single case antibody activity of the IgG paraprotein against hypophyseal cells supported the idea that this syndrome might have an autoimmune origin.

Leprosy

No book on inflammatory neuropathy would be complete without a consideration of leprosy, arguably the commonest cause of peripheral neuropathy in the world. In parts of India, the prevalence of leprosy remains as high as 6% of the population (Chatterjee et al. 1976), and leprosy remains a serious public health problem in the Far East, Africa and South and Central America. The disease is caused by chronic infection with *Mycobacterium leprae*, an intracellular bacterium 1–7 μm long and 0.25 μm wide which resembles *M. tuberculosis*. The infection produces a spectrum of disease which ranges from tuberculoid, through borderline (dimorphous), to lepromatous forms, according to the vigour of the host response. The cell-mediated immune response is marked in tuberculoid and absent in lepromatous leprosy (Ridley and Jopling 1966). The reader is referred to standard texts for detailed accounts of leprosy and its treatment (Waters et al. 1987) but the main forms of leprosy and relevant nerve pathology will be briefly mentioned.

In tuberculoid leprosy there are only one or two well-circumscribed skin lesions with thin raised red margins and flattened hypopigmented anaesthetic and analgesic centres. The rest of the skin remains normal and nerves are not thickened apart from those supplying the skin lesions. Biopsy of the margin of the tuberculoid lesions shows a non-caseating tuberculoid chronic granulomatous reaction with epitheloid cells, giant cells and lymphocytes. Further evidence of cell-mediated immunity is provided by a positive lepromin test. This involves injecting a suspension of autoclaved *M. leprae* into the skin. Induration and erythema may be detected 48 hours later. This early (Fernandez) reaction is similar to the tuberculin test. More specific is the late Mitsuda reaction, which is read after three to five weeks and is positive in tuberculoid but negative in lepromatous leprosy. There is, however, some cross-reactivity with tuberculin, and the lepromin test is not specific for leprosy.

At the other end of the spectrum, lepromatous leprosy is a much more serious condition in which there are at first numerous small hypopigmented macules or papules and congestion of the nasal mucosa with a bloody discharge. As the disease advances, the skin becomes thickened especially on the lips, face and pinnae. The eyebrows and eyelashes are lost. There is progressive impairment of cutaneous sensation, especially of those modalities subserved by non- and small myelinated fibres and the peripheral nerves become thickened. This causes a more or less glove and stocking pattern of sensory loss and impairment of sweating. Position sensation and tendon reflexes are preserved. The profound analgesia permits recurrent trauma which leads to resorption of the fingers and

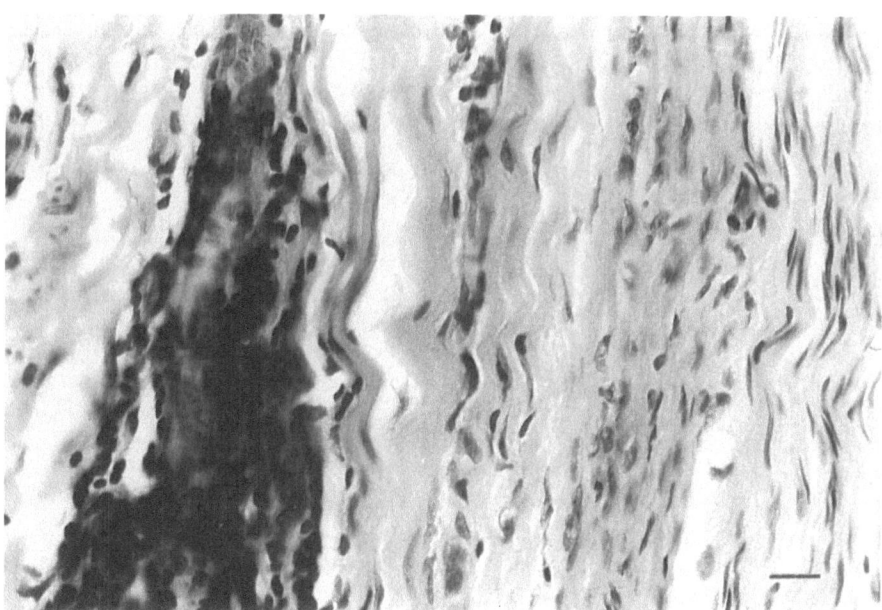

Fig. 12.6a. Pure neural leprosy. Chronic inflammatory infiltrate around an epineurial vessel. Bar = 20 μm.

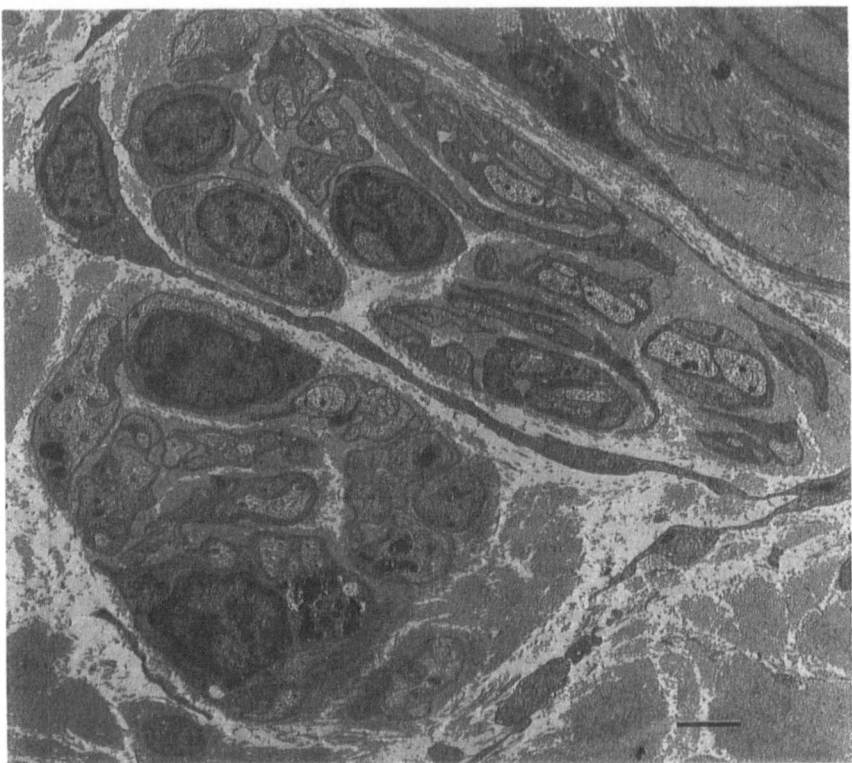

Fig. 12.6b. Electron micrograph from the same nerve showing debris in a non-myelinating Schwann cell. Bar = 2 μm.

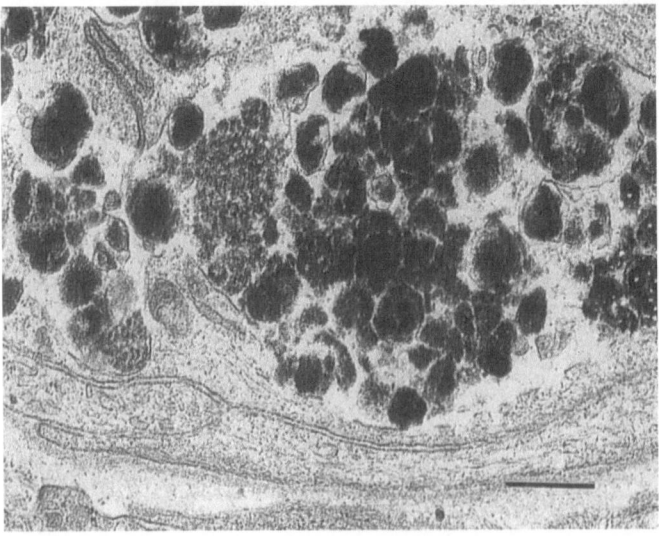

Fig. 12.6c. Higher power micrograph of debris. Bar = 0.5 μm.

toes and severe deformity. The gross deformities attracted the unfortunate stigma to the disease. Between the tuberculoid and lepromatous poles have been segregated borderline tuberculoid, borderline and borderline lepromatous forms. Fine distinctions between these are difficult for the non-leprologist but the principles are clear. In borderline leprosy the skin lesions are variable in size. Some are hypo-, and some hyperpigmented. They characteristically have broad irregular rims and anaesthetic centres. Nerves at sites of predilection may be asymmetrically thickened. These include ulnar above the medial epicondyle, superficial radial and median at the wrist, greater auricular, common peroneal at the fibula head and posterior tibial at the medial malleolus. Of particular interest to the neurologist are rare cases of pure neural leprosy in which granulomas are confined to the nerve and skin lesions are not seen (Fig. 12.6).

During drug treatment of any of the borderline forms of leprosy reactions occur in which the skin lesions become swollen and red and nerve damage may ensue. These are considered to be due to a reversal or upgrading of immunity to *M. leprae*. At the lepromatous end of the spectrum, especially during treatment, a severe illness may occur with fever, malaise and often painful nerve lesions. These reactions, called erythema nodosum leprosum, have been shown to be due to immune complex vasculitis. They are sometimes accompanied by other manifestations of vasculitis including iritis and rarely nephritis and arthritis. Central American patients with untreated lepromatous leprosy may develop a different form of reaction causing scars on the skin of the lower limbs which are due to vasculitis.

Treatment of leprosy has moved on from the days when long-continued dapsone was the mainstay to multi-drug therapy with dapsone, rifampicin and clofazimine which provides the prospect of cure and eradication of the disease (Editorial 1988). Erythema nodosum leprosum can be treated with thalidomide, prednisolone or clofazimine. Other reactions may require aspirin or prednisolone depending on their severity. The details of drug treatment and problems of detection of disease and delivering treatment and health education to the affected populations are dealt with in standard texts (Waters et al. 1987; Bryceson and Pfaltzgraff 1979).

Despite years and volumes of work the precise mode of infection and spread of leprosy have not been conclusively established. In epidemiological studies the frequency of leprosy in family members was higher when the index case was an infant than when the index case was an adolescent or adult, which suggests that spread is via droplet infection. This is understandable when it is realised that the nasal mucosa and secretions of an untreated patient with lepromatous leprosy contain large numbers with *M. leprae*. The portal of entry might be the skin and spread via the nerves is usually postulated. *M. leprae* has a predilection for growing in non-myelinating Schwann cells (Shetty et al. 1988) in which it can be seen in nerve biopsies (Fig. 12.7) and grown in tissue culture. The bacilli can be found in axons but this is rare in human material and seems unlikely to be a major route of spread. In multibacillary (lepromatous) cases *M. leprae* are also present in endothelial cells, raising the alternative possibility that blood-borne spread is important. The reason why some people develop lepromatous and others tuberculoid leprosy has not been discussed. Its basis may be partly genetic since patients with tuberculoid leprosy are more likely to have the HLA DR2+ major histocompatibility antigen than those who do not (De Vries et al. 1980; van Eden et al. 1982).

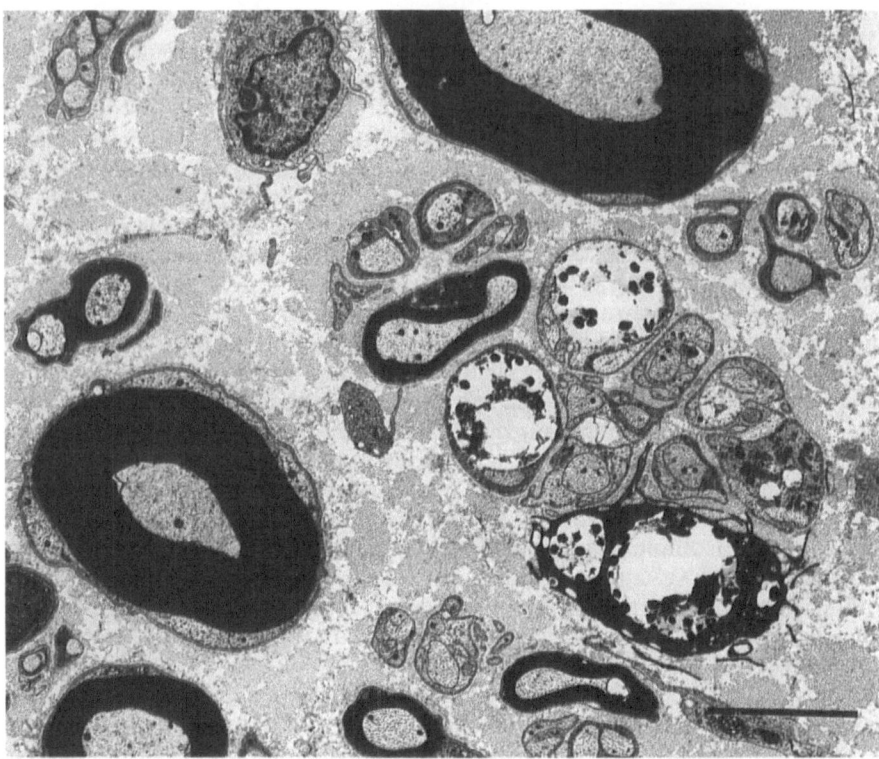

Fig. 12.7. Lepromatous leprosy. Electron micrograph showing *M. leprae* in a macrophage and in Schwann cells of unmyelinated nerve fibres. The large myelinated fibre has a thick myelin sheath for the axon size. Bar = 5 μm. (From Shetty et al. (1988), with permission.)

The nerve lesion in tuberculoid leprosy consists of the same type of granuloma as in the skin. Indeed the granuloma may merge into that in the skin with thickening of the epineurium and perineurium and infiltration of these and the endoneurium by granuloma. Both axons and myelin are destroyed by the granulomatous process and the architecture of the fascicles may be lost. In some cases the granuloma may be so aggressive that caseation occurs and nerve abscesses are formed (Sabin and Swift 1984). It is extremely difficult to find acid-fast bacilli in tuberculoid lesions.

In lepromatous leprosy nerve trunks are grossly enlarged particularly at the sites of predilection mentioned. Sabin and Swift (1984) argue that this distribution is because the sites of predilection are where long nerve trunks come near the surface so that the endoneurium is one or two degrees cooler than elsewhere. *M. leprae* grows preferentially at a temperature slightly cooler than body temperature. However, there are alternative explanations and objections to this hypothesis. Most, but not all, of the sites of predilection are also sites of compression where Schwann cell proliferation is probably more common than elsewhere. If temperature were an important factor it is surprising that leprosy should be more common in hotter parts of the world. The thickening of the nerves is due to the presence of masses of foamy macrophages in the epineurium and perineurium. The macrophages contain masses of acid-fast bacilli. In the endoneurium *M. leprae* is also present in the non-myelinating Schwann cells, in macrophages and in endothelial cells but rarely in myelinating Schwann cells or axons.

Earlier confusion about reports of demyelination occurring in leprous neuropathy has been resolved by recent teased fibre and ultrastructural studies. In most forms of leprous neuropathy a combination of axon loss and demyelination may occur. Teased fibre studies have confirmed that the demyelination is paranodal or segmental. Ultrastructurally there is no associated macrophage stripping of myelin such as is seen in GBS or EAN. There is, however, evidence of axonal shrinkage resulting in axons which are thinner than is appropriate for their myelin sheaths in transverse section and in wrinkling of the myelin sheath in longitudinal section (Shetty et al. 1988). It is surprising that the granulomatous lesion within the nerves of patients at the tuberculoid end of the spectrum does not induce autoimmunity to neural antigens and an experimental allergic neuritis-like illness, as occurs in cauda equina neuritis in horses (Chapter 11). Although patients with leprosy have increased titres of antibody to intermediate filaments they do not have demonstrable antibodies to myelin P_2 protein, the major antigen responsible for producing EAN, nor do they exhibit positive lymphocyte transformation tests to that antigen (Mshana et al. 1983a,b). There is no evidence so far that *M. leprae* directly or indirectly stimulates damaging autoimmune reactions to nerve antigens. The demyelination observed in the studies by Jacobs et al. (1987) could be explained as a secondary consequence of axonal atrophy. Axonal damage could be the consequence of release of lymphokines from activated macrophages but might merely be caused by compression and fibrosis, either as a direct pressure effect or indirectly by causing endoneurial ischaemia.

Human Immunodeficiency Virus Infection and Neuropathy

Human immunodeficiency virus (HIV) infection is associated with a wide range of neuromuscular complications which differ in nature according to the stage of disease. The commonest neuropathy is a distal sensory polyneuropathy occurring in the context of fully developed acquired immunodeficiency syndrome (AIDS). At or soon after the acquisition of HIV infection both GBS and CIDP are well recognised. Herpes zoster and vasculitis multiple mononeuropathy occur at any stage of the infection. The neuromuscular complications of HIV infection (Dalakas and Pezeshkpour 1988; De la Monte et al. 1988; Leger et al. 1989; Parry 1988) are summarised in Table 12.3. GBS and CIDP occurring in the context of HIV infection have already been considered in Chapters 5 and 11.

When GBS occurs with HIV infection, the picture may not differ clinically or electrophysiologically from GBS occurring without HIV infection. However, the CSF consistently contains an increased number of lymphocytes (Dalakas and Pezeshkpour 1988) and sural nerve biopsies show more extensive mononuclear cell infiltration than is usual in biopsies from the common form of GBS (Cornblath et al. 1987). The prognosis is variable. The author has had two cases which

Table 12.3. Neuromuscular complications of infection with human immunodeficiency virus

Clinical syndrome	Pathology	Relative frequency at different stages of infection			
		Seroconversion	HIV antibody positive only	ARC	AIDS
Distal sensory neuropathy	Axonopathy				+++
GBS	Inflammatory demyelinating	++	++	+	+
CIDP	Inflammatory demyelinating		++	++	+
Multiple mononeuropathy	Vasculitis		+	+	+
Acute sensory neuronopathy	Dorsal root ganglionitis	+			
Cauda equina syndrome	Dorsal root cytomegalovirus infection		+		
Herpes zoster			++	++	++
Polymyositis			++	+	+

+++, common (about 15%); ++, well recognised; +, rare cases reported.

have recovered without treatment. Favourable responses to plasma exchange have been suggested but it is not possible to deduce from single cases whether this represents a real therapeutic effect. Other cases are reported to have a fulminating course leading to death (Dalakas and Pezeshkpour 1988). It is not yet possible to discern whether the occurrence of GBS has any bearing on the likelihood of developing ARC (AIDS-related complex) or AIDS later.

Chronic idiopathic demyelinating polyradiculoneuropathy (CIDP) is a more common manifestation of the early stages of HIV infection. The clinical and electrophysiological features of CIDP with HIV antibodies in the serum are the same as in seronegative CIDP. Cerebrospinal fluid pleocytosis and inflammatory infiltrates in the nerves are common whereas they are usually absent in sero-negative CIDP (Cornblath et al. 1987). HIV antigen has been identified in the CSF and HIV has been isolated from nerve tissue (Dalakas and Pezeshkpour 1988; De la Monte et al. 1988).

Multiple mononeuropathy has been reported at most stages of HIV infection (Dalakas and Pezeshkpour 1988; Parry 1988; Said et al. 1988) and has sometimes been documented as due to necrotising vasculitis. Alternative causes are herpes zoster and lymphomatous infiltration of the meninges, roots, plexuses or nerves. The clinical distinction between an asymmetrical demyelinating polyradiculo-neuropathy and a vasculitic multiple mononeuropathy is sometimes blurred. Furthermore, the intensity of the inflammation in some cases of CIDP is so great that the histological distinction from vasculitis is difficult and the picture of vasculitic multiple mononeuropathy may evolve into CIDP (Parry 1988).

A personal case of painful sensory neuropathy in a man with HIV antibodies and generalised lymphadenopathy but no immunodeficiency was slightly dif-ferent from any of these clinical pictures and demonstrates this difficulty of distinguishing inflammatory from vasculitic neuropathy. Although the symptoms

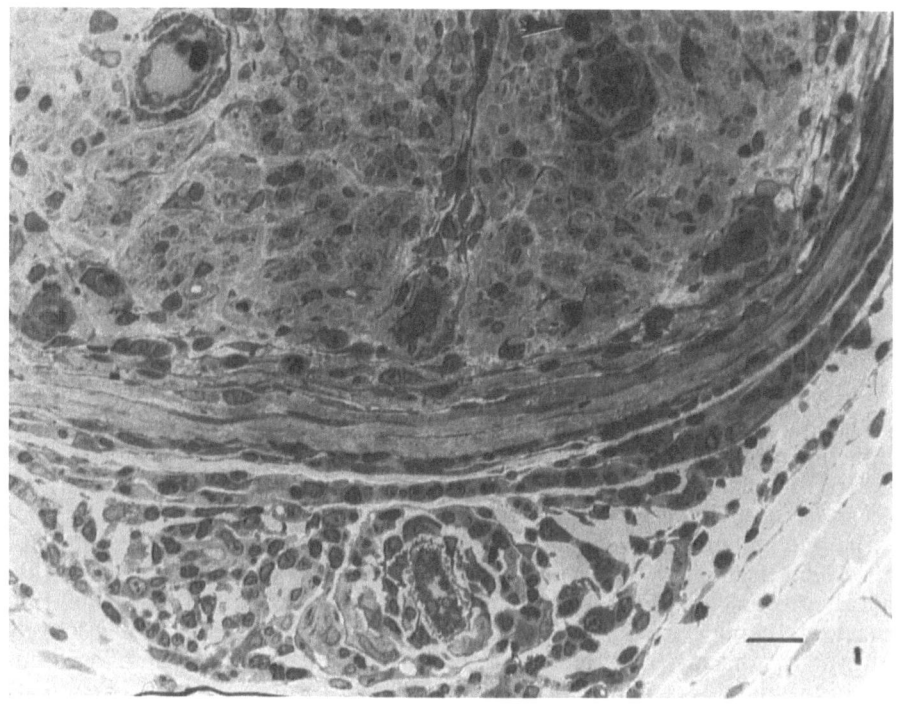

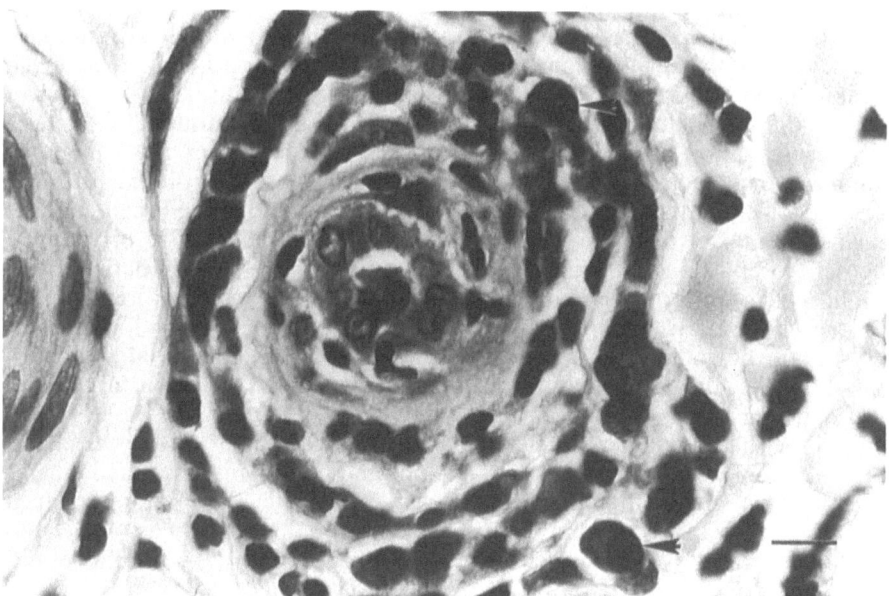

Fig. 12.8. Painful sensory neuropathy in a patient with HIV antibodies but without AIDS. **a** Thionin and acridine orange-stained resin section showing almost complete loss of myelinated nerve fibres, ongoing Wallerian degeneration (*arrowhead*) and perivascular infiltration around an epineurial vessel. Bar = 40 μm. **b** Haematoxylin and eosin-stained paraffin section showing perivascular infiltration around an epineurial vessel. Note plasma cells (*arrowheads*). Bar = 10 μm.

and signs were predominantly sensory there was electrophysiological evidence of slowing of motor nerve conduction indicating demyelination. The predominant change in a sural nerve biopsy was of nerve fibre depletion with very extensive intense endoneurial inflammatory cell infiltration including large numbers of plasma cells (Fig. 12.8). Although fibrinoid necrosis was absent, the inflammation involved not only the perivascular space but also the vessel wall so as to suggest an ongoing vasculitis.

A single case of acute sensory neuronopathy has been reported to occur after a febrile illness probably due to HIV seroconversion (Elder et al. 1986).

An unusual syndrome of progressive lumbosacral polyradiculopathy, unlike any clinical picture commonly encountered in the absence of HIV infection, has been reported in eight patients with AIDS (Jeantils et al. 1986; Eidelberg et al. 1986; Behar et al. 1987) and also in asymptomatic patients with HIV antibodies (Crawford et al. 1987). There is intense inflammation of affected nerves and nerve roots with typical intranuclear and intracytoplasmic inclusions in Schwann cells and endothelial cells indicating infection by cytomegalovirus. Small endoneurial and epineurial vessels exhibited focal necrosis, inflammatory cell infiltration and segmental occlusion, clear evidence of vasculitis. Although the clinical syndrome predominantly affects the lumbar and sacral segments, the thoracic and cervical roots were also shown to be affected at autopsy. It is not yet clear whether this rare syndrome represents a separate disorder, or is a localised form of a generalised multiple vasculitic mononeuropathy. It seems likely that the nerve damage is caused by an inflammatory response to endoneurial CMV infection. Since CMV is present in endothelial cells and Schwann cells it is scarcely surprising that both are damaged.

Sensory polyneuropathy affects 10%–30% of patients with AIDS and also occurs with AIDS-related complex (Snider et al. 1983; Lipkin et al. 1985; Cornblath and McArthur 1988). It causes pain and paraesthesiae in the soles of the feet so that walking becomes uncomfortable. Distal impairment of pain and vibration sensation with diminished or absent ankle reflexes are commonly found but power is preserved. Electrophysiologically absent or diminished sural nerve action potentials indicate an axonal neuropathy. The CSF cell count is usually normal but the protein concentration is increased in 50% of patients. There is little published information on the pathology of nerves in patients with painful sensory neuropathy associated with AIDS. The one sural nerve in the series of Snider et al. (1983) was normal. In two patients studied by Lipkin et al. (1985) both showed axonal neuropathy, in one case with epineurial and endoneurial perivascular inflammation. Nine patients with AIDS and distal symmetrical polyneuropathy (not necessarily purely sensory) studied by De la Monte et al. (1988) had evidence at autopsy of patchy involvement of the peripheral nervous system with loss of myelin and axons and only slight associated inflammation. From the paraffin-embedded material available the authors deduced that moderate or severe myelin loss was present, but this was usually associated with axonal degeneration to which it was probably secondary. It seems likely and is generally presumed that the AIDS-related sensory neuropathy is an axonopathy caused by dying back of the sensory axons. There is post-mortem evidence of gracile tract degeneration in some patients with the syndrome (Rance et al. 1988), indicating that the dying back also affects the central process of the dorsal root ganglion cell. The dying back could be due to infection of the dorsal root ganglion neuron by HIV but this has yet to be demonstrated.

Summary

All forms of vasculitis may affect the PNS causing either multiple mononeuro-pathy or symmetrical polyneuropathy. Nerve fibres are damaged causing axonal degeneration and a non-uniform pattern of nerve fibre loss affecting the centres of nerve fascicles, and affecting some fascicles more than others. Generalised peripheral neuropathy in hypereosinophilic syndrome, rheumatoid arthritis and SLE is usually due to vasculitis. Rarely, neuropathy may be due to vasculitis apparently confined to peripheral nerves.

Sarcoidosis usually causes a multiple mononeuropathy, affecting the cranial more than the spinal nerves. The underlying pathology has sometimes been shown to be sarcoid granulomas in the endoneurium and epineurium.

Peripheral neuropathy may be associated with paraproteinaemia in several ways, including a syndrome of progressive sensory and motor demyelinating neuropathy with postural tremor. This is associated with IgM paraprotein of undetermined significance (benign monoclonal gammopathy). The IgM para-protein is an antibody to an epitope on myelin-associated glycoprotein, and on acidic glycolipids which are present in PNS but not CNS myelin.

Patients with multifocal demyelinating motor neuropathy sometimes have antibodies to gangliosides, and may respond to immunosuppressive treatment.

In lepromatous and tuberculoid leprosy there is loss of nerve fibres in affected nerves. Associated demyelination is due to axonal shrinkage and autoimmune mechanisms have not been demonstrated.

It is sad to conclude this book with an account of a form and cause of neuropathy which was not heard of eight years ago. Nevertheless the occurrence of GBS and CIDP in the early stages of HIV infection raises fresh questions about the relationship between infection of endoneurial tissue, autoimmunity to neural antigens and mechanisms of demyelination. The answers may shed further light on the cause of peripheral and, perhaps, central nervous system demyelinating disease.

References

Adams CWM, Buk SJA, Hughes RAC, Leibowitz S, Sinclair E (1989) Perl's ferrocyanide test for iron in the diagnosis of vasculitic neuropathy. Neuropathol Appl Neurobiol 15:443–440

Bailey AA, Sayre GP, Clarke EC (1956) Neuritis associated with SLE. Arch Neurol Psychiatry 75:251–259

Behar R, Wiley C, McCutchan JA (1987) Cytomegalovirus polyradiculoneuropathy in acquired immune deficiency syndrome. Neurology 37:557–561

Bloch SL, Jarret MP, Swerdlow M, Grayzel AI (1979) Brachial plexus neuropathy as the initial presentation of systematic lupus erythematosus. Neurology 29:1633–1634

Bosch EP (1982) Peripheral neuropathy associated with monoclonal gammopathy. Studies of intraneural injections of monoclonal immunoglobulin sera. J Neuropathol Exp Neurol 41:446–459

Bryceson A, Pfaltzgraff RF (1979) Leprosy, 2nd edn. Churchill Livingstone, Edinburgh

Caselli RJ, Daube JR, Hunder GG, Whisnant JP (1988) Peripheral neuropathic syndromes in giant cell (temporal) arteritis. Neurology 38:685–690

Case records of the Massachussetts General Hospital (1987) Periarteritis nodosa with involvement of peripheral nerves. N Engl J Med 316:1139–1147

Chatterjee BR, Taylor CE, Thomas J, Naidu GN (1976) Acid fast bacillary positivity in asymptomatic individuals in leprosy endemic villages around Jhalda in West Bengal. Lepr India 480: 119–131

Chazot G, Berger B, Carrier H (1976) Manifestations neurologiques des gammopathies monoclonales. Rev Neurol 132:193–202

Churg J, Strauss L (1951) Allergic granulomatosis, allergic angitis and periarteritis nodosa. Am J Pathol 27:277–294

Chusid MJ, Dale DC, West BC, Wolff SM (1975) The hypereosinophilic syndrome: analysis of 14 cases with review of the literature. Medicine 54:1–27

Clamp JR (1967) Some aspects of the first recorded case of multiple myeloma. Lancet ii:1354–1356

Conn DL, Dyck PJ (1984) Angiopathic neuropathy in connective tissue diseases. In: Dyck PJ, Thomas PK, Lambert EH, Bunge RH (eds) Peripheral neuropathy, 2nd edn. WB Saunders, Philadelphia

Contamin F, Singer B, Mignot R, Eloffet M, Kazatchkine M (1976) Polyneuropathie à rechutes évolvuant depuis 19 ans, associée à une gammopathie monoclonale IgG benigne. Effet favorable de la corticothérapie. Rev Neurol 132:741–762

Copeman WSC (1964) Textbook of the rheumatic diseases, 3rd edn. Livingstone, Edinburgh, pp 174–239

Cornblath DR, McArthur JC (1988) Predominantly sensory neuropathy in patients with AIDS and AIDS-related complex. Neurology 38:794–796

Cornblath DR, McArthur JC, Kennedy PGE, Witte AS, Griffin JW (1987) Inflammatory demyelinating peripheral neuropathies associated with human T-cell lymphotropic virus type III infection. Ann Neurol 21:32–40

Crawford EJP, Baird PRE, Clark AL (1987) Cauda equina and lumbar root compression in patients with AIDS. J Bone Jt Surg 69:36–37

Cream JJ, Hern JEC, Hughes RAC, MacKenzie ICK (1974) Mixed or immune complex cryoglobulinaemia and neuropathy. J Neurol Neurosurg Psychiatry 37:82–97

Dalakas MC, Engel WK (1981) Polyneuropathy with monoclonal gammopathy. Ann Neurol 10:45–52

Dalakas MC, Pezeshkpour GH (1988) Neuromuscular diseases associated with human immunodeficiency virus infection. Ann Neurol 23:s38–s48

De la Monte SM, Gabuzda DH, Ho DH et al. (1988) Peripheral neuropathy in the acquired immunodeficiency syndrome. Ann Neurol 23:485–492

De Vries RRP, Mehra NK, Vaidya MC, Gupta MD, Meera Kahn P, Van Rood JJ (1980) HLA-linked control of susceptibility to tuberculoid leprosy and association with HLA-DR types. Tissue Antigens 294–304

Dellagi K, Dupoue Y, Brouet JC et al. (1983) Waldenstrom's macroglobulinaemia and peripheral neuropathy: a clinical and immunologic study of 25 patients. Blood 62:280–285

Dorfman LJ, Ransom BR, Forno LS, Kelts A (1983) Neuropathy in the hypereosinophilic syndrome. Muscle Nerve 6:291–298

Drachman DA (1963) Neurological complications of Wegener's Granulomatosis. Arch Neurol 8:145–155

Dubois EL (1974) Lupus erythematosus, 2nd edn. University of South California Press, Los Angeles

Dyck PJ, Conn DL, Okazaki H (1972) Three-dimensional morphology of fiber degeneration related to sites of occluded vessels. Mayo Clin Proc 47:461–475

Dyck PJ, Berstead TJ, Conn DL, Stevens JC, Windebank AJ, Low PA (1987) Non-systemic vasculitic neuropathy. Brain 110:843–854

Editorial (1972) Wegener's granulomatosis. Lancet ii:519

Editorial (1988) Chemotherapy of leprosy. Lancet ii:487–488

Eidelberg D, Sotrel A, Vogel H, Walker P, Keefield J, Crumpacker CS (1986) Progressive polyradiculopathy in AIDS. Neurology 36:912–916

Elder G, Dalakas M, Pezeshkpour G, Sever J (1986) Ataxic neuropathy due to ganglioneuronitis after probable acute human immunodeficiency virus infection. Lancet ii:1275–1276

Falk RJ, Jennette JC (1988) Anti-neutrophil cytoplasmic auto-antibodies with specificity for myeloperoxidase in patients with systemic vasculitis and idiopathic necrotizing crescentic glomerulonephritis. N Engl J Med 318:1651–1657

Fauci AS, Haynes BF, Katz P (1978) The spectrum of vasculitis: clinical, pathologic, immunological and therapeutic considerations. Ann Intern Med 89:660–676

Fauci AS, Katz P, Haynes BF, Wolff SM (1979) Cyclophosphamide therapy of severe systemic necrotizing vasculitis. N Engl J Med 301:235–238

Feinglass EJ, Arnett FC, Dorsch CA, Thomas MZ, Stevens MB (1976) Neuropsychiatric manifestations of SLE: diagnosis, clinical spectrum and relationship to other features of the disease. Medicine 55:323–339

Fukase M, Tsunematsu T, Nishitani H et al. (1969) Report of a case of solitary plasmacytoma in the abdomen presenting with polyneuropathy and endocrinological disorders. Clin Neurol 9:657

Gainsborough N, Hall SM, Hughes RAC, Leibowitz S (1990) Sarcoid neuropathy. J Neurol (in press)

Galassi G, Gibertoni M, Mancini A et al. (1984) Sarcoidosis of the peripheral nerve: clinical electrophysiological and histological study of two cases. Eur Neurol 23:459–465

Gardner-Thorpe C, Harriman DGF, Parsons M, Rudge P (1971) Löffler's eosinophilic endocarditis with Balint's syndrome (optic ataxia and paralysis of visual fixation). Q J Med 40:249–260

Garland HG, Thompson JG (1953) Uveoparotid tuberculosis (Febris uveoparotidea of Heerfordt). Q J Med 26:157–160

Gibson T, Myers AR (1976) Nervous system involvement in SLE. Ann Rheum Dis 35:398–406

Gocke DJ, Hso K, Morgan C, Bombardien S, Lockskin M, Christian CL (1971) Vasculitis in association with Australia antigen. J Exp Med 134 suppl:330–336

Goetz CG (1980) Polyarteritis nodosa. Handbook Clin Neurol 39:295–311

Grisold W, Jellinger K (1985) Multifocal neuropathy with vasculitis in hypereosinophilic syndrome. An entity or drug induced effect. J Neurol 231:301–306

Haas DC, Tatum AH (1988) Plasmapheresis alleviates neuropathy accompanying IgM anti-MAG paraproteinaemia. Ann Neurol 23:394–396

Harati Y, Niakan E (1986) The clinical spectrum of inflammatory-angiopathic neuropathy. J Neurol Neurosurg Psychiatry 49:1313–1318

Hays AP, Latov N, Takatsu M, Sherman WH (1987) Experimental demyelination of nerve induced by serum of patients with neuropathy and an anti MAG M protein. Neurology 37:242–246

Heerfordt CF (1909) Ueber eine Febris uveo-parotidea subchronica. Graefes Arch Ophthalmol 70:254–273

Heptinstall RH, Sowry GSC (1952) Peripheral neuritis in systemic lupus erythematosus. Br Med J i:525–527

Hughes RAC, Cameron JS, Hall SM, Heaton J, Payan JA, Teoh R (1982) Multiple mononeuropathy as the initial presentation of systemic lupus erythematosus – nerve biopsy and response to plasma exchange. J Neurol 228:239–247

Ito H, Latov N (1988) Monoclonal IgM in two patients with MND bind to the carbohydrate antigens Gal(beta 1–3)GalNAc and Gal(beta 1–3)GlcNAc. J Neuroimmunol 19:245–254

Jacobs JM, Shetty VP, Antia NH (1987) Teased fibre studies in leprous neuropathy. J Neurol Sci 79:310–314

Jeantils V, Lemaitre MO, Robert J et al. (1986) Subacute polyneuropathy with encephalopathy in AIDS with human cytomegalovirus pathogenicity. Lancet ii:1039

Johnson RT, Richardson EP (1968) The neurological manifestations of systemic lupus erythematosus. A clinico-pathological study of 24 cases and review of the literature. Medicine 47:337–369

Kahn SN, Riches PG, Kohn J (1980) Paraproteinaemia in neurological disease: incidence, associations and classification of monoclonal immunoglobulins. J Clin Pathol 33:617–621

Kaltreider HB, Talal N (1969) The neuropathy of Sjogren's syndrome: trigeminal nerve involvement. Ann Intern Med 70:751–762

Kauffmann RH, Herrmann WA, Meijer CJLM, Daha MR, Van Es LA (1980) Circulating and tissue bound immune complexes in allergic vasculitis: relationship between immunoglobulin class and clinical features. Clin Exp Immunol 41:459–471

Kelly JJ, Kyle RA, Miles JM, O'Brien PC, Dyck PJ (1981a) The spectrum of peripheral neuropathy in myeloma. Neurology 31:24–31

Kelly JJ, Kyle RA, O'Brien PC, Dyck PJ (1981b) The prevalence of monoclonal gammopathy in peripheral neuropathy. Neurology 31:1480–1483

Kelly JJ, Kyle RA, Miles JM, Dyck PJ (1983) Osteosclerotic myeloma and peripheral neuropathy. Neurology 33:202–210

Kelly JJ, Kyle RA, Latov N (1987) Polyneuropathies associated with plasma cell dyscrasias. Martinus Nijhoff, Boston

King RHM, Thomas PK (1984) The occurrence and significance of myelin with unusually large periodicity. Acta Neuropathol 63:319–329

Kissel JT, Slivka AP, Warmolts JR, Mendell JR (1985) The clinical spectrum of necrotising angiopathy of the peripheral nervous system. Ann Neurol 18:251–257

Klemperer P (1947) Diseases of the collagen system. NY Acad Med Bull 23:581–594

Kussmaul A, Maier R (1866) Ueber eine bisher nicht beschriebene eigentumliche Arterienerkrankung (Periarteritis nodosa) die mit Morbus Brightii und rapid fortschreitender allgemeiner Muskellahmung einhergeht. Dtsch Arch Klin Med 1:484-517

Lamaria J, Casquero P, Pou A (1987) Mononeuritis multiplex in Waldenstrom's macroglobulinaemia. Ann Neurol 22:268-272

Latov N (1981) Plasma cell dyscrasia and peripheral neuropathy: identification of the myelin antigens that react with human paraproteins. Proc Natl Acad Sci USA 78:7139-7142

Latov N, Sherman WH, Nemni R et al. (1980) Plasma cell dyscrasia and peripheral neuropathy with a monoclonal antibody to peripheral nerve myelin. N Engl J Med 303:618-621

Latov N, Hays AP, Donofrio PD et al. (1988) Monoclonal IgM with unique specificity to gangliosides GM_1 and GD_{1b} and to lacto-N-tetraose associated with human motor neuron disease. Neurology 38:763-768

Lecky BRF, Hughes RAC, Murray NMF (1987) Trigeminal sensory neuropathy. A study of 22 cases. Brain 110:1463-1485

Leger JM, Bouche P, Bolgert F et al. (1989) The spectrum of polyneuropathies in patients infected with HIV. J Neurol Neurosurg Psychiatry 52:1369-1374

Leibowitz S, Hughes RAC (1983) Immunology of the nervous system. Edward Arnold, London, pp 1-304

Leibowitz S, Gregson NA, Kennedy M, Kahn SN (1983) IgM paraproteins with immunological specificity for a Schwann cell component and peripheral nerve myelin in patients with polyneuropathy. J Neurol Sci 59:153-165

Leluyer B, Devaux AM, Dailly R, Ensel P (1983) Polyradiculonévrite révélatrice d'une sarcoidose de l'enfant. Arch Fr Pediatr 40:175-178

Lipkin WI, Parry G, Kiprov D, Abrams D (1985) Inflammatory neuropathy in homosexual men with lymphadenopathy. Neurology 35:1479-1483

Logothetis J, Silverstein P, Coe J (1960) Neurologic aspects of Waldenstrom's macroglobulinaemia. Arch Neurol 3:564-573

MacFadyen DJ (1960) Wegener's granulomatosis with discrete lung lesions and peripheral neuritis. Can Med Assoc J 83:760-763

Matthews WB (1984) Sarcoid neuropathy. In: Dyck PJ, Thomas PK, Lambert EH (eds) Peripheral neuropathy. Saunders, Philadelphia, pp 2018-2026

Matthews WB (1987) Neuropathies. In: Matthews WB (ed) Handbook of clinical neurology. Elsevier, Amsterdam, pp 195-198

McGinnis S, Kohriyama T, Yu RK, Pesce MA, Latov N (1988) Antibodies to sulfated glucuronic acid containing glycosphingolipids in neuropathy associated with anti-MAG antibodies and in normal subjects. J Neuroimmunol 17:119-126

Meier C, Vandevelde M, Steck A, Zurbriggen A (1983) Demyelinating polyneuropathy associated with monoclonal IgM paraproteinaemia. J Neurol Sci 63:353-367

Meier C (1985) Polyneuropathy in paraproteinaemia. J Neurol 232:202-214

Mellgren SI, Conn DL, Stevens JC, Dyck PJ (1989) Peripheral neuropathy in primary Sjogren's syndrome. Neurology 39:390-394

Mendell JR, Sahenk Z, Whitaker JN et al. (1985) Polyneuropathy and IgM monoclonal gammopathy: studies on the pathogenic role of anti-MAG antibody. Ann Neurol 17:243-254

Milanese C, Procaccia S, La Mantia L, Guzzetti E, Corridori F (1982) Peripheral neuropathy and solitary myeloma: analysis of serum and CSF IgG in two cases. J Neurol Neurosurg Psychiatry 45:468-470

Mshana RN, Humber DP, Harboe M, Belehu A (1983a) Immune responses to bovine neural antigens in leprosy patients. II. Absence of in vitro lymphocyte stimulation to peripheral nerve myelin proteins. Lepr Rev 54:217-227

Mshana RN, Harboe M, Stoner GL, Hughes RAC, Kadlubowski M, Belehu A (1983b) Immune responses to bovine neural antigens in leprosy patients. Absence of antibodies to an isolated myelin protein. Int J Lepr 51:33-40

Nakanski T, Sobue I, Toyokura Y et al. (1984) The Crow-Fukase syndrome: a study of 102 cases in Japan. Neurology 34:712-720

Nardelli E, Steck AJ, Barkas T, Schleup M, Jerusalem F (1988) Motor neuron syndrome and monoclonal IgM with antibody activity against gangliosides GM_1 and GD_{1b}. Ann Neurol 23: 524-528

Neild GH, Williams DG (1987) Polyarteritis and related syndromes. In: Weatherall DJ, Ledingham JGG, Warrell DA (eds) Oxford textbook of medicine, 2nd edn. Oxford University Press, Oxford, pp 16.28-16.34

Nemni R, Galassi G, Cohen M et al. (1981) Symmetric sarcoid polyneuropathy: analysis of a sural nerve biopsy. Neurology 31:1217-1223

Noring L, Kjellin KG, Siden A (1980) Neuropathies associated with disorders of plasmacytes. Eur

Neurol 19:224–230

Oh SJ (1979) Sarcoid polyneuropathy: a histologically proved case. Ann Neurol 7:178–181

Ohnishi A, Hikano A (1981) Uncompacted myelin lamellae in dysglobulinaemic neuropathy. J Neurol Sci 51:131–140

Osler LD, Sidell AD (1960) The Guillain–Barré syndrome: the need for exact diagnostic criteria. N Engl J Med 262:964–969

Pallis CA, Scott JT (1965) Peripheral neuropathy in rheumatoid arthritis. Br Med J i:1141–1147

Parry GJ (1988) Peripheral neuropathies associated with human immunodeficiency virus infection. Ann Neurol 23:s49–s53

Pestronk A, Adams RN, Clawson L et al. (1988a) Serum antibodies to GM_1 ganglioside in amyotrophic lateral sclerosis. Neurology 38:1457–1461

Pestronk A, Cornblath DR, Ilyas AA et al. (1988b) A treatable multifocal motor neuropathy with antibodies to GM_1 ganglioside. Ann Neurol 24:73–78

Pollock M, Nukada H, Taylor P (1983) Comparison between fascicular and whole nerve biopsy. Ann Neurol 13:65–68

Powell HC, Rodriguez M, Hughes RAC (1984) Microangiopathy in dysglobulinaemic neuropathy. Ann Neurol 15:386–394

Rance NE, McArthur JC, Cornblath DR, Landstrom DL, Griffin JW, Price DL (1988) Gracile tract degeneration in patients with sensory neuropathy and AIDS. Neurology 38:265–271

Read DJ, Banhegan RI, Matthews WB (1978) Peripheral neuropathy and benign IgG paraproteinaemia. J Neurol Neurosurg Psychiatry 41:215–219

Rechthand E, Cornblath DR, Stern BJ, Meyerhoff JO (1984) Chronic demyelinating polyneuropathy in systematic lupus erythematosus. Neurology 34:1375–1377

Richardson EP (1980) Systemic lupus erythematosus. In: Vinken PJ, Bruyn GW, Klawans HL (eds) Handbook of clinical neurology, 39. Neurological manifestations of systemic diseases. North-Holland, Amsterdam, pp 273–293

Ridley DS, Jopling WH (1966) Classification of leprosy according to immunity: a five-group system. Int J Lepr 34:255–273

Rinne UK (1967) Neurologische Manifestationen der Sarkoidose. Dtsch Z Nervenheilk 191:245

Rowland LP, Defendini R, Sherman WH et al. (1982) Macroglobulinaemia with peripheral neuropathy simulating motor neuron disease. Ann Neurol 11:532–536

Rudnicki S, Chad DA, Drachman DA, Smith TW, Anwer HE, Levitan N (1987) Motor neuron disease and paraproteinaemia. Neurology 37:335–337

Sabin TD, Swift TR (1984) Leprosy. In: Dyck PJ, Thomas PK, Lambert EH, Bunge R (eds) Peripheral neuropathy, 2nd edn. Saunders, Philadelphia, pp 958–987

Sadeh M, Sarova-Pinhas I, Ohry A (1980) Sensory ataxia as an initial symptom of SLE. J Rheumatol 3:420–421

Sahenk Z, Mendell JR, Rossio JL, Hurtubise P (1977) Polyradiculoneuropathy accompanying procainamide-induced lupus erythematosus: evidence for drug-induced enhanced sensitization to peripheral nerve myelin. Ann Neurol 1:378–384

Said G, Lacroix-ciaudo C, Fujimura H, Blas C, Faux N (1988) The peripheral neuropathy of necrotising arteritis: a clinicopathological study. Ann Neurol 23:461–465

Scadding JG, Mitchell DN (1985) Sarcoidosis. Chapman and Hall, London, pp 309–310

Schott B, Michel D, Lejeune E et al. (1968) Polyradiculonévrites au cours de la maladie de Besnier–Boeck–Schaumann. J Méd Lyon 49:931

Sewell HF, Matthews JB, Gooch E et al. (1981) Autoantibody to nerve tissue in a patient with peripheral neuropathy and an IgG paraprotein. J Clin Pathol 34:1163–1166

Sherman WH, Latov N, Hays AP et al. (1983) Monoclonal IgM^k antibody precipitating with chondroitin sulfate C from patients with axonal polyneuropathy and epidermolysis. Neurology 33:192–201

Sherman WH, Olarte MR, McKieran G, Sweeney K, Latov N, Hays AP (1984) Plasma exchange treatment of peripheral neuropathy associated with plasma cell dyscrasia. J Neurol Neurosurg Psychiatry 47:813–819

Shetty VP, Antia NH, Jacobs JM (1988) The pathology of early leprous neuropathy. J Neurol Sci 88:115–132

Shy ME, Heiman-Patterson T, Parry GJ, Tahmoush AJ, Evans VA, Schick PK (1988) Motor neuronopathy in a patient with autoantibodies against gangliosides GM_1 and GD_{1b}. Improvement following immunotherapy. Neurology 38 suppl:352

Sidiq M, Kirsner AB, Sheon RP (1972) Carpal tunnel syndrome: first manifestation of SLE. JAMA 222:1416–1417

Silverstein A, Doniger DE (1963) Neurologic complications of myeloma. Arch Neurol 9:534–544

Smith IS, Kahn SN, Lacey BW et al. (1983) Chronic demyelinating neuropathy associated with benign IgM paraproteinaemia. Brain 106:169–195

Snider WD, Simpson DM, Nielson G et al. (1983) Neurological complications of AIDS: analysis of 50 patients. Ann Neurol 14:403–418

Solomons REB (1982) Plasma cell dyscrasia with polyneuropathy, organomegaly, endocrinopathy, monoclonal gammopathy and skin changes; the POEMS syndrome. J R Soc Med 75:553–555

Stern G (1980) Wegener's granulomatosis. Handbook Clin Neurol 39:343–345

Strickland GT, Moser KM (1967) Sarcoidosis with a Landry–Guillain–Barré syndrome and clinical response to corticosteroids. Am J Med 43:131–135

van Eden W, De Vies RRP, D'Amaro J, Schrender I, Lecker DL, Van Rood JJ (1982) HLA DR associated genetic control of the type of leprosy in a population from Surinam. Hum Immunol 4:343–350

Vaughan JH, Barnett EV, Leadley PJ (1967) Serum sickness: evidence in man of antigen–antibody complexes and free light chains in the circulation during the acute reaction. Ann Intern Med 67:596–602

Vital C, Aubertin J, Ragnault JM, Aimgues H, Mouton L, Bellance R (1982) Sarcoidosis of the peripheral nerve: a histological study of two cases. Acta Neuropathol (Berl) 58:111–114

Vital C, Brechenmacher C, Reiffers J et al. (1983) Uncompacted myelin lamellae in two cases of peripheral neuropathy. Acta Neuropathol 60:252–256

Waters MFR, Ledingham JGG, Warrell DA (1987) Leprosy (Hansen's Disease Hanseniasis). In: Weatherall DJ (ed) Oxford textbook of medicine. Oxford University Press, Oxford

Wegener F (1936) Ueber generalisierte septische Gefässerkrankungen. Verh Dtsch Ges Pathol 29:202–231

Weller RO, Bruckner FE, Chamberlain MA (1970) Rheumatoid neuropathy: a histological and electrophysiological study. J Neurol Neurosurg Psychiatry 33:592–604

Wiederholt WC, Siekert RG (1965) Neurological manifestations of sarcoidosis. Neurology 15:1147

Zeek PM (1953) Periarteritis nodosa and other forms of necrotizing angiitis. N Engl J Med 248: 764–772

Guillain–Barré Syndrome: A Short Guide for the Patient, Relative and Friend*

Introduction

This guide is written by a doctor for patients who have been told that they may have Guillain–Barré syndrome and their relatives and friends. It has to be honest and is meant to be reassuring. Unfortunately words cannot talk back and the writer cannot respond to the concern which this information may unintentionally arouse. Consequently if you do not understand or are worried by the information offered here, do ask your doctor to explain.

What is Guillain–Barré syndrome?

Guillain–Barré syndrome (GBS) is an uncommon illness causing weakness and loss of sensation which usually recovers completely after a few weeks or months. It is named after two French physicians, Guillain (pronounced Ghee-Yun) and Barré (pronounced Bah-Rray), who described it in 1916 in two soldiers who got better! It affects one person in 2000 each year, i.e. 1000 altogether each year in the United Kingdom. It can occur at any age from infancy onwards but is slightly more common in the old. It is more common in men than women. It is not hereditary: it is not passed on to your children. It is not infectious: it is not caught from or transmitted to anybody else. However, it does often develop a week or two after a throat or intestinal infection.

Early Symptoms

The first symptoms are usually either tingling (pins and needles) or loss of feeling (numbness) beginning in the toes and fingers or weakness so that legs feel heavy and wooden, arms feel limp and hands cannot grip or turn things properly. These symptoms may remain mild and clear up within a week or two without need for hospital admission but most people need to be admitted to hospital. At the earliest stage it may be difficult for the patient to persuade the doctor

---

* Reproduced with permission of The British Guillain–Barré Syndrome Support Group.

that there is anything physical wrong. Within a few days it is all too obvious that something has gone wrong because legs simply will not bear weight, and arms become terribly weak and the doctor finds that the tendon reflexes have disappeared.

Investigations

The diagnosis of GBS is made on clinical grounds, not laboratory tests. This means that the doctor has to rely on the history and clinical examination fitting into the pattern of GBS. The doctor will particularly want to know of any recent possible toxin exposure (insecticides, solvents), alcohol intake, tick bites, family history of nerve disease, or symptoms of any coincidental illnesses, such as diabetes (thirst, frequent urination, weight loss): any of these *might* lead to a different diagnosis. Investigations will include blood tests and usually a lumbar puncture and electromyogram (EMG). The lumbar puncture involves lying on one side and having a needle inserted under local anaesthesia between the vertebrae into the sac of cerebrospinal fluid which surrounds the nerve roots. The idea is worse than the procedure really is and it does not usually hurt (the writer has had one). The cerebrospinal fluid often contains much more protein than usual while the cell content remains normal. If different changes are found the doctor has to review the diagnosis with even more care. The electromyogram or EMG is an electrical recording of the muscle activity. If a nerve is stimulated with a brief electrical pulse (felt like a sharp tap or jolt), muscle activity can be recorded and the speed of nerve conduction worked out. Often in GBS nerve conduction is slowed or even blocked. The test usually lasts about half an hour. It is only slightly uncomfortable and quite harmless.

Plateau Phase

The worst is usually reached within two or three weeks. After this begins the plateau phase, which usually lasts a few days or weeks, during which the course of the disease seems stationary. Most people are so weak during this stage that they are confined to bed and rest is probably a good thing. However, it is very important to keep all the joints moving through a full range to stop them stiffening up. The physiotherapist is in charge of this physical therapy and may be pleased to supervise relatives and friends in what they can do to help. Unfortunately some patients get a lot of pain during GBS. This may come from the inflammation of the nerves themselves, from the muscles which have temporarily lost their nerve supply, from stiff joints or simply because the patient is lying in an uncomfortable posture and is too weak to move into a more comfortable position. To combat the pain, the doctors will prescribe pain-killers and the nurses and physiotherapists will help with repositioning and physical therapy. It helps to know that some pain is common in GBS, that the pain will disappear as the condition improves, and that the occurrence of pain does not mean anything else is going wrong.

The Intensive Care Unit

This section is directed only towards the few patients that are so severely affected that they are transferred to an intensive care unit. About 20% of patients have weakness of the breathing muscles and have to be placed on a machine called a ventilator which will take over their breathing. This is less worrying than it sounds because it is absolutely routine for all intensive care units to take over breathing with an artificial ventilator. The connection to the ventilator is made under a short general anaesthetic to a tube in the windpipe (trachea) via the nose or mouth. This tube, called an endotracheal tube, can be left in place for a week or two. If artificial ventilation is required for longer a surgeon may make a small opening, called a tracheostomy, into the windpipe at the base of the neck which is much more comfortable and permits artificial ventilation for as long as necessary. The tracheostomy is also performed under a general anaesthetic. Fortunately in GBS artificial ventilation is rarely necessary for more than a few weeks, and most people, 80%, do not need artificial ventilation at all. When ventilation is no longer needed, the tracheostomy tube can be removed quite painlessly. The wound closes in a few days eventually leaving a faint scar below the line of a collar.

Intensive care has become a very sophisticated part of medicine which has enormously improved the care of severe GBS. To make this possible pulse, blood pressure, temperature and blood chemistry have to be measured frequently. The pulse will be recorded by monitoring the heart beat, with an electrocardiogram, on a television screen to detect abnormalities which may need treatment. Patients may need infusions into veins to provide fluids and give drugs and a tube called a catheter in the bladder to drain the urine. A tube, called a nasogastric tube, may be passed from the nose into the stomach because swallowing may be difficult or impossible and food is still necessary. Constipation can be a troublesome problem at first, but eventually nurses and patients invariably work out a regime of laxatives and suppositories which works. Communication can be a problem for a patient who is unable to talk but with blinks, nods, communication boards and above all patience it is almost always possible to get the message across. If the intensive care regime seems tedious, it is worth remembering that modern intensive care has made death from GBS a very rare event indeed.

Recovery Phase

Eventually the numbness begins to recede and strength begins to come back. Once it is clear that this is a genuine improvement rather than wishful thinking, there is cause for general rejoicing because improvement is likely to continue steadily and two-thirds of patients recover completely. The time taken for recovery to occur is very variable. Sometimes it is only a week or two but most people remain affected for between three and six months.

The one-third of patients who do not recover completely may be left with minor degrees of weakness, numbness and sometimes discomfort which do not usually seriously interfere with their lives. Some, however, are left so disabled that they cannot resume their former occupation. This is usually because of residual weakness of their hands and ankles so that strenuous manual work and walking are impaired. It is very rare to be left dependent on a wheelchair. Improvement is fastest during the first few months but some patients report continued gradual improvement even after a year has elapsed.

What Causes Guillain–Barré Syndrome?

The disease is due to inflammation of the peripheral nerves or "neuritis". The peripheral nerves connect the central nervous system to the skin and muscles. Because *many* nerves are inflamed GBS is called a "*poly*neuritis". Because it often occurs after an infection it is sometimes called "post-infective polyneuritis". The most likely explanation for the inflammation is that the immune cells, called lymphocytes, make a mistake and attack the nerves. They may have been tricked into making this mistake by the previous infection. Eventually the immune cells realise their mistake and stop attacking the nerves. A disease in which the immune system attacks its own body is called an *auto* (self) *immune* disease. Fortunately the part of the nerve attacked, which is the insulating or myelin sheath, can be replaced by the myelin-forming cells, named Schwann cells after the doctor who described them. Usually the conducting core of the nerve, called the axon, is not damaged. If the axons are damaged too, as sometimes occurs, they can regrow but recovery takes longer.

The way in which infections trick the lymphocytes into attacking the body and producing such an autoimmune disease is still the subject of research. The lymphocytes may cause the formation of chemicals called antibodies which circulate in the blood and latch onto and damage the myelin. The lymphocytes also attract macrophages, cells which can attack and digest myelin: this digestion is not altogether bad because the damaged myelin has to be removed before new myelin can be formed. Eventually special suppressor lymphocytes are formed which are able to suppress the action of the harmful lymphocytes. We need to know more about what induces the formation of these suppressor lymphocytes.

How Can We Cure Guillain–Barré Syndrome?

Since GBS usually gets better on its own the most important treatment is ordinary nursing and medical care with physiotherapy, and if necessary intensive care. No *drugs* have been proven to make any difference to the speed of recovery. Some doctors prescribe steroid hormones in the acute stage but others do not. Several recent trials have demonstrated that on average *plasma exchange* is helpful for severely affected patients in the first week or two of the illness. Plasma exchange involves being connected to a machine which can separate the blood cells from the fluid or plasma: about 250 ml of blood is removed at a time, the plasma is discarded and the blood cells are returned to the patient with clean plasma. The procedure is repeated several times on each of about five days until sufficient plasma has been exchanged. The risks of the procedure are extremely small and modern sterilisation has, for practical purposes, eliminated the risk of transmitting unpleasant infections in the clean plasma. Plasma exchange is probably not worthwhile in mildly affected patients or in acute GBS after the first couple of weeks. Although plasma exchange does seem to shorten the duration of the illness, particularly the time on a ventilator and the time to walk unaided, it is a help rather than a cure and improved treatments are constantly being sought.

Chronic Idiopathic Demyelinating Polyradiculoneuropathy

Chronic idiopathic demyelinating polyradiculoneuropathy, or CIDP for short, is much less common than GBS and most people reading this leaflet need not bother with this section. Less than 3% of patients with GBS have a second attack. It is usually clear from soon after the onset that a patient has CIDP because of its *chronic*, i.e. gradually progressive, course evolving over months or years rather than days or weeks as in GBS. Some patients with CIDP do have periods of worsening and then improvement and individual relapses are sometimes rather confusingly like GBS. The name CIDP is derived as follows: *C* stands for chronic which refers to the gradual course. *I*diopathic means "of its own cause" and is an admission that we cannot identify a specific cause such as diabetes or toxin exposure. Some doctors call CIDP "chronic inflammatory demyelinating polyradiculoneuropathy" or "chronic idiopathic polyneuritis" names which imply that, like GBS, the nerves are damaged by inflammation. *D*emyelinating means that the damage is to the insulating myelin sheaths round each nerve fibre and the central "wire" or axon is more or less spared. "Poly" means many, "radiculo" means root, "neuro" means peripheral nerve and "opathy" means disease: so polyradiculoneuropathy means disease of many peripheral nerves and the spinal roots (which are the points of origin of the peripheral nerves from the spinal cord).

CIDP is probably an autoimmune disease similar to GBS. Although CIDP is a chronic condition several different treatments are thought to be helpful. They all act by suppressing the damaging autoimmune response. Examples are steroids, azathioprine and plasma exchange. Obviously suppressing the immune response cannot be undertaken lightly because it runs the risk of suppressing normal immune responses to infections. The decision whether to try these treatments has to be tailored by the doctor to the individual needs of each patient. However, it is reassuring to know that demyelination can be repaired, that treatment is available and that some patients get better without treatment.

Where to go for help

Guillain–Barré Syndrome Support Group International
PO Box 262
Wynnewood,
Pa 19096, USA Tel: (215) 642 6855

Mrs G Sanders
Guillain–Barré Syndrome Support Group
Foxley, Holdingham
Sleaford, Lincs NG34 8NR, UK Tel: 0529 304615

In the UK the following may be useful:

Department of Employment. The Disablement Resettlement Officer (DRO) can be extremely helpful where there are employment problems. He (or she) can usually be contacted through a local Job Centre or Employment Office.

Department of Health & Social Security. Local offices deal with National Insurance Benefits. National Insurance Benefits may include Sickness Benefit, Invalidity Benefit, Attendance Allowance and Mobility Allowance.

Disablement Income Group and DIG Charitable Trust, Millmead Business Centre, Millmead Road, London N17 9QU. Tel: 081-801 8013. Founded to fight for adequate income to meet the extra costs of disability, the charitable trust has an advisory service to help disabled people with problems regarding obtaining allowances for which they may be eligible.

Disabled Living Foundation,
346 Kensington High Street, London W14 8NS

Occupational Therapists give advice on all types of aids and equipment to help with the tasks of everyday living. They can also advise on adaptations to a dwelling. They can be contacted at Social Services Departments in the Community, or through referral at a hospital Occupational Therapy Department.

Royal Association for Disability and Rehabilitation (RADAR) 25 Mortimer Street, London W1N 8AB.

Sexual and Personal Relationships of the Disabled (SPOD), 286 Camden Road, London N7 OBJ. Tel: 071-607 8851. Advisory leaflets available.

Social Services Department of the local authority is empowered to provide services for the handicapped when necessary. The services available vary from area to area. However, home helps, meals on wheels and advice from an occupational therapist are generally available. A social worker will also visit on request to discuss any problems.

Further Reading

Hughes RAC, Winer JB (1984) Guillain–Barré syndrome. In: Matthews WB, Glaser G (eds) Recent advances in clinical neurology. Blackwells, Oxford, pp 19–49

Editorial (1986) Plasma exchange for neurological disorders. Lancet ii:1313–1314
Guillain–Barré Syndrome Study Group (1985) Plasmapheresis and acute Guillain–Barré syndrome. Neurology 35:1096–1104
Heller J, Vogel S (1986) No laughing matter. Jonathan Cape, London
Hughes RAC (1990) Guillain–Barré syndrome. Springer-Verlag, Berlin, London
Ropper AH, Shahani BT (1984) Diagnosis and management of acute areflexic paralysis with emphasis on Guillain–Barré syndrome. In: Asbury AK, Gilliatt RW (eds) Peripheral nerve disorders: a practical approach. Butterworths, London, pp 21–45

Subject Index